Dentine Hypersensitivity

Dentine Hypersensitivity

Developing a person-centred approach to oral health

Edited by

***Peter G. Robinson** BDS, MSc, PhD, FRACDS, FDSRCS, FHEA, FFPH*

School of Clinical Dentistry, Claremont Crescent, University of Sheffield, Sheffield, UK

AMSTERDAM • BOSTON • HEIDELBERG • LONDON
NEW YORK • OXFORD • PARIS • SAN DIEGO
SAN FRANCISCO • SINGAPORE • SYDNEY • TOKYO
Academic Press is an imprint of Elsevier

Academic Press is an imprint of Elsevier
32 Jamestown Road, London NW1 7BY, UK
225 Wyman Street, Waltham, MA 02451, USA
525 B Street, Suite 1800, San Diego, CA 92101-4495, USA
The Boulevard, Langford Lane, Kidlington, Oxford OX5 1GB, UK

Library of Congress Cataloging-in-Publication Data
A catalog record for this book is available from the Library of Congress

British Library Cataloguing-in-Publication Data
A catalogue record for this book is available from the British Library

ISBN: 978-0-12-801631-2

For information on all Elsevier publications
visit our web site at http://store.elsevier.com

Dedication

David Locker would laugh at the idea of a book about dentine hypersensitivity being dedicated in his name. This is for him then.

Contents

List of Contributors

Finbarr Allen Cork Dental School & Hospital, Wilton, Cork, Ireland

Sarah R. Baker School of Clinical Dentistry, Claremont Crescent, University of Sheffield, Sheffield, UK

Ashley P.S. Barlow GlaxoSmithKline Consumer Healthcare, Weybridge, UK

Katrin Bekes Department of Operative Dentistry and Periodontology, University School of Dental Medicine, Martin-Luther-University Halle-Wittenberg, Halle, Germany

Olga V. Boiko School of Clinical Dentistry, Claremont Crescent, University of Sheffield, Sheffield, UK

Susan E. Coldwell Department of Oral Health Sciences, University of Washington, Seattle, WA

Joana Cunha-Cruz Department of Oral Health Sciences, School of Dentistry, University of Washington, Seattle, WA

Barry J. Gibson School of Clinical Dentistry, Claremont Crescent, University of Sheffield, Sheffield, UK

David Gillam Centre for Adult Oral Health, Institute of Dentistry, Barts and The London School of Medicine and Dentistry, QMUL, London

Melanie Hall School of Clinical Dentistry, Claremont Crescent, University of Sheffield, Sheffield, UK

S.L. He Chongqing Key Laboratory for Oral Diseases and Biomedical Sciences; Department of Pediatric Dentistry, The Affiliated Hospital of Stomatology, Chongqing Medical University, Chongqing, China

Lisa J. Heaton Department of Oral Health Sciences, University of Washington, Seattle, WA

Marta Krasuska School of Clinical Dentistry, Claremont Crescent, University of Sheffield, Sheffield, UK

David Locker Faculty of Dentistry, University of Toronto, Toronto, Canada

Carolina Machuca School of Clinical Dentistry, Claremont Crescent, University of Sheffield, Sheffield, UK

Steve Mason GlaxoSmithKline Consumer Healthcare, Weybridge, UK

Ninu R. Paul School of Clinical Dentistry, Claremont Crescent, University of Sheffield, Sheffield, UK

Tess Player GlaxoSmithKline Consumer Healthcare, Weybridge, UK

Jenny M. Porritt School of Clinical Dentistry, Claremont Crescent, University of Sheffield, Sheffield, UK

Peter G. Robinson School of Clinical Dentistry, Claremont Crescent, University of Sheffield, Sheffield, UK

Farzana Sufi GlaxoSmithKline Consumer Healthcare, Weybridge, UK

Elena Talioti Centre for Adult Oral Health, Institute of Dentistry, Barts and The London School of Medicine and Dentistry, QMUL, London

J.H. Wang Chongqing Key Laboratory for Oral Diseases and Biomedical Sciences; Department of Pediatric Dentistry, The Affiliated Hospital of Stomatology, Chongqing Medical University, Chongqing, China

John C. Wataha Department of Restorative Dentistry, School of Dentistry, University of Washington, Seattle, WA

Part One

Introduction and Background

Introduction

1

Peter G. Robinson, Sarah R. Baker and Barry J. Gibson
School of Clinical Dentistry, Claremont Crescent, University of Sheffield, Sheffield, UK

Diseases, people, and society

The purpose of this book is to present a case for adopting a person-centered approach in oral health care and oral health research. We have used dentine hypersensitivity (DH) as a case study, because in many different ways, it exemplifies the interaction between the person and the disease, the part of that person's body affected by the disease, and the society in which that person lives.

The current definition tells us that "Dentine hypersensitivity is characterized by short, sharp pain arising from exposed dentine in response to stimuli, typically thermal, evaporative, tactile, osmotic, or chemical and which cannot be ascribed to any other dental defect or pathology."[1] This definition reveals that the dental view immediately focuses on pain through abnormal loss of tissue that exposes the underlying dentine. Thus, the definition also tells us something about dentistry; there is no mention of the person who has the condition.

The omission of the person undermines the definition considerably. First, it encourages the mistaken belief that the diagnosis of DH is objective. The definition requires there to be pain in the absence of any other cause. This means that the person with the condition must identify the pain for the condition to be present. That person's perception of pain is based on his or her experiences, interpretations, and beliefs. That is to say, it is subjective. Consequently, the entire existence of DH in a tooth is, of necessity, based on a subjective opinion, and no matter how much one may wish to, it is impossible to ignore the person. The "person" is central to the diagnosis of the condition. In DH, despite this key role of the person, little research has studied what it is like for a person to live with it.

It might also be worth thinking about the name of the condition. It tells us the dentine is overly sensitive. However, shouldn't exposed dentine be sensitive? Does the name imply that the person is too sensitive, too? Put another way, does the name reflect professional views on an acceptable level of sensitivity?

There is also the question of why the dentine is exposed. Recession of the gingivae (gums) may be a manifestation of a more severe disease. In which case, why does this person have that disease? Recession often exposes dentine if the person brushes too aggressively or uses a hard toothbrush or abrasive toothpaste. Perhaps the social pressures to keep the mouth clean and fresh and worrying about the appearance of the teeth have led to brushing ferociously or using gritty toothpaste. In all these cases, things happening beyond the person influence the cause of the condition.

Dentine Hypersensitivity. DOI: http://dx.doi.org/10.1016/B978-0-12-801631-2.00001-4

The existence of consumer products for DH also reveals how the condition is more than merely dental. It is people, and not teeth or tubules, who buy products. Television advertisements for those products also convey meanings beyond exposed dentine. They show people wincing in pain, whose enjoyment of food or drink or social occasions is spoiled. Some of those advertisements feature dentists in surgeries, whereas others involve an anonymous (but usually decorative) narrator in a public place. The narrator advocates the use of a product that apparently brings immediate and powerful relief. During our research, we discovered that these two advertising styles reflect whether products were conceived as medicaments or cosmetics. Thus, the way a product is placed in a legal framework directly influences the messages received by the public about an oral condition.

The influence of these advertisements on people's purchasing also shows how the consumer products industry (as part of wider society) affects our personal knowledge and behavior related to DH. If the products reduce pain, then we can congratulate the industry on creating and disseminating effective products. And yet, this industry also carries a danger. If the advertisements draw viewers' attention to a condition they did not know they had, if they *sensitize* subjective opinions to sensations that they hadn't noticed, then they will encourage people to identify the pain. In this way, the advertisements will be making people ill!

These examples all illustrate the role of factors outside the mouth regarding the causes, diagnosis, and consequences of DH, and all involve the person. In doing so, they widen the idea of what oral health is. They demonstrate the role of the mouth and oral health, the way it is viewed, and its effect on everyday life, not simply in terms of the consequences of toothache, but what the mouth means, and what it communicates. One very direct result of thinking about the mouth in this way is considering the effect of oral conditions on the everyday life of the person affected.

The operation was a success, but the patient died

It is hardly surprising that dentists and oral health researchers focus so much on disease and the technical aspects of dental treatment. A strong image we all share of dentistry involves someone looking down at us, working on our teeth. The work is clearly very intricate, highly skilled, and demanding of enormous concentration. It is even very difficult for people to communicate with their dentist during these procedures! Young people for whom this kind of work resonates will therefore be attracted to dentistry. At dental school, students must spend a huge amount of time acquiring these necessary and exacting technical skills. Even after graduation, dentists have been paid according to the number of these treatment procedures they undertake. Cumulatively, these processes select and reinforce a biomedical focus.

In contrast, many of us have encountered a clinician, either as a teacher or as someone caring for us, who showed a gift for seeing beyond the teeth and seeing the patient as a person. Clinicians like this know what it is that is bothering their patients, and they regard treatment success as when those problems have been overcome. This difference between concentrating on pathology and the technical aspects of dentistry as opposed to thinking about the person reflects the distinction between two contrasting ideas of health.

The biomedical model of health defines health as the absence of disease. This perspective has been useful in health care, because it directly links clinical signs to the mechanisms of disease, therefore guiding diagnosis and treatment. The model evolved from the premise that diseases are organ-specific pathological processes that affect the function of cells within the organs. Its focus is on clinical, physiological, and biochemical outcomes, and its foundations are in the physical and biological sciences.

In many respects, this approach has served us well. The dominance of the basic sciences of genetics, biology, pathology, physiology, biochemistry, and molecular biology in clinical practice and medical research (including dentistry) has provided the understanding that has underpinned huge advances in health care over the centuries.[2] Nevertheless, the model has limitations. Its core problem is that it restricts the way we think about health and health care, because it is reductionist.

The term "reductionist" refers to the reduction of health and disease to their smallest common denominators and the exclusion of "peripheral" or complicating factors. One aspect of reductionism is mind–body dualism, which treats the mind and body as discrete and unrelated objects. Physicians have been known to argue that their responsibility is to treat only "real diseases" rather than to be concerned with psychological and social problems.[3] Thus, the physical and biological sciences are seen in isolation from their personal and social etiologies and consequences.

The definition of DH gives us a perfect example of reductionism, where the disease is seen purely as a problem of specific organs (the teeth or the mouth). It exemplifies how the biomedical model characterizes specific diseases when their etiologic and pathogenic processes are obvious, and we have already seen how treatments are specific to the disease. And yet, we also saw how DH, like so many other human diseases, is not a specific disease with a specific etiology. The condition can only be diagnosed when all other diseases have been ruled out, therefore rendering DH "a diagnosis of exclusion." We also began to see how there could be many etiological factors for the condition. Furthermore, the focus on disease and its treatment means that prevention or upstream health promotion becomes only an afterthought.

A key aspect of DH is that exposed dentine is ubiquitous, and yet not everybody with exposed dentine has hypersensitivity. Therefore, even in this relatively minor condition, the biomedical model gives insufficient attention to psychological and behavioral factors that might be responsible for the condition or that might be the consequences of it. Research in this area has largely bypassed the experiences of people and concentrated much more on microscopy, laboratory tests, and other technical processes. Indeed, a recent review explains DH in almost entirely biomedical terms.[4] Clinical examinations, radiographs, and results of special tests may indicate the possibility, but not the actuality, of the disease. You need people for that, and yet the biomedical model omits them. Later chapters in this book will discuss the need to counsel people with DH that the treatments for it are not always immediately or totally effective, which again indicates that the biomedical approach to DH is inadequate.

There are a number of other consequences of the biomedical model's narrow focus on the clinical and the technical. Clinicians assess oral health with indicators such as caries and gingival and plaque indices. The impacts of oral diseases and oral conditions are not recorded formally, even though they may denote substantial effects on individual daily functioning, including the ability to eat, smile, and talk to other people.[5]

Clinicians and scientists like to work with objective data. "Objective" implies something that is valid and free from random measurement error or even bias. In contrast, subjective assessments may give the impression of being undependable, less tangible, less reliable, and seemingly prone to all manner of influences. In fact, there are countless examples of unreliable clinical assessments in dentistry, ranging from patients receiving different treatment plans from different dentists to the formal study of diagnostic agreement.[6,7] An overreliance on the accuracy of clinical assessments is particularly ironic in the case of DH. First, hypersensitivity is often a consequence of periodontal conditions, the assessments of which are notoriously unreliable.[8] Second, there simply is no objective test for DH, because it is reliant on the subjective assessment of pain.

This latter point can be applied more broadly as the whole notion of objectivity in the identification of disease is dubious. For example, the value for the threshold at which diastolic blood pressure is said to be diagnostic of hypertension is arbitrary. There is nothing special about the value of 100 mmHg; it merely falls at a particular point on a distribution and is associated with slightly more adverse outcomes than slightly lower values. It is a socially constructed threshold, yet it has the power to alarm people with higher values and falsely reassure those with low values.[9] Social constructionists also claim that arbitrary values are the result of power relations and bias and can be changed.[10] Clinicians will also recognize lack of objectivity in prescribing patterns. When presented with the same patient, dentists are more likely to prefer treatment options within their own area of expertise.[11]

Yet another critique of the biomedical model is its tendency to see patients and people as passive objects rather than active participants in health and health care. The biomedical expertise and language required to be a clinician to some extent excludes lay people from involvement in their health care. In addition, as we have seen, some practitioners ignore people's psychological, social, and cultural contexts, even though those contexts shape their perception and experiences of illness. A classic example of this is women's experiences of childbirth, which was removed from homes to hospitals in the medical approach to obstetrics.[12] Childbirth came to be treated as a medical problem, despite being a normal aspect of life. Illich saw this medicalization as iatrogenic.[13] He argued that medicine contributed to illness by distracting from the real causes of health problems. Moreover, the relationship between clinician and patient strongly influences the quality of therapeutic outcomes (see Chapter 4). The success of treatment is restricted by the clinician's ability to modify patient behaviors; therefore, the exclusion of lay people from health care is at least limiting and may be harmful.

Table 1.1 Problems with the biomedical model

Problems with the biomedical model
Subjective, unreliable, and value laden
Exclusive
Questionable ethics and consumer participation
Normative needs are often high
The norms of dentists do not correspond to the functional norms or social needs of people
Based on absence of disease rather than health
Ignores the functioning of the oral cavity or person
Ignores social and motivational factors
Excludes alternate treatments
Ignores health promotion and prevention

Source: Derived from Sheiham et al.[16]

At a public health level, ignoring people and their contexts gives little attention to who gets a disease, and why. This failure to consider the distribution of disease therefore also fails to account for social inequalities in health. It is clear that health and illness are socially patterned, being influenced by gender, ethnicity, age, and social position.[14,15]

One of the most strident criticisms of the biomedical model in dentistry was written by Sheiham et al.,[16] who framed the argument in terms of using clinical (normative) assessments to measure the treatment needs of population groups. Their specific application of the critiques can be applied generally and serves as an excellent summary of the critiques of the model (Table 1.1).

Biopsychosocial model of health

One antidote to the biomedical model of health is the biopsychosocial perspective, which considers people and their health in their physical and social milieu. The biopsychosocial model is clarified in the World Health Organization (WHO) definition of health as "*a complete state of physical, mental, and social well-being and not merely the absence of disease or infirmity*."[17] Interestingly, this conception of health recalls the old English derivation of "hoelth," which means "wholeness or being sound or well."[18]

The WHO definition explicitly goes beyond clinical assessments of disease to incorporate the notion of well-being. Well-being is a subjective perception; therefore, immediately the person is at the heart of this definition. Moreover, well-being calls up the image of feeling good and of being able to do the things one needs or wants to do. Social well-being is related to getting along with other people. Health is therefore shaped by the relationship between the person and his or her physical and social environment. Dubos (a microbiologist) saw health as the manner in which one responds to the challenges of one's environment.[19]

Placing people and health in a physical and social context has profound implications for the work of dentists and researchers. It helps to explain health, its determinants, and its consequences. It demands consideration of psychological and social factors such as lay beliefs, coping strategies, and emotions such as depression and fear. Consequently, the model sees a role for health care in homes and communities (although in this respect, the biopsychosocial model may encourage further medicalization).

Recognizing the psychological and behavioral factors responsible for ill-health or that form its consequences focuses attentions on these lay experiences. It means we should think about the effects of disease on everyday life, including the ability to eat, smile, and talk to other people.[20] It means that clinicians should analyze the meaning of patients' reports of their illness in psychological, social, and cultural terms to complement their physiological and biological assessments.[21] Allowing patients and people to become active participants in health and health care also helps to overcome the mismatch between lay and normative assessments of health and militates against the problems of clinical assessments of disease by considering the role and social functioning of lay people. However, the perspective does not come without a cost. Clinicians must supplement their clinical abilities with psychological insights and communication skills to enhance treatment. Finally, greater emphasis on the social determinants of health encourages us to think about alternate ways of dealing with disease, including prevention and population-based health promotion. Such actions are more likely to overcome social inequalities in health.

In summary, for the purpose of this book, adopting the biopsychosocial model of health carries four huge implications. The first is that a definition of health that incorporates the person means that we, in turn, will be person centered. Second, a new definition of health gives health care a new goal. It becomes less concerned with the treatment of disease and can be reoriented toward care that demonstrably improves health. That, in turn, brings the third consequence, the need for better ways to measure health. Finally, any research that addresses the personal and social causes and consequences of disease must inevitably be multidisciplinary.

Health-related quality of life

Incorporating personal and social aspects to broaden our conception of health requires much broader assessments of health. One concept that has been used is that of health-related quality of life (HQoL). HQoL forms part of overall quality of life but is restricted to aspects of life that are health related. Of course, the two are not entirely discrete as they are linked by environmental factors. For example, low income is a facet of non-HQoL but can influence health.

As might be imagined, there are many definitions of HQoL (Table 1.2). The definitions differ but are mutually compatible, and all reach past the merely clinical, extending to include all those aspects of daily life that might be influenced by health conditions. This definition will be used for the purposes of this book.

Table 1.2 Definitions of HQoL

- Physical, psychological, and social domains of health, seen as distinct areas that are influenced by a person's experiences, beliefs, expectations, and perceptions.[22]
- The impact of a perceived health state on the ability to live a fulfilling life.[23]
- The impact of disease and treatment on disability and daily functioning.[24]
- Optimum levels of mental, physical, role, and social functioning, including relationships, and perceptions of health, fitness, life satisfaction, and well-being. It should also include some assessment of the patient's level of satisfaction with treatment, outcome and health status, and future prospects.[25]

HQoL is multidimensional, combining physical health, perceptions of health, and/or disability.[26] It may include psychological and emotional well-being measured with indicators of anxiety, depression, social networks, and support. Many measures incorporate functioning in social roles and indicators of life satisfaction and self-esteem.[27]

Only those directly affected can describe their effects on everyday life.[28] Therefore, HQoL must, as much as possible, be recorded by the person affected. By systematically considering the lay perspective, such subjective measures can be seen as the first step to incorporating lay people into clinical practice and research.

Assessments of HQoL take the form of validated and standardized questionnaires or interview schedules that measure subjective perceptions of functional status and well-being. Such assessments may be generic (capturing impacts on health in general) or may be condition specific (focusing closely on the impacts of specific conditions).[29] For example, SF-36 assesses generic health status, whereas the child perceptions questionnaire specifically relates to oral HQoL (OHQoL).[30]

Oral health-related quality of life

In just the same way that aspects of quality of life were selected to include only those related to health, we can specify further and select aspects only relevant to oral health. Almost 40 years ago, Cohen and Jago[31] suggested the development of "sociodental indicators" to record the psychosocial impact of oral problems. This proposal reflected the aim of dentistry, to maintain a functional, pain-free, and esthetically and socially acceptable dentition. Consequently, it became necessary to account for disruptions in physical, psychological, and social functioning caused by oral diseases.

The starting point for this task required the adoption of psychosocial definitions of oral health akin to the WHO definition. Dolan defined oral health as "a comfortable and functional dentition that allows individuals to continue their social role," whereas the UK Department of Health adopted "a standard of health of the oral and related tissues which enables an individual to eat, speak, and socialize without active disease, discomfort, or embarrassment and which contributes to general well-being."[32,33] As was the case with general health, these definitions incorporate daily activities and the effects of oral conditions on life overall. Thus, assessing the ability to perform these activities and effects is as important as identifying oral disease itself. OHQoL is a useful means to measure these diverse aspects of health, and it is therefore multidimensional.

Definitions of OHQoL have evolved over the decades.[20,34–38] Early attempts were vague but became more clearly specified with time. In 1989, Locker defined OHQoL as "the functioning of the oral cavity and the person as a whole and with subjectively perceived symptoms such as pain and discomfort."[20] His subsequent definitions emphasized the effects of the mouth on general health and well-being, going beyond the oral cavity to the individual and the way in which oral disorders, diseases, and conditions threaten health, well-being, and quality of life.[34] Notably, the ability to perform normal social roles was adopted as a benchmark against which impacts might be measured. However, this approach has been criticized as being negative and therefore for ignoring positive aspects of health.[39]

Most recently, Locker and Allen distinguished between subjective oral health and OHQoL. Subjective health status describes the person's current health state, while the concept of OHQoL also includes a subjective evaluation of that status. This book will use this most current definition throughout: *the impact of oral disease and disorders on aspects of everyday life that a patient or person values, that are of sufficient magnitude, in terms of frequency, severity, or duration to affect their experience and perception of their life overall.*[38]

Like assessments of HQoL, OHQoL measures are also highly structured but enquire about the symptoms and impacts of oral conditions on everyday life. The best-designed comprise a number of closely related questions (items) to form scales. There are often several subscales, with each recording a different dimension or domain of OHQoL. Participants' responses to the items are assigned values that are summarized in some way to create a score.

Unlike the false assumption of accuracy in clinical measurement, psychometrics recognizes that participant responses are subject to error.[40] Surprisingly, few of these sources of error are specific to subjective assessments, as the list includes different forms of misunderstanding of the question and biases in responses, all of which are present in clinical assessments. However, by making this assumption explicit, psychometric assessment attempts to minimize error in many ways. There are dedicated procedures to screen individual items, determine the consistency of responses to items, and assess the whole scale by correlating scores against other related variables.

Applications of OHQoL

So far, we have stressed the need to think about health more comprehensively in oral health care and research. That perspective is the rationale for being person centered and leads us to measure health in new ways. The question now becomes one about the utility of doing so. Potential applications of OHQoL can be summarized as political, clinical, research, public health, and for theoretical purposes (Table 1.3).

Table 1.3 Potential uses of oral health-related quality of life measures[41]

Field of work	Potential uses in health field/oral health-related quality of life
Political	• Planning public health policy
	• Planning in resource allocation
Clinical uses	• Communication tools
	• Commissioning programs of care
	• Evaluating interventions
	• Assessing the outcomes of new treatments
	• Aiding understanding of the patient's point of view
	• Screening
	• Identifying and prioritizing patient problems and preferences
	• Monitoring and evaluating individual patient care
	• Identifying which patients have more benefit from treatment
	• Involving patients perspectives in decision making and self-care
	• Predicting outcomes in order to provide appropriate care
	• Clinical audit
Public health	• Describing and monitoring illness in populations
	• Planning, monitoring, and evaluating services
	• Needs assessment and prioritization
	• Encouraging greater lay participation in health care
Research	• Evaluating outcomes of health care interventions
	• Elucidating the relationships between different aspects of health
Theoretical	• Exploring models of health
	• Describing factors influential to health

Politically, OHQoL data can be used to plan public health policies and inform resource allocation. In countries where health care is planned according to need, personal and social impacts of oral disease may prove valuable indicators. Typically, clinical data have been used to capture the imagination of politicians with regard to oral health, but OHQoL data might have more direct meaning to policymakers. Rather than lobbying with specialist data on mean DMFT values, information on the number of children kept awake with toothache may be a much more powerful lever.

There is clearly a role for the use of OHQoL data in screening and monitoring oral health in individual patient care. OHQoL data can act as outcomes to evaluate oral care, for both individual patients and organizations providing care. Public health applications represent an upscaling of clinical applications and can describe illness in populations to guide needs assessment and prioritization and to plan and evaluate services. Colleagues new to this field should take caution, though; more stringent measurement properties are required for measures used in individuals than in groups of people. Measures for individuals need to be more reliable and therefore often have more items.

The field of OHQoL research has expanded, although not without its criticisms. We know more about the impact of oral disorders and what determines those impacts.

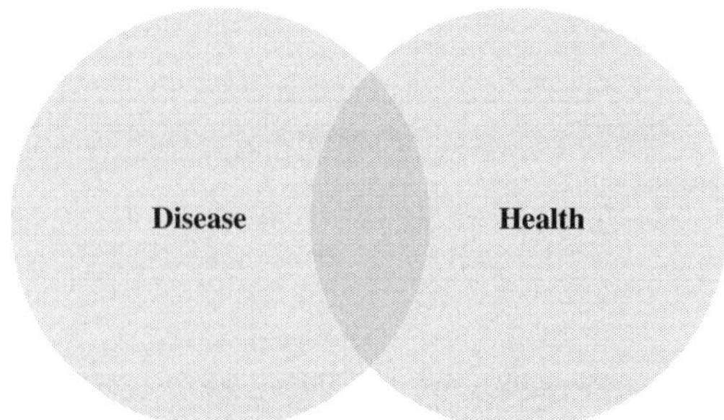

Figure 1.1 The relationship between health and disease.[34]

There is also a growing body of research regarding the effects of oral health care interventions on OHQoL. However, much research in this area has simply been to evaluate measures in different settings. Less work has used the measures for purposes that can be applied to improve health of either individuals or populations.

Adopting the concept of OHQoL has important ramifications for the development of theory. Naturally, there is a tendency among clinicians to see the links between the state of the mouth and OHQoL as paramount, and to assume that if they repair people's teeth, improvements in OHQoL will inevitably follow. This perspective sees clinical status and OHQoL existing on a logical continuum. Innumerable data from within oral and general health challenge this view by showing only weak relationships between clinical and HQoL data.[42,43] In the most extreme cases, there is a "disability paradox," whereby people with profound health problems (such as cancer or chronic kidney disease) report quality of life that is as good as that of the general population.[44] These findings suggest that clinical status and OHQoL are distinct but related ideas, which Locker depicted as a Venn diagram rather than being linearly related (Figure 1.1).[34]

The notion that OHQoL is not a simple consequence of the state of the mouth means that to some extent, OHQoL has a life of its own and suggests that other factors intervene in the relationship between the two. For instance, many of us would feel embarrassment or loss of self-esteem if one of our front teeth was removed. Conversely, someone living in an area or social group where tooth loss is common may be less affected. In this case, social and psychological factors are modifying the effects of tooth loss. One result of recognizing the influence of other factors on the impact of oral conditions is that identifying those factors will be useful if we want to improve OHQoL. However, this requires us to bring together many different ideas from clinical, behavioral, and social science. This multidisciplinary work is exciting but not easy.

The value of theoretical models

The reader will have realized by now that the different aspects of health can create difficulties when trying to measure them. In addition, these dimensions and the causes and subjective consequences of disease are interrelated, leaving the possibility for an almost infinite mess of hypotheses from which it will be difficult to distinguish the real from the spurious relationships. We overcome these problems from a position of conceptual clarity, defining our terms and theorizing in advance how the different concepts are related.

Although often unfamiliar to clinicians, theoretical models are fundamental to the development of science, because they are a practical method of matching theory to reality. They take the form of organized theories constructed to describe natural phenomena and can be applied to predict measurable material consequences.[45] For example, in medicine, biological models can shape the practice of medicine and are commonly used to explain natural processes of diseases. Much oral and dental research has been criticized for being atheoretical, so it is worth considering the merits of theoretical models in a book that is going to try to address the relationships between people, diseases, and the environment.[46]

First, theoretical models serve as explanatory tools, explaining natural phenomena as relationships among variables. As we have seen, by explaining diseases in terms of etiology and pathological processes, the biomedical model aids understanding by connecting causes, pathological changes, and clinical features.[47]

Second, theoretical models also predict specific natural processes and logical material consequences that can be measured. For example, Janket et al.[48] theorized that the presence of dental conditions with inflammatory mediators (pericoronitis, dental caries, root remnants, and gingivitis) would predict chronic heart disease.

Third, theoretical models guide further actions. The model that guided much of the work in this book had already been used to evaluate the relationships among clinical variables, symptoms, psychosocial factors, functioning and general health perceptions, and overall quality of life.[49] With this understanding, this model guided the development of interventions to improve function and HQoL.[50]

Fourth, theory can be used to guide systematic data analysis by restricting analyses to hypothesized relationships, rather than allowing researchers to embark on a "fishing trip" of cross-tabulations.[51] In an extension of this, theoretical models can be developed into statistical models that can be used to estimate the magnitude of specified effects.

Two theoretical models have been used in OHQoL research: Locker's model of oral health and Wilson and Cleary's model linking clinical variables to quality of life.[5,49] Locker's study of chronic illnesses led him to consider the adverse consequences of oral conditions for aspects of life such as pain, discomfort, speaking, eating, and smiling in line with the WHO's International Classification of Functioning, Disability, and Health.[5,52] For instance, tooth decay may damage a tooth (disease, impairment, and functional limitation of the organ), which might restrict eating and sleeping (a disability of the individual) and might affect school performance (social handicap, a disadvantage experienced by impaired and disabled people, because they do not meet the expectations of society) (Figure 1.2).

The development of this model marked a change in dentistry, from the purely biomedical to a biopsychosocial perspective. For example, the two most widely used measures of OHQoL in adults, the Oral Health Impact Profile and the Oral

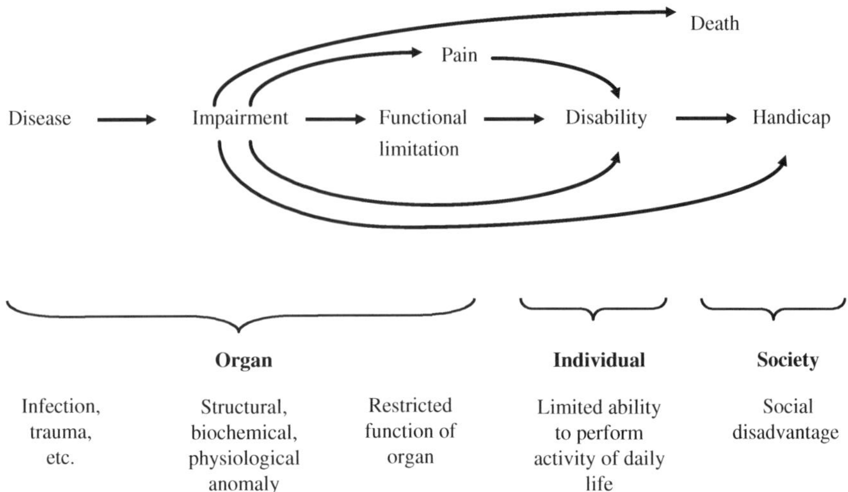

Figure 1.2 Locker's conceptual model of oral health.

Impacts on Daily Performance, were based on the Locker model and used the different concepts within them as frameworks for questions and scoring.[53,54]

The Wilson and Cleary model operationalizes the biopsychosocial approach by linking clinical factors with quality of life in a continuum of five levels: biological and physiological factors, symptoms, functioning, general health perceptions, and overall quality of life (Figure 1.3).

Biological and physiological factors form the biomedical pole of the spectrum and are measured in laboratory tests and clinical practice. Symptoms represent a shift in attention to the person. They are the subjective perceptions of abnormal physical, emotional, or cognitive states and the things that form the basis of complaints. The person's ability to perform particular tasks is referred to as functioning, and those functions may be physical, psychological, or social, or involve performing one's role at work or among one's family. General health perceptions represent the summation of how the person judges his or her health. This level integrates those that precede it in a subjective rating of self-rated health. In oral health, general health perceptions are analogous to global oral health ratings. Overall quality of life refers to subjective well-being, happiness, and life satisfaction. As described, overall quality of life is determined by both health and nonmedical factors.

We have already seen how individual and environmental characteristics might influence HQoL. Factors unique to the individual might include personal preferences, behaviors, and emotional or psychological factors such as stress or self-esteem. Environmental factors include socioeconomic status, culture, and living conditions.

Wilson and Cleary[55] hypothesized what they believed to be the principal relationships in the model, with linear relationships operating from the biomedical to influence overall quality of life, mediated by symptoms, functioning, and general health perceptions. For example, receiving new dentures changed the clinical status and

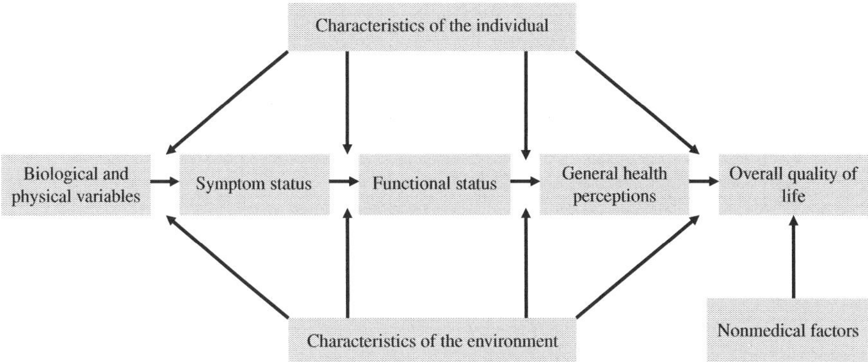

Figure 1.3 The Wilson and Cleary model. Linking clinical variables with health-related quality of life. A conceptual model of patient outcomes.[55]

consequently improved oral functioning and global oral health in housebound elders.[55] The linear pathways may be modified by individual and environmental factors at each stage. For instance, low income or feelings of stress could exacerbate the impact of symptoms on emotional and social function. In relation to the mouth, Gregory[56] showed that the effects of visibly missing front teeth on symptoms and functional impairment were modified by the relevance of oral health. Relationships within the model were also hypothesized to be bidirectional, so that difficulties brushing one's teeth (a functional limitation) might allow the progression of periodontitis, therefore leading to greater symptoms of pain from hypersensitive dentine.

The Wilson and Cleary model has been used increasingly in oral health, and that use consistently supports its utility. Research has tested direct and mediated pathways among people with xerostomia and among housebound older people with no teeth.[55,57] Subsequent prospective cohort studies with lagged analysis and structural equation modeling have confirmed its validity as a framework for studying OHQoL in children and have even guided the design and analysis of a randomized controlled trial of an intervention to improve OHQoL by enhancing an individual factor.[43,50,58] In summary, theory forms a framework for understanding how factors of interest are related; it guides the analysis and interpretation of data and can suggest the design and evaluation of interventions.

This book

In making the case for person-centered oral health care and research, our objective is to provide a detailed and integrated account of multidisciplinary research of DH. We present the findings of a series of studies of DH, drawing on the research traditions of epidemiology, sociology, psychology, laboratory science, clinical dentistry, and dental public health, all of which have been integrated as part of the study of a single oral condition.

DH provides an excellent case study for person-centered work for both direct and more symbolic reasons. Directly, despite being a very common condition (affecting about 10% of adults), it has not received the attention it deserves from clinical dentists and academics. More symbolically, DH is an example of how the different levels of human health interact (person, disease, organ, and society). We have argued that a biomedical focus on disease and specific parts of the body diverts attention from the principal concern in dentistry: the oral health of people. This breadth of perspective has required us to work as a multidisciplinary team. The authors are dentists, psychologists, sociologists, and industry scientists working together in the field of DH. The work described in this book shows how social and behavioral science can bring new insights into the experience, treatment, and fundamental knowledge of an important dental condition.

This book comprises four parts. The first part introduces the book. It provides biomedical background regarding the clinical presentation and physiological mechanisms of DH as a basis for understanding the technologies to treat it and starts to show how a purely biomedical perspective might be inadequate (Chapter 2). Knowledge of the epidemiology in Chapter 3 demonstrates the scale of the problem of DH, reveals the implications of a diagnosis of exclusion, and illustrates the need for dental professionals to diagnose DH in research studies. Significantly, the review of contemporary treatments for DH describes the mechanisms of action of individual therapies but does not lose sight of its behavioral causes or of the need to counsel patients on the effectiveness of treatment (Chapter 4).

We hope that any biomedical emphasis in the first part is counterbalanced by this introductory chapter, which has provided a rationale for a person-centered approach to oral health and explained the implications of such an approach. This case is supplemented by Chapter 5, which provides more details of the assessment of HQoL and the properties of a good measure.

Part 2 brings together our work on the subjective experience of DH and is focused on the development of a condition-specific OHQoL measure: the dentine hypersensitivity experience questionnaire (DHEQ). Our person-centered approach made it essential that the DHEQ is relevant to people with DH and grounded in their experiences. Chapter 6 describes the conduct and results of an exhaustive qualitative study exploring the range of personal and functional experiences of DH in everyday life. This knowledge gave us the framework and language to use within the DHEQ, although that development required some difficult decisions. The careful thought behind those decisions is described in Chapter 8, which is a narrative of the construction of the questionnaire. This type of material is not normally published, because it is rather like a laboratory scientist typing up their notes. It is included because we felt it would be useful for others attempting similar tasks. Chapter 7 is a more formal account of the construction and preliminary validation of the DHEQ. It describes the analyses and data used to select the items and the psychometric performance of the measure in two cross-sectional studies.

The original DHEQ performed very well as a research tool but was too long, placing too much of a burden for use with patients in clinics or participants in field epidemiology. Chapter 9 tells how several short forms were derived and tested, resulting in a 15-item version (DHEQ-15) with outstanding measurement properties and that appears to be suitable for use in clinical settings with individual patients. A comprehensive analysis of data from three randomized controlled trials extends the validation of the DHEQ as an evaluative measure and confirms that DHEQ is stable, valid, and responsive, and discriminates between treatments of different efficacy (Chapter 10).

This introduction has illustrated how health can be seen as the manner in which one interacts with one's physical and social environment. Consequently, the same clinical problem may have different impacts in other cultural settings. These differences pose special challenges for the translation of HQoL measures. As well as ensuring linguistic translation, questionnaires must be culturally relevant for use in new settings. The adaptation of the DHEQ for use in China is outlined in Chapter 11. The measure has been translated to other languages, but this version provides an excellent case study.

Chapter 12 applies the DHEQ to a pressing problem in HQoL research. In a phenomenon known as response shift, subjective assessments vary if the person making the assessment changes over time. The person may adapt to a new health condition, so that it no longer impacts on his or her life. For example, people with DH may reconcile themselves to never having ice cream again, to the point where they never consider eating ice cream and do not think their diet is restricted. Such changes in internal standards might undermine evaluative research. Chapter 12 introduces the field of response shift and discusses two applications of DHEQ to identify and account for it.

Part 3 contains two chapters written from a more psychological position to describe how an understanding of psychological theory and method has informed the measurement of pain in DH. One difficulty of research of DH relates to its relatively minor nature, with only fleeting episodes of pain. DH might barely register on a scale used to measure the pain of childbirth, for instance. Psychologists in Washington have devised elegant solutions for this problem in the form of labeled magnitude scales, which act like a magnifying glass at the less severe end of the scale, allowing people to record quite subtle differences in pain. Chapter 13 describes the development and testing of these scales.

The existence of an accurate measure of the impact of DH and the availability of new technologies allow very sophisticated research of the psychological processes involved in living with the condition. The research presented in Chapter 14 used daily diaries to study the role of illness beliefs and coping in the adjustment to DH. Once again, this is an example of how high-quality research in oral health can increase the understanding of wider and more fundamental concepts.

The emphasis in the fourth part of the book shifts to the meaning of DH and again reveals the applicability of oral health research to add to knowledge beyond the mouth. It begins with the claim that when we talk about something, we draw on language to make ourselves understood. The language we use is socially produced and is constantly being modified over time. It enables us to make sense out of the world. This pool of references has become known as a "pool of meaning." When we speak of health or illness, then, we are drawing on this pool of meaning, which is socially produced and constructed over time. Chapter 15 reports on explorations of the structure and origins of accounts of illness associated with DH. Chapter 16 analyses the meaning of the language used in accounts of dentine hypersensitivity. A specific language was developed, and its history shows how dentine hypersensitivity became separated from dentine sensitivity and how these meanings reveal how the condition came to be regarded in the medical, scientific and economic spheres. We then go on to a detailed study of how DH is communicated in advertising. Chapter 17 explores how advertisements "animate" inanimate toothpastes to produce productive patterns of meaning and enable these substances to develop meaning and become incorporated into everyday life.

Finally, in Chapter 18, we draw together the different parts of the book to reach conclusions about the utility of DHEQ and how DHEQ can or might be used. We also identify contributions to the understanding of DH and OHQoL from the social and behavioral sciences. We then close by emphasizing the value of a multidisciplinary approach to oral health care research. The long and short forms of the DHEQ are presented as appendices.

The introduction to this book would not be complete without a clear acknowledgement of the very generous support of GlaxoSmithKline Consumer Healthcare Research and Development. The book is an output of a long and fruitful partnership with them, for which we are very grateful. Colleagues from GlaxoSmithKline have collaborated on some of the chapters, but editorial control has rested entirely with PGR.

References

1. Canadian Advisory Board on Dentin Hypersensitivity. Consensus-based recommendations for the diagnosis and management of dentin hypersensitivity. *J Can Dent Assoc* 2003;**69**:221–6.
2. Younossi ZM, Guyatt GH. Quality of life assessment and chronic liver disease. *Am J Gastroenterol* 1998;**93**:1037–41.
3. Engel GL. The need for a new medical model: a challenge for biomedicine. *Science* 1977;**196**:129–36.
4. Cartwright RB. Dentinal hypersensitivity: a review. *Community Dent Health* 2014;**31**:15–20.
5. Locker D. Measuring oral health: a conceptual framework. *Community Dent Health* 1988;**5**:3–18.
6. Elderton RJ, Nuttall NM. Variation among dentists in planning treatment. *Br Dent J* 1983;**154**:201–6.
7. Winters D. Getting drilled: cavity opinions differ by dentist. WHOtv.com <http://whotv.com/2013/05/06/getting-drilled-dentists-cavity-opinions-can-differ/> [accessed 18.03.14].
8. Hefti AF, Preshaw PM. Examiner alignment and assessment in clinical periodontal research. *Periodontol 2000* 2012;**59**:41–60. Available from: http://dx.doi.org/doi:10.1111/j.1600-0757.2011.00436.x.
9. Rose G. *The strategy of preventive medicine*. Oxford: Oxford University Press; 1992.
10. Lupton D. *Medicine as culture: illness, disease and the body in western societies*. London: Sage Publications; 2003.
11. Bartlett D, Preiskel A, Shah P, Ahmed A, Moazzez R. An audit of prosthodontics undertaken in general dental practice in the south east of England. *Br Dent J* 2009;**207**:E15 Available from: http://dx.doi.org/doi:10.1038/sj.bdj.2009.908. Epub 2009 Oct 16.

12. Donnison J. *Midwives and medical men: a history of inter-professional rivalries and women's right*. London: Heinemann; 1977.
13. Illich I. *Limits to medicine*. London: Marion Boyars; 1976.
14. Engel GL. The clinical application of the biopsychosocial model. *Am J Psychiatry* 1980;**137**:535–44.
15. NICE. *NICE public health guidance 6: behaviour change at population, community and individual levels*. London: National Institute for Health and Clinical Excellence; 2007.
16. Sheiham A, Maizels JE, Cushing AM. The concept of need in dental care. *Int Dent J* 1982;**32**:265–70.
17. WHO. *Constitution*. Geneva: World Health Organization; 1948.
18. Dolfman ML. The concept of health: an historic and analytic examination. *J Sch Health* 1973;**43**:491–7.
19. Dubos R. *The mirage of health: utopia, progress, and biological change*. London: Harper Collins Publishers; 1959.
20. Locker D. *An introduction to behavioural science & dentistry*. London: Tavistock/ Routledge; 1989.
21. Engel GL. Enduring attributes of medicine relevant for the education of the physician. *Ann Intern Med* 1973;**78**:587–93.
22. Testa MA, Simonson DC. Assessment of quality-of-life outcomes. *N Engl J Med* 1996;**334**:835–40.
23. Bullinger M, Anderson R, Cella D, Aaronson N. Developing and evaluating cross-cultural instruments from minimum requirements to optimal models. *Qual Life Res* 1993;**2**:451–9.
24. Kaplan RM, Atkins CJ, Timms R. Validity of a quality of well-being scale as an outcome measure in chronic obstructive pulmonary disease. *J Chronic Dis* 1984;**37**:85–95.
25. Bowling A. *Measuring disease*. Buckingham: Open University Press; 1995.
26. Gift HC, Atchison KA, Dayton CM. Conceptualizing oral health and oral health-related quality of life. *Soc Sci Med* 1997;**44**:601–8.
27. Bowling A. *Research methods in health: investigating health and health services*. 2nd ed. Buckingham: Open University Press; 2004.
28. Black N, Jenkinson C. How can patients' views of their care enhance quality improvement? *Br Med J* 2009;**339**:202–5.
29. Dawson J, Doll H, Fitzpatrick R, Jenkinson C, Carr AJ. The routine use of patient reported outcome measures in health care settings. *Br Med J* 2010;**340**:c186.
30. Jokovic A, Locker D, Stephenson M, Kenny D, Tompson B, Guyatt GH. Validity and reliability of a questionnaire for measuring child oral health related quality of life. *J Dent Res* 2002;**81**:459–63.
31. Cohen LK, Jago JD. Toward the formulation of sociodental indicators. *Int J Health Serv* 1976;**6**:681–98.
32. Dolan TA. Identification of appropriate outcomes for an aging population. *Spec Care Dentist* 1993;**13**:35–9.
33. Department of Health. *An oral health strategy for England*. London: Department of Health; 1994.
34. Locker D. Concepts of oral health. Disease and the quality of life. In: Slade GD, editor. *Measuring oral health and quality of life*. Chapel Hill: University of North Carolina; 1997. p. 11–24.
35. Kressin NR. The oral health related quality of life measure (OHQOL). In: Slade GD, editor. *Measuring oral health and quality of life*. Chapel Hill: University of North Carolina; 1997. p. 113–20.

36. US Department of Health and Human Services. *Oral health in America: A report of the surgeon general*. Rockville, MD: US Department of Health and Human Services; 2000.
37. Gift HC, Atchison KA. Oral health, health, and health-related quality of life. *Med Care* 1995;**33**:NS57–77.
38. Locker D, Allen F. What do measures of "oral health-related quality of life" measure?. *Community Dent Oral Epidemiol* 2007;**35**:401–11.
39. McGrath C, Bedi R. A study of the impact of oral health on the quality of life of older people in the UK-findings from a national survey. *Gerodontology* 1998;**15**:93–8.
40. Streiner DL, Norman GR. *Health measurement scales*. 3rd ed. Oxford: Oxford University Press; 2003.
41. Robinson PG, Carr AJ, Higginson IJ. How to choose a quality of life measure. In: Carr AJ, Higginson IJ, Robinson PG, editors. *Quality of life*. London: BMJ Books; 2003. p. 88–100.
42. Locker D, Slade G. Association between clinical and subjective indicators of oral health status in an older adult population. *Gerodontology* 1994;**11**:108–14.
43. Baker SR, Mat A, Robinson PG. What psychosocial factors influence adolescents' oral health? *J Dent Res* 2010;**89**:1230–5.
44. Albrecht GL, Devlieger PJ. The disability paradox: high quality of life against all odds. *Soc Sci Med* 1999;**48**:977–88.
45. McLaren N. A critical review of the biopsychosocial model. *Aust N Z J Psychiatry* 1998;**32**:86–92.
46. Baker SR, Gibson BJ. Social Oral Epidemi(olog)2y Where next: one small step or one giant leap? *Community Dentistry Oral Epidemiology*. In Press.
47. Quintner JL, Cohen ML, Buchanan D, Katz JD, Williamson OD. Pain medicine and its models: helping or hindering? *Pain Med* 2008;**9**:824–34.
48. Janket S, Qvarnström M, Meurman J, Baird A, Nuutinen P, Jones J. Asymptotic dental score and prevalent coronary heart disease. *Circulation* 2004;**109**:1095–100.
49. Wilson IB, Cleary PD. Linking clinical variables with health-related quality of life: conceptual model of patient outcomes. *JAMA* 1995;**273**:59–65.
50. Nammontri O, Robinson PG, Baker SR. Enhancing oral health via sense of coherence: a cluster-randomized trial. *J Dent Res* 2013;**92**:26–31. Available from: http://dx.doi.org/10.1177/0022034512459757.
51. Boorse C. Health as a theoretical concept. *Philos Sci* 1997;**44**:542–73.
52. WHO. *International classification of impairments, disabilities, and handicaps. A manual of classification relating to the consequences of disease*. Geneva: World Health Organisation; 1980.
53. Slade GD, Spencer AJ. Development and evaluation of the oral health impact profile. *Community Dent Health* 1994;**11**:3–11.
54. Adulyanon S, Sheiham A. Oral impacts on daily performances. In: Slade GD, editor. *Measuring oral health and quality of life*. Chapel Hill: University of North Carolina; 1997. p. 151–60.
55. Baker SR, Pearson NK, Robinson PG. Testing the applicability of a conceptual model of oral health in housebound edentulous older people. *Community Dent Oral Epidemiol* 2008;**36**:237–48.
56. Gregory J, Gibson B, Robinson PG. Variation and change in the meaning of oral health related quality of life: a "grounded" systems approach. *Soc Sci Med* 2005;**60**:1859–68.
57. Baker SR, Pankhurst CL, Robinson PG. Testing relationships between clinical and non-clinical variables in xerostomia: a structural equation model of oral health-related quality of life. *Qual Life Res* 2007;**16**:297–308.
58. Gururatana O, Baker SR, Robinson PG. Determinants of children's oral health related quality of life over time. *Community Dentistry Oral Epidmiol* 2014;**42**:206–215.

Clinical presentation and physiological mechanisms of dentine hypersensitivity

2

Katrin Bekes
Department of Operative Dentistry and Periodontology, University School of Dental Medicine, Martin-Luther-University Halle-Wittenberg, Halle, Germany.

Introduction

Dentine hypersensitivity (DH) is a significant global clinical oral health problem in the adult population.[1] It is a painful symptom of the exposed and innervated cervical pulp–dentine complex that may disturb a person during eating, drinking, toothbrushing, and sometimes even breathing.[2] The quality of life of people with DH is often altered, because the pain experienced with DH causes tangible and frequent discomfort.[3] Recent studies indicate that members of the Western population retain their functional natural dentition longer than previous generations, resulting in continued tooth wear.[4] Given these demographic and health trends, it is likely that DH will become a more frequent dental finding in the future. Thus, the condition needs to be diagnosed and addressed at an early stage, or indeed prevented, to reduce lifelong oral pain symptoms associated with DH. Extensive research efforts have been invested in understanding the processes leading to DH and developing effective treatments to alleviate and prevent this painful condition.[1,5–7]

The first part of this chapter outlines the clinical features of DH. Although numerous reports on all topics related to DH have been published, its diagnosis, etiology, and predisposing factors may not be well-understood by all dental professionals. The second part of this chapter focuses on the pain mechanisms causing DH, describing three distinct hypotheses that have been proposed (as well as the results of investigations of each), and discussing the nature of the pain experienced with DH.

Clinical presentation of DH

Definition

DH is defined as a "short, sharp pain arising from exposed dentine in response to stimuli, typically thermal, evaporative, tactile, osmotic, or chemical, which cannot be ascribed to any other form of dental defect or pathology."[8,9] A modification to

Dentine Hypersensitivity. DOI: http://dx.doi.org/10.1016/B978-0-12-801631-2.00002-6

this definition was suggested by the Canadian Advisory Board on Dentine Hypersensitivity in 2003, who proposed that "disease" should be substituted for "pathology."[10] Classically, the pain experienced with DH is of rapid onset, short and sharp in character, and of a duration equal to that of the applied stimuli, although it can persist as a dull throbbing ache for variable periods.[11] It may be localized or generalized, affecting one or more tooth surfaces simultaneously.[1,12]

The definition of DH has two aspects. Whereas the first is a clinical description of the condition,[13] the second, perhaps more importantly, identifies DH as a distinct clinical entity and therefore invites the clinician to consider a differential diagnosis, given that other conditions may have identical symptoms but require different management strategies.[14]

Differential diagnosis

Great care must be taken in the diagnosis of DH to exclude all other dental defects and pathology. A number of clinical conditions may elicit the short, sharp tooth sensitivity that is characteristic of DH (summarized in Table 2.1).[9,10,12,15,16] These conditions require treatment options that are usually quite different from those used for DH. Therefore, it is necessary to take the proper time to make a correct diagnosis, because DH is always a diagnosis of exclusion; it can only be definitively confirmed after all other possible conditions have been diagnostically eliminated. A proper history of the nature of the pain, clinical evaluation, and radiographic examination, as well as the use of diagnostic tests (such as percussion, palpation, and pulp-vitality testing) will allow the clinician to confirm DH by excluding other conditions.[13]

Table 2.1 Possible causes of tooth sensitivity that do not represent DH

Dental caries
Chipped teeth
Fractured teeth
Fractured restorations
Cracked tooth syndrome
Postoperative sensitivity
Pulpal response to caries and to restorative treatment
Pulpitis or other endodontic problems
Ditching of margins of amalgam restorations and surface wear on composites
Improperly insulated metallic restorations
Incorrect placement of dentine adhesives in restorative dentistry leading to nanoleakage
Palatogingival groove
Vital bleaching procedures
Acute hyperfunction of teeth
Atypical facial odontalgia
Hypoplastic enamel
Congenitally open cementum–enamel junction

Prevalence

Tooth sensitivity is often reported in clinical dental practice. From the relatively few studies dealing with the prevalence of tooth sensitivity, it can be concluded that DH is a frequent condition.[17] Nevertheless, epidemiological studies of the prevalence of DH have produced conflicting data, with figures ranging from 1.34% to 98%.[18–20] The heterogeneity of prevalence data may be related to assessment methods, ranging from questionnaires to clinical detection, and could possibly be related to study location as well.[21] In addition, most studies of DH have examined highly selected populations, such as patients at periodontal offices,[22] students,[23] or hospitalized patients.[24] Several studies indicate that even though high percentages of a population may report having sensitive teeth, a much smaller proportion actually has DH diagnosed on the basis of defined clinical diagnostic criteria.[2,16,17]

Another factor that may explain the huge range in self-reported prevalence figures is the subjective nature of pain. The experience of a sensory event is highly subjective, can vary substantially between individuals, and is related to individual tolerance of pain as well as to physical and emotional factors.[25,26] It is known that although many individuals do not seek treatment to desensitize their teeth because they do not perceive DH to be a severe oral health problem,[27] a substantial number of people experience discomfort to the extent that it interferes with their eating, drinking, oral hygiene habits, and sometimes even breathing.[28,29] It appears that the incidence of true DH in most general populations ranges from 10% to 30%.[2] (See Chapter 3 for an exhaustive review of the prevalence of DH).

Distribution

In general, a slightly higher prevalence of DH has been reported in women than in men,[30] although the observed difference often fails to reach the level of statistical significance.[17,22,31] The reasons for any differences are not yet clear, but they have been presumed to be possibly related to womens' better overall health care and oral hygiene awareness, which would make them more aware of DH.[8] Furthermore, women have been found to be more sensitive to pain, and this physiological phenomenon may be another possible reason for any gender difference in DH prevalence.[32]

DH may affect people of any age.[15] It can present at any time from the early teenage years to older ages. It can be assumed that with increased life expectancy and with more people keeping their own teeth longer, the prevalence of DH will increase because of the concomitant increased prevalence of periodontal disease, periodontal treatment, gingival recession, and erosive tooth wear, all of which can expose dentine to external stimuli. However, epidemiological data suggest that DH is most often diagnosed in individuals from 20 to 40 years of age, with the peak occurrence found at the end of the third decade.[33] This observation may be attributable to the natural processes of aging. After the age of approximately 40 years, secondary or reparative dentine develops and subsequent sclerosis of the dentinal tubules is accompanied by reduced sensitivity.[15]

In terms of intraoral distribution, DH occurs most frequently on the buccal cervical zones of permanent teeth. In general, maxillary teeth seem to be more affected than mandibular teeth but, again, this difference often fails to reach statistical significance.[17,31] The teeth with the most common occurrence are premolars and canines.[9,17,30,34] Teeth with lower plaque scores are associated with DH, suggesting a connection between regular (possibly overzealous) toothbrushing and the onset of sensitivity.[9] The most common stimuli that cause DH are, in order, cold drinks, hot drinks, toothbrushing, and sour substances.[35]

Etiology and risk factors

Although many risk factors have been identified that lead to the exposure of dentine, dentinal tubular openings, and subsequent pain have been identified, it is not clear which are key to the development of DH.[20] It seems that interactions between several factors (both stimuli as well as predisposing factors) appear to play an important role in initiating DH.[36] DH can only arise when two processes occur. First, the dentine surface of the tooth has to become exposed (lesion localization); second, a number of dentinal tubules in close proximity to each other have to be opened and must be patent from the pulp to the oral environment (lesion initialization).[37]

Exposure of dentine may be the result of hard (enamel) or soft tissue loss (gingival recession). Loss of enamel can occur via attrition, abrasion, erosion, or abfraction, with increased dentine wear and tubule exposure often being the result of the synergistic effects of erosion and abrasion (Figure 2.1).[38]

Attrition describes the wearing away of tooth substance as a result of tooth-to-tooth contact during normal or parafunctional masticatory activity (Figure 2.2).[39] Abrasion is considered the pathological wear of tooth substance from factors other than tooth contact through biomechanical frictional processes (e.g., toothbrushing).[40] The use of abrasive toothpastes has been identified as possibly being responsible for lesion development, as it may abrade dentine.[41] Conversely, such abrasiveness may also produce a smear layer, thereby reducing sensitivity.[42] Further work is needed to investigate the relationship between DH and toothbrushing habits, force, duration of brushing, frequency, toothbrush filament characteristics, and toothpaste.[36]

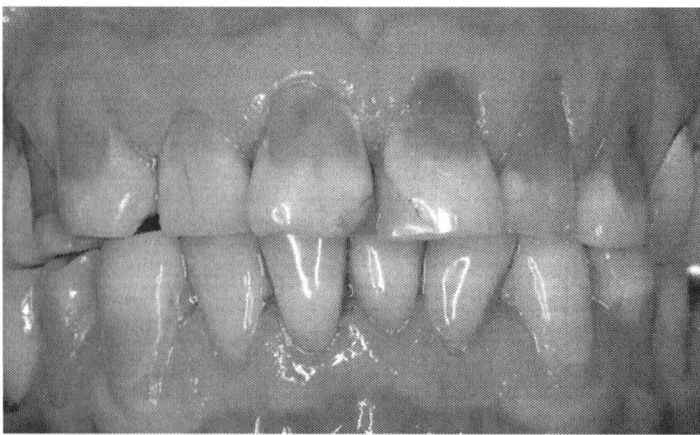

Figure 2.1 Enamel loss attributable to abrasion and erosion, exposing dentine.

Dental erosion is defined as chemical wear as the result of extrinsic or intrinsic acid or chelators acting on plaque-free tooth surfaces.[43] It is characterized by initial softening of the enamel surface and is followed by continuous layer-by-layer dissolution, leading to permanent loss of tooth volume and leaving a softened layer at the surface of the remaining tissue. In advanced stages, dentine becomes increasingly exposed. Erosion is further classified, according to the source of the acid, as either intrinsic or extrinsic.[40] Extrinsic acid exposure is associated with dietary acids, such as citrus fruits, pickled food, fruit juices, carbonated drinks, wines and ciders, and others. Intrinsic acids are associated with eating disorders and mainly comprise gastric acid, which moves to the oral cavity as a result of gastroesophageal reflux,[23] vomiting syndromes (such as bulimia), or from vomiting caused by drugs that act as irritants to the gastric mucosa (Figure 2.3). When erosion is caused by gastric regurgitation, the palatal aspects of the upper incisors and the occlusal and buccal aspects of lower posterior teeth are primarily affected.[40]

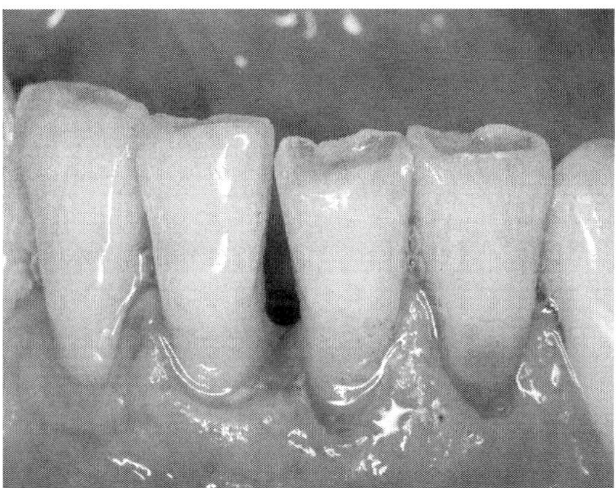

Figure 2.2 Enamel loss attributable to attrition, exposing dentine.

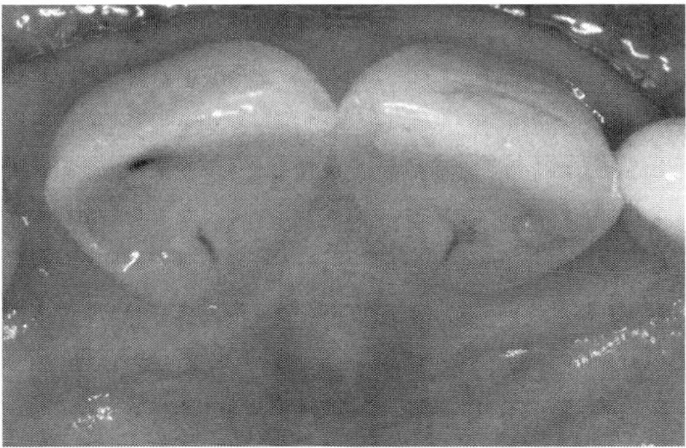

Figure 2.3 Enamel loss on the palatal surfaces as a result of persistent gastroesophageal reflux, exposing dentine.

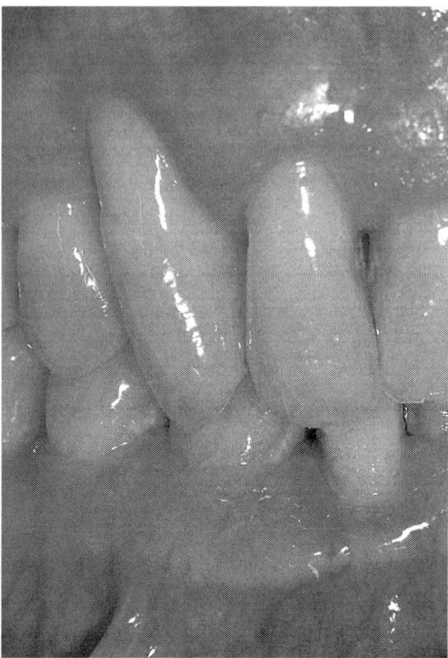

Figure 2.4 Gingival recession with exposed root surfaces.
Source: Acknowledgment to J. Feldmann.

Abfraction is the microstructural loss of tooth substance in areas of stress concentration.[43] This loss occurs most commonly in the cementoenamel region of teeth, where flexure may lead to a breaking away of parts of the thin layer of enamel rods, as well as microfracture of cementum and dentine.[43] Such lesions, when observed on a single tooth or on nonadjacent teeth, are hypothesized to be the result of eccentrically applied occlusal forces (e.g., during grinding, clenching, temporomandibular disorders (TMD), trauma) that lead to tooth flexure rather than to be the result of abrasion alone.[39,44]

The most common cause of exposing radicular dentine is recession of the gingival marginal tissues (Figure 2.4). Clinical evidence indicates that gingival recession accounts for a much greater dentine area of exposure than cervical enamel loss.[12] This process is characterized by the displacement of the gingival margin apical to the cementoenamel junction, thereby exposing visible cementum of the root surface.[36] It allows for rapid and extensive exposure of dentinal tubules, because the cementum layer is thin and easily removed.[45] Factors causing gingival recession are thin alveolar cortex, periodontal disease, and management of the condition, buccal, or lingual dehiscence and fenestration of alveolar bone, trauma, orthodontic therapy, oral piercing, self-inflicted injury, prosthodontic treatment traumatizing the keratinized gingiva, and traumatic toothbrushing. Because more than one of these factors could be present simultaneously, the etiology of gingival recessions appears to be multifactorial.[36,45]

Not all exposed dentine is sensitive.[1,12] To induce DH, a localized lesion has to be initiated in a second step.[12] This initiation occurs when the smear layer or tubular plugs are removed and the outer ends of the dentinal tubules are opened.[33] Abrasive or erosive agents could produce this opening. Evidence suggests that erosion is the more dominant factor in this initiation process, but it can be potentiated by abrasion.[1,36]

The influence of plaque on DH remains controversial. Plaque accumulation on root surfaces may lead to demineralization of tooth structures, which in turn leads to the opening of the dentinal tubules and, therefore, to pain. It has been demonstrated that people with poor plaque control report more problems with DH and vice versa. That is, people who maintain good levels of plaque control are less likely to report DH.[46,47] Conversely, some clinical studies have shown more gingival recession with improper or aggressive oral hygiene practices.[48] The most brushed teeth and, therefore, the ones with the lowest plaque scores exhibited the most gingival recession and the most DH, despite having no plaque present.[48]

Physiological mechanisms of DH

Dentine

Dentine is sensitive because of its anatomy and physiology. It is a porous, mineralized connective tissue with an organic matrix of collagenous proteins and an inorganic component, hydroxyapatite. Dentine is highly permeable, mainly because of the presence of numerous dentinal tubules that extend from the pulp to the dentine–enamel junction and are surrounded by hypermineralized tissue (known as peritubular dentine). The dentinal tubule contains serum-like fluid and an odontoblast cell process.[21]

Dentine exhibits regional differences in tubule diameter and density. It has been reported that dentinal tubule diameter can vary from 2.5 μm at the pulp to 0.9 μm peripherally.[49] Scanning electron microscopic investigations of human dentinal tubules have demonstrated approximately 45,000 tubules per square millimeter at the pulp, 29,500 tubules per square millimeter in the middle dentine, and 20,000 tubules per square millimeter peripherally.[49] Interestingly, odontoblast processes were seen only in the tubules near the pulp.[14,49]

Mechanisms of DH

Although the innervation of pulp and dentine are well-understood, the mechanisms of DH pain remain poorly elucidated. Three theories have been used to explain the mechanisms of DH: the dentinal receptor theory; the odontoblastic transduction theory; and the hydrodynamic theory. All three are intimately related to the anatomy and histology of the dentine–pulpal complex.

The dentinal receptor theory was one of the early hypotheses about the mechanisms of DH. It implied that DH is caused by the direct stimulation of the nerve endings in dentine. However, based on experimental studies, it seems unlikely that neural cells exist in the sensory portion of the outer dentine,[50] discounting this early theory.

The odontoblastic transduction theory proposed by Rapp[51] assumed that odontoblasts extend to the peripheral dentine, with any odontoblastic processes exposed at the dentine surface being susceptible to excitation by chemical and mechanical stimuli.[52,53] However, microscopic experiments failed to confirm this concept. It was observed that the odontoblast process was restricted to the inner third of the dentinal tubules. The outer part was not observed to contain any cellular elements and appeared only to be filled with dentinal fluid. Moreover, no synapses have been demonstrated between odontoblasts and nerve terminals.[54] Thus, the odontoblast transduction mechanism theory lacks supporting evidence.

By far the most widely accepted theory regarding the mechanisms underlying DH is the hydrodynamic theory, which was first proposed by Gysi[55] in 1900 and was validated by Brännstrom and coworkers six decades later.[56,57] The hydrodynamic theory indicates that when an appropriate stimulus (temperature, physical, or osmotic change) is applied to dentine, a change in the rate of fluid movement within the dentinal tubules occurs, creating a pressure change in the dentine and triggering a response in the pulp nerves, ultimately causing pain for the person.[9] Intradental myelinated A-β and some A-δ nerve fibers are thought to respond to stimuli, resulting in the characteristic short, sharp pain of DH.[36] *In vivo* studies have revealed that the response of the pulpal nerves was proportional to the magnitude of the pressure change and, therefore, the rate of fluid flow.[58] It has been demonstrated that stimuli that cause fluid to flow away from the pulp (such as cold) produced more rapid and greater pulp nerve responses than those that caused an inward flow (such as heat).[58] This finding explains the rapid and severe response to cold stimuli compared with the slow, dull response to heat most typically associated with DH. In addition to dentine pressure changes, another process may be involved. When the rate of fluid flow changes in a dentinal tubule, an electrical discharge called "streaming potential" occurs across the dentine. This discharge may be able to stimulate nerves electrically.[59] Nevertheless, to date, the exact mechanism by which the fluid flow stimulates pulpal nerves is not known with certainty. However, an understanding of the hydrodynamic mechanisms of DH provides a basis for developing desensitizing therapies.[12]

"Sensitive" versus "nonsensitive" dentine

When seeking to understand DH, it should be noted that not all exposed dentine is sensitive. The number of open tubules and their diameters are factors likely to distinguish sensitive from nonsensitive dentine. Teeth affected by DH have almost eight times more patent tubules per unit area compared with nonsensitive teeth.[60] Tubule diameter has been found to be two times greater in teeth affected by DH versus nonsensitive teeth.[61] In addition, scanning electron microscopy of replica models of hypersensitive and nonsensitive dentine showed that the smear layer was thinner, different in structure, and likely undercalcified in hypersensitive dentine.[62]

Although the number and diameter of open dentinal tubules are relevant to dentinal tubule fluid flow and, therefore, sensitivity, tubule diameter is likely a more important characteristic because fluid flow is proportional to the fourth power of the radius (hence, if the diameter is doubled, then the tubule fluid flow increases by 16-fold).[63] This relationship explains why tubular occlusion, of whatever nature, may reduce DH pain.[36]

Pain

The pain of DH is classically short, sharp, of rapid onset, and of duration equal to that of the applied stimuli. The pain-producing stimuli can be thermal, tactile, osmotic, chemical, or evaporative,[9] but cold stimulus has been reported to cause the greatest problems for people affected by DH.[30] Cold stimulus increases the outward flow of fluid in the dentinal tubules, whereas a thermal (hot) stimulus produces contraction of the fluid in the tubules.[58]

Nevertheless, pain has extremely variable characteristics, ranging from discrete discomfort to extreme severity.[13,25,26] Not all people experiencing tooth sensitivity seek treatment to desensitize their teeth, because some do not perceive DH to be a severe oral health problem.[27] The level of pain associated with DH varies among different teeth[30,34] and different individuals.[17] Perception of, expression of, and reaction to pain are influenced by genetic, developmental, familial, psychological, social, and cultural variables. Psychological factors, such as the situational and emotional factors that exist when pain is experienced, can profoundly alter the strength of these perceptions. Some people are concerned whenever they experience dental pain; for some, the first time experiencing DH creates the fear that something more serious is occurring.[26] However, the influence of pain can be considerable and can affect daily activities depending on the source of discomfort [such as pain associated with consuming hot or cold foods and beverages (coffee, ice cream), during toothbrushing, or sometimes even while breathing].[28,29] In these instances, these symptoms are highly relevant from the person's point of view and often have a considerable adverse impact on daily quality of life.[64] Accordingly, DH has been reported to have a negative impact on people's oral health-related quality of life,[3] although only a few studies have been performed to date.[65,66] People with DH experience more impacts on oral health-related quality of life and have poorer oral health than the general population.[3]

Summary

In dental practice, DH is a commonly presenting condition. It may disturb people during eating, drinking, and oral hygiene habits. The etiology of DH, which is directly connected with dentine exposure, is multifactorial; however, interactions between several factors, including stimuli as well as predisposing factors, may play an important role in initiating this condition. The most current theory regarding the physiological mechanism responsible for the pain associated with DH is the hydrodynamic theory. This theory suggests that fluids within the dentinal tubules become disturbed by temperature, physical, or osmotic changes, subsequently triggering a response in the pulp nerves that leads to a neural pain signal. The pain experienced by people with DH can cause such discomfort that it interferes with daily activities.

References

1. Mantzourani M, Sharma D. Dentine sensitivity: past, present and future. *J Dent* 2013;**41** (Suppl. 4):S3–17.
2. Bartold PM. Dentinal hypersensitivity: a review. *Aust Dent J* 2006;**51**:212–8 [quiz 76].
3. Bekes K, John MT, Schaller HG, Hirsch C. Oral health-related quality of life in patients seeking care for dentin hypersensitivity. *J Oral Rehabil* 2009;**36**:45–51.
4. Nuttall N, Steele JG, Nunn J, Pine C, Treasure E, Bradnock G, et al. *A guide to the UK Adult Dental Health Survey 1998*. London: British Dental Association; 2001.
5. Gillam D, Chesters R, Attrill D, Brunton P, Slater M, Strand P, et al. Dentine hypersensitivity—guidelines for the management of a common oral health problem. *Dent Update* 2013;**40**:514–6 18–20, 23–4
6. Martens LC. A decision tree for the management of exposed cervical dentin (ECD) and dentin hypersensitivity (DHS). *Clin Oral Investig* 2013;**17**(Suppl. 1):S77–83.
7. Schmidlin PR, Sahrmann P. Current management of dentin hypersensitivity. *Clin Oral Investig* 2013;**17**(Suppl. 1):S55–9.
8. Addy M. Etiology and clinical implications of dentine hypersensitivity. *Dent Clin North Am* 1990;**34**:503–14.
9. Addy M, Smith SR. Dentin hypersensitivity: an overview on which to base tubule occlusion as a management concept. *J Clin Dent* 2010;**21**:25–30.
10. Canadian Advisory Board on Dentin Hypersensitivity. Consensus-based recommendations for the diagnosis and management of dentin hypersensitivity. *J Can Dent Assoc (Tor)* 2003;**69**:221–6.
11. Trowbridge HO. Mechanism of pain induction in hypersensitive teeth. In: Rowe NH, editor. *Proceedings of symposium on hypersensitive dentine origin and management*. Ann Arbor, MI: University of Michigan, School of Dentistry; 1985. p. 1–6.
12. Orchardson R, Gillam DG. Managing dentin hypersensitivity. *J Am Dent Assoc* 2006;**137**:990–8 [quiz 1028–9].
13. Porto IC, Andrade AK, Montes MA. Diagnosis and treatment of dentinal hypersensitivity. *J Oral Sci* 2009;**51**:323–32.
14. Dowell P, Addy M, Dummer P. Dentine hypersensitivity: aetiology, differential diagnosis and management. *Br Dent J* 1985;**158**:92–6.
15. West NX. Dentine hypersensitivity. In: Lussi A, editor. *Monographs in oral science: dental erosion*. Basel: Karger; 2006. p. 173–89.
16. Gillam DG. Current diagnosis of dentin hypersensitivity in the dental office: an overview. *Clin Oral Investig* 2013;**17**(Suppl. 1):S21–9.
17. Splieth CH, Tachou A. Epidemiology of dentin hypersensitivity. *Clin Oral Investig* 2013;**17**(Suppl. 1):S3–8.
18. Bamise CT, Olusile AO, Oginni AO, Dosumu OO. The prevalence of dentine hypersensitivity among adult patients attending a Nigerian teaching hospital. *Oral Health Prev Dent* 2007;**5**:49–53.
19. Chabanski MB, Gillam DG, Bulman JS, Newman HN. Clinical evaluation of cervical dentine sensitivity in a population of patients referred to a specialist periodontology department: a pilot study. *J Oral Rehabil* 1997;**24**:666–72.
20. West NX, Sanz M, Lussi A, Bartlett D, Bouchard P, Bourgeois D. Prevalence of dentine hypersensitivity and study of associated factors: a European population-based cross-sectional study. *J Dent* 2013;**41**:841–51.
21. Shiau HJ. Dentin hypersensitivity. *J Evid Based Dent Pract* 2012;**12**:220–8.

22. Chabanski MB, Gillam DG, Bulman JS, Newman HN. Prevalence of cervical dentine sensitivity in a population of patients referred to a specialist Periodontology Department. *J Clin Periodontol* 1996;**23**:989–92.
23. Bamise CT, Kolawole KA, Oloyede EO, Esan TA. Tooth sensitivity experience among residential university students. *Int J Dent Hyg* 2010;**8**:95–100.
24. Micheelis W, Bauch J. Oral health of representative samples of Germans examined in 1989 and 1992. *Community Dent Oral Epidemiol* 1996;**24**:62–7.
25. McGrath PA. The measurement of human pain. *Endod Dent Traumatol* 1986;**2**:124–9.
26. McGrath PA. Psychological aspects of pain perception. *Arch Oral Biol* 1994;**39** (Suppl.):55S–62S.
27. Gillam DG, Seo HS, Bulman JS, Newman HN. Perceptions of dentine hypersensitivity in a general practice population. *J Oral Rehabil* 1999;**26**:710–4.
28. Gillam DG, Bulman JS, Eijkman MA, Newman HN. Dentists' perceptions of dentine hypersensitivity and knowledge of its treatment. *J Oral Rehabil* 2002;**29**:219–25.
29. Schuurs AH, Wesselink PR, Eijkman MA, Duivenvoorden HJ. Dentists' views on cervical hypersensitivity and their knowledge of its treatment. *Endod Dent Traumatol* 1995;**11**:240–4.
30. Orchardson R, Collins WJ. Clinical features of hypersensitive teeth. *Br Dent J* 1987;**162**:253–6.
31. Amarasena N, Spencer J, Ou Y, Brennan D. Dentine hypersensitivity in a private practice patient population in Australia. *J Oral Rehabil* 2011;**38**:52–60.
32. Que K, Ruan J, Fan X, Liang X, Hu D. A multi-centre and cross-sectional study of dentine hypersensitivity in China. *J Clin Periodontol* 2010;**37**:631–7.
33. Dababneh RH, Khouri AT, Addy M. Dentine hypersensitivity—an enigma? A review of terminology, mechanisms, aetiology and management. *Br Dent J* 1999;**187**:606–11 [discussion 03].
34. Oyama T, Matsumoto K. A clinical and morphological study of cervical hypersensitivity. *J Endod* 1991;**17**:500–2.
35. Chrysanthakopoulos NA. Prevalence of dentine hypersensitivity in a general dental practice in Greece. *J Clin Exp Dent* 2011;**3**:e445–51.
36. West NX, Lussi A, Seong J, Hellwig E. Dentin hypersensitivity: pain mechanisms and aetiology of exposed cervical dentin. *Clin Oral Investig* 2013;**17**(Suppl. 1):S9–19.
37. Addy M. Dentine hypersensitivity: definition, prevalence distribution and aetiology. In: Addy M, Embery G, Edgar WM, Orchardson R, editors. *Tooth wear and sensitivity: clinical advances in restorative dentistry*. London: Martin Dunitz; 2000. p. 239–48.
38. Barbour ME, Rees GD. The role of erosion, abrasion and attrition in tooth wear. *J Clin Dent* 2006;**17**:88–93.
39. Bartlett D, Smith BGN. Definition, classification and clinical assessment of attrition, erosion and abrasion of enamel and dentine. In: Addy M, Embery G, Edgar WM, Orchardson R, editors. *Tooth wear and sensitivity: clinical advances in restorative dentistry*. London: Taylor and Francis; 2000. p. 87–92.
40. Bartlett DW, Shah P. A critical review of non-carious cervical (wear) lesions and the role of abfraction, erosion, and abrasion. *J Dent Res* 2006;**85**:306–12.
41. West NX, Hooper SM, O'Sullivan D, Hughes N, North M, Macdonald EL, et al. In situ randomised trial investigating abrasive effects of two desensitising toothpastes on dentine with acidic challenge prior to brushing. *J Dent* 2012;**40**:77–85.
42. Adams D, Addy M, Absi EG. Abrasive and chemical effects of dentifrices. In: Embry G, Rolla G, editors. *Clinical and biological aspects of dentifrices*. Oxford: Oxford University Press; 1992. p. 345–55.

43. Ganss C. Definition of erosion and links to tooth wear. In: Lussi A, editor. *Monographs in oral science: dental erosion*. Basel: Karger; 2006. p. 9–16.
44. Grippo JO. Abfractions: a new classification of hard tissue lesions of teeth. *J Esthet Dent* 1991;**3**:14–9.
45. Smith RG. Gingival recession. Reappraisal of an enigmatic condition and a new index for monitoring. *J Clin Periodontol* 1997;**24**:201–5.
46. Hiatt WH, Johansen E. Root preparation. I. Obturation of dentinal tubules in treatment of root hypersensitivity. *J Periodontol* 1972;**43**:373–80.
47. Trowbridge HO, Silver DR. A review of current approaches to in-office management of tooth hypersensitivity. *Dent Clin North Am* 1990;**34**:561–81.
48. Addy M, Mostafa P, Newcombe RG. Dentine hypersensitivity: the distribution of recession, sensitivity and plaque. *J Dent* 1987;**15**:242–8.
49. Garberoglio R, Brännstrom M. Scanning electron microscopic investigation of human dentinal tubules. *Arch Oral Biol* 1976;**21**:355–62.
50. Irvine JH. Root surface sensitivity: a review of aetiology and management. *J N Z Soc Periodontol* 1988;15–8.
51. Rapp R, Avery JK, Strachen DS. Possible role of the acetylcholinesterase in neural conduction within the dental pulp. In: Finn SB, editor. *Biology of dental pulp organ*. Birmingham: University Alabama Press; 1968. p. 309.
52. Bernick S. Innervation of the human tooth. *Anat Rec* 1948;**101**:81–107.
53. Frank RM. Attachment sites between the odontoblast process and the intradentinal nerve fibre. *Arch Oral Biol* 1968;**13**:833–4.
54. Pashley DH. Dynamics of the pulpo-dentin complex. *Crit Rev Oral Biol Med* 1996;**7**:104–33.
55. Gysi A. An attempt to explain the sensitiveness of dentin. *Br J Dent Sci* 1900;**43**:865–8.
56. Brännstrom M. A hydrodynamic mechanism in the transmission of pain producing stimuli through the dentine. In: Anderson DJ, editor. *Sensory mechanisms in dentine*. Oxford: Pergamon Press; 1962. p. 12–9.
57. Brännstrom M. Sensitivity of dentine. *Oral Surg Oral Med Oral Pathol* 1966;**21**:517–26.
58. Matthews B, Vongsavan N. Interactions between neural and hydrodynamic mechanisms in dentine and pulp. *Arch Oral Biol* 1994;**39**(Suppl.):87S–95S.
59. Mumford JM, Newton AV. Transduction of hydrostatic pressure to electric potential in human dentin. *J Dent Res* 1969;**48**:226–9.
60. Absi EG, Addy M, Adams D. Dentine hypersensitivity. A study of the patency of dentinal tubules in sensitive and non-sensitive cervical dentine. *J Clin Periodontol* 1987;**14**:280–4.
61. Pashley DH. Mechanisms of dentin sensitivity. *Dent Clin North Am* 1990;**34**:449–73.
62. Rimondini L, Baroni C, Carrassi A. Ultrastructure of hypersensitive and non-sensitive dentine. a study on replica models. *J Clin Periodontol* 1995;**22**:899–902.
63. Guyton A, Hall JE. *Textbook of medical physiology*. 12th ed. Philadelphia, London: W B Saunders; 2010.
64. Locker D. Measuring oral health: a conceptual framework. *Community Dent Health* 1988;**5**:3–18.
65. Boiko OV, Baker SR, Gibson BJ, Locker D, Sufi F, Barlow AP, et al. Construction and validation of the quality of life measure for dentine hypersensitivity (DHEQ). *J Clin Periodontol* 2010;**37**:973–80.
66. Bekes K, Hirsch C. What is known about the influence of dentine hypersensitivity on oral health-related quality of life? *Clin Oral Investig* 2013;**17**(Suppl. 1):S45–51.

The burden of dentine hypersensitivity

Joana Cunha-Cruz[1] *and John C. Wataha*[2]
[1]Department of Oral Health Sciences, School of Dentistry, University of Washington, Seattle, WA, [2]Department of Restorative Dentistry, School of Dentistry, University of Washington, Seattle, WA

Introduction

Dentine hypersensitivity is a significant problem for patients and practitioners alike. For patients, the chronic pain of this condition may lead to changes in diet, speaking, and social well-being (see Chapters 6 and 7). Furthermore, the condition fills most practitioners with uncertainty about diagnosis, about treatment, and about how to best help their patients (Chapter 4). The key for both groups lies in understanding how to accurately diagnose dentine hypersensitivity and its true prevalence, both of which are the focuses of this chapter.

Diagnosis of dentine hypersensitivity

Dentine hypersensitivity is diagnosed by a person's self-report of pain, the evaluation of the person's response to a stimulus, and by excluding other dental and periodontal conditions.[1] Diagnosis also may include a subjective evaluation of the impact of sensitivity on daily life, as well as pain intensity, severity, and frequency.

A self-report of pain may be elicited or spontaneous, that is, a person is queried about pain in the mouth or spontaneously reports pain during a dental office visit. Several questions can be used to enquire about dentine hypersensitivity, ranging from general questions such as "Do you have pain or discomfort in your mouth, teeth, or gums?" to specific questions such as "Do you have sensitive teeth?" (Table 3.1). Other options include asking about pain during routine activities (eating, drinking cold liquids). When using a nonspecific question such as "Have you recently had any pain, sensitivity, or discomfort in your teeth or gums," a dentist would rely on the patient's own pain threshold as a trigger for a positive response. A spontaneous report of a self-perceived problem usually indicates a significant problem.

Because dentine hypersensitivity symptoms are often intermittent, the time frame of the question becomes important. Studies have shown that use of a specific time frame when enquiring will influence the person's response, and thus the apparent prevalence of dentine hypersensitivity.

Dentine Hypersensitivity. DOI: http://dx.doi.org/10.1016/B978-0-12-801631-2.00003-8

Table 3.1 Examples of questions for self-reporting of dentine hypersensitivity

Do you have...	...any pain, sensitivity or discomfort in your teeth or gums?
Have you recently had...	...pain or discomfort when eating?
During the last month, have you experienced...	...pain or discomfort when drinking cold liquids?
Have you ever experienced...	...pain or discomfort when brushing your teeth?
	...sensitive teeth?

In a survey of general dentists in the northwestern region of the United States,[2] dentists reported that they relied primarily on patient reports, either spontaneous or via query, to identify patients with dentine hypersensitivity. The most reported diagnostic method was a spontaneous patient report (97% of practitioners), which was frequently used by 70% of practitioners. Patient reports after query were also common (96%) but not as frequently used (38%) as spontaneous reports (Figure 3.1).

By definition, dentine hypersensitivity should be elicited by a stimulus. With a positive response to the self-report of sensitivity, a clinical examination generally follows to identify the sensitive teeth and to exclude other causes of pain to reach a formal diagnosis. The clinician stimulates suspected teeth to try to trigger pain and record the patient's reaction. Different stimuli can be applied, including thermoevaporative (air blast), thermal (cold water), tactile or mechanical (force-sensitive probe, explorer, or curette), and osmotic (glucose and sodium chloride). In both research studies and clinical practice, the most common stimuli are thermoevaporative and tactile. For the thermoevaporative stimulus, the clinician applies a 1-second cold air blast from an air–water syringe to the sensitive tooth from a distance of 1 cm away from the vestibular or buccal surface while covering the adjacent teeth with gloved fingers. Whereas dentists in clinical practice more frequently use an explorer or curette as a tactile stimulus,[2] investigators in research studies use a force-sensitive probe (e.g., Jay sensitivity sensor probe and the Yeaple electronic force-sensing probe).[3–5] The clinician places the force-sensitive probe on each tooth perpendicular to the vestibular or buccal surface and the pressure is gradually increased until the pain threshold is reached or a preestablished maximum force is applied.

General dentists in the northwestern United States used a variety of means to clinically confirm and assess dentine hypersensitivity[2] (Figure 3.1). Explorer and air blast (90%) usage was the most common means of clinically, but cold water (83%) or a curette (67%) also were common. Less common methods of assessment were the electric pulp tester (37%), application of glucose or sodium chloride (22% and 9%, respectively), or use of a force-sensitive electronic probe (11%). Although these methods were diverse, most practitioners relied on mechanical, thermal, or osmotic stimulation, suggesting that they are aware of and subscribe to the current state of knowledge on hydrodynamic mechanisms[6] (Chapter 2). An exception was

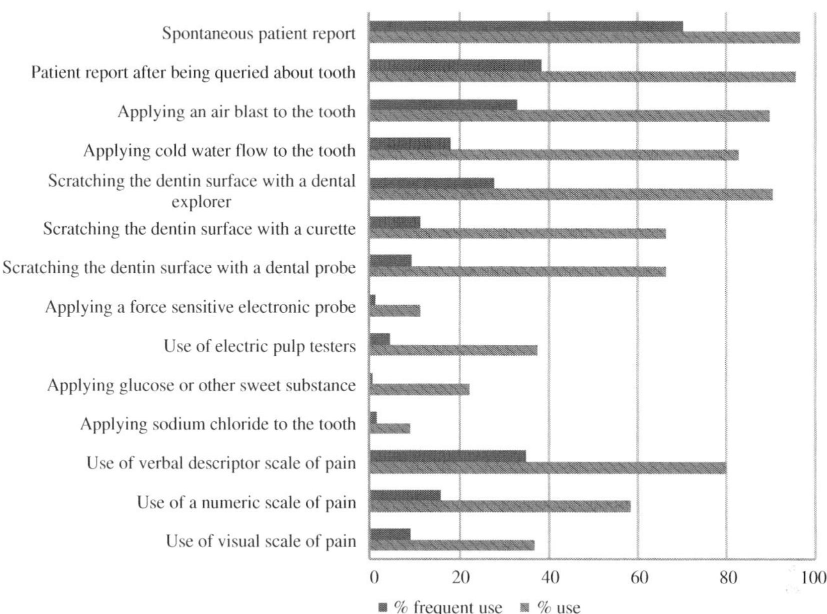

Figure 3.1 Methods to diagnose dentine hypersensitivity used by general dentists ($n = 209$) in the northwestern United States, 2009.[2]

the use of the electric pulp tester, which was used by nearly 40% of respondents, but was used frequently by fewer than 5% of dentists.

A subjective response to tooth stimulation may be measured by self-report or indirectly by an observer. After application of the stimulus, the person is asked to rate the pain felt during the application. Several pain questionnaires and scales have been used to assess the quality and severity of pain experienced, such as the McGill pain questionnaire, visual analog scales (VAS), numerical rating scales, verbal description scales, faces pain scales, and labeled magnitude scales[7] (Table 3.2). Pain-rating labeled magnitude scales have been developed specifically for dentine hypersensitivity (see Chapter 13).[8]

An observer can also judge the person's pain-related behavior. Specific semisubjective scales have been developed for dentine hypersensitivity using categorical ordinal scales, such as the Schiff's scale[9] and Cumulative Hypersensitivity Index (CHI), a novel Schiff Index sextant cumulative score.[10] For the Schiff's scale, a single examiner isolates each tooth individually using a gloved finger and applies the stimulus sequentially to each tooth surface. An observer records based on his or her judgment and on a scale of 0–3, the person's pain reaction to the stimulus, including whether the person requested discontinuation of the stimulus and whether the person expressed pain-related behaviors. The CHI is calculated as the total cumulative score of the highest Schiff's score in each sextant.

Table 3.2 Description of pain scales used for the assessment of dentine hypersensitivity

Pain scale	Description
Visual analog scale	Continuous scale comprised of a horizontal or vertical line, usually 10 cm (100 mm) in length, anchored by two verbal descriptors, one for each symptom extreme (e.g., "no pain" and "pain as bad as it could be" or "worst imaginable pain"). The distance measured from the "No pain" end to the patient's mark is the score.
Numerical rating scale	Segmented numeric version of the VAS comprised of a bar or line numbered with integers (0–10 or 1–10), anchored by two verbal descriptors, one for each symptom extreme.
Verbal description scale	Ordered categorical scale commonly of four to seven categories, with each response option consisting of adjectives (e.g., "no pain," "mild pain," "moderate pain," "severe pain," "extreme pain," and the "most intense pain imaginable"). The adjectives are scored by assigning numbers (e.g., 0–6) to each response option.
Face interval scale	Graphical scale commonly of six or seven expressions of faces increasing in pain intensity at approximately equal intervals. The faces are scored by assigning numbers (e.g., 0–5) to each response option.
Labeled magnitude scale	Semantically labeled scale consisting of a bar or line labeled with verbal descriptors spaced according to their semantic magnitude (e.g., "no pain", "dim", "dull", "sharp", "stabbing"). The distance measured from the "no pain" end to the patient's mark is the score.

General dentists in the northwestern United States[2] frequently requested verbal (80%) or numeric (58%) descriptions of dentine hypersensitivity pain or asked patients to rate the pain level on a visual scale of some type (37%) (Figure 3.1). In addition to providing an indication of the severity and quality of pain in dentine hypersensitivity, the pain scales and the CHI (cumulative Schiff Index score per subject) also are used as guides to clinical management and improvement over time.

Because dentine hypersensitivity is a diagnosis of exclusion,[1] during the clinical examination other conditions that could also cause the pain or discomfort must be ruled out. These conditions include dental caries, pulpitis, fractured teeth, fractured restorations, postrestorative sensitivity, marginal leakage, chipped teeth, and gingival inflammation[11] (Chapter 4).

The assessment of dentine hypersensitivity may also consider the impact of pain on daily life. Oral health-related quality of life questionnaires such as the short form of the Oral Health Impacts Profile (OHIP-14)[12] and Oral Impacts on Daily Performances (OIDP)[13] can be used, but they do not discriminate well between persons with dentine hypersensitivity and the general population. For this reason, the Dentine Hypersensitivity Experience Questionnaire (DHEQ) has been developed specifically to capture the daily experiences of people with dentine hypersensitivity and its impacts on different dimensions of life, such as eating, drinking, talking, toothbrushing, and social interaction[14] (Chapters 7, 9, and 10).

Prevalence of dentine hypersensitivity

Knowledge of the prevalence of a condition is used to guide diagnosis under the maxim "common things commonly occur." Therefore, uncertainty about the prevalence of dentine hypersensitivity has significant consequences for patients and dental practitioners. With vague knowledge about prevalence comes uncertainty about diagnosis, the appropriate time to treat, and how aggressive treatment should be. These difficulties are exacerbated by uncertainty over factors associated with hypersensitivity that often make diagnosis elusive. Development of new treatments, assessing their effectiveness, choosing the appropriate treatment, and understanding the mechanisms causing hypersensitivity all ultimately depend on a clear understanding of the prevalence of the condition. We therefore systematically reviewed prevalence studies of dentine hypersensitivity to ascertain a precise estimate of its frequency and the factors that might influence its apparent prevalence.

A systematic search identified 44 published studies reporting on 56 cross-sectional studies of the prevalence of dentine hypersensitivity in different populations[15–58]. Most studies were conducted in Europe (43%) and Asia (30%) during or after the year 2000 (66%). Studies were funded by universities, governments, and other nonprofit organizations (30%), industries (21%), or funding was not reported (48%). Participants were patients from general dental practices (either at university clinics or private practices) (43%) or volunteers from the general population (38%). Two-thirds of studies included only adults (64%); only adolescents and young adults (23%) and the remaining studies included adults and adolescents with or without children (13%). Participants were mostly sampled in one site (63%) out of convenience for the investigators (43%), but more appropriate methods of selection such as random (36%) and consecutive (21%) samples also were used.

Investigators in the majority of the studies reported only the study sample actually enrolled in the study (70%) without mentioning the number of people in the eligible population (intended sample) or the target (external) population (to which results may be generalized). Of the 17 studies (30%) that reported some type of nonresponse rate, 8 exceeded the widely accepted rate of less than 20% of nonrespondents. This lack of reporting of the eligible and target populations (and nonresponse rates) prevented an assessment of selection bias attributable to sampling or nonparticipation, giving rise to external validity uncertainties of the studies. Thus, in most studies the prevalence of dentine hypersensitivity cannot be considered representative of the population.

To diagnose dentine hypersensitivity, more than half of published studies have relied on self-reports from questionnaires and response to stimulation during clinical examination (55%), and the remaining relied on questionnaire only (36%) or clinical examination only (9%). Whereas 40% of the studies relying on self-reports did not mention the questions used, questions that were used to assess dentine hypersensitivity ranged from nonspecific questions such as presence of pain or discomfort to more specific questions mentioning the presence of sensitive teeth. Of the studies using both self-reports and clinical examinations (55%), 35% conducted the clinical examinations to assess response to stimulation only on a positive response to the questionnaire. The majority of studies used only a thermoevaporative stimulus (air blast) to assess dentine hypersensitivity (59%). A tactile stimulus using a sharp probe ($n = 3$), dental explorer ($n = 2$), or a Yeaple probe ($n = 1$) was also used alone (3%) or in combination with a thermoevaporative stimulus (19%) or a thermal stimulus (3%). Diagnosis was confirmed by excluding other causes of pain or sensitivity during clinical examination in 55% of the studies. Measurement bias may have been present in several studies, particularly those that did not exclude other causes of pain or sensitivity (45%) or examined all teeth through stimulation regardless of the participant reports of pain or sensitivity (45% of all studies or 69% of clinical examination studies).

The 56 studies included 73,669 participants.[15–58] The median number of participants was 734 (range: 51–12,692). The reported prevalence of dentine hypersensitivity varied from 1.3% to 84%. Fixed-effects and random-effects meta-analyses were conducted by combining the results of all studies (ignoring the clustering of studies within a publication). When studies reported the prevalence of dentine hypersensitivity using different methods, one prevalence estimate from each study was selected, giving preference to those based on clinical examinations rather than questionnaires and, among the former, those based on thermoevaporative stimulus to elicit pain (Figure 3.2).

Fixed-effects meta-analysis resulted in a summary estimate of the prevalence of dentine hypersensitivity of 9.9% [95% confidence interval (CI): 9.7–10.1%]. The random-effects meta-analysis resulted in a summary estimate of 33% (95% CI: 29.3–36.6). The lower summary prevalence from the fixed-effects model can be interpreted as the "best estimate" or the "best guess" for the prevalence in the absence of heterogeneity, whereas the higher summary prevalence from the random-effects model can be interpreted as the average prevalence from all studies.[59] Because there was a high degree of heterogeneity ($I^2 = 99.7\%$), we calculated the predictive interval for a future study and further explored the heterogeneity using meta-regression. The predictive interval applicable for a future study based on past experience was wide, ranging from 5% to 61%. This interval includes the possibility that a new study would observe a prevalence of dentine hypersensitivity as low as 5%, and also the possibility that a prevalence much greater than the expected average of 33% would be observed.

Random-effects meta-regression models with study characteristics as covariates were fitted to understand the impact of such characteristics as modifiers of the study effect size. No statistically significant differences in prevalence were observed by

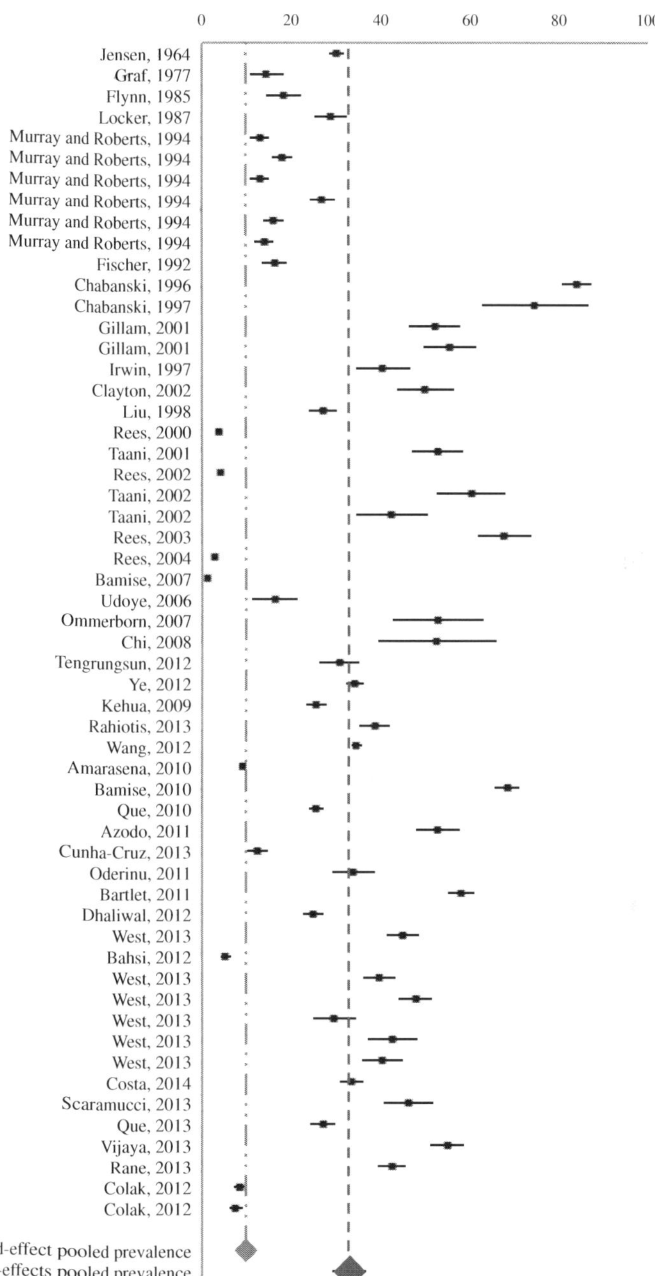

Figure 3.2 Forest plot of studies on prevalence of dentine hypersensitivity.[15–58] Prevalence (95% CI) for individual studies and fixed-effects (narrow dash) and random-effects (wide dash) summary prevalence. (For interpretation of the references to color in this figure legend, the reader is referred to the web version of this book.)

study decades, continents, multiple-site or single-site studies, funding sources, or nonresponse rates. Surprisingly, characteristics of the diagnostic methods used did not explain the variability in prevalence estimates among the studies. There were no differences in prevalences among studies when compared by the method of diagnosis (self-reports versus clinical examination), the method of stimulation during clinical examination (none versus thermoevaporative versus other), or performance of a clinical examination to exclude other causes of sensitivity. A tentative explanation for this finding is the wide variability in both methods of measurement of dentine hypersensitivity. The methods for querying participants varied from specific questions regarding "sensitive teeth" to general questions regarding "pain or discomfort in teeth," and the methods for clinical examination varied from eliciting pain in all teeth present to eliciting pain only after a positive self-report.

Effect modifiers that were significant included the source and age of participants and recruitment strategy. Studies among specialty practice patients (mean prevalence: 61%; 95% CI: 31–91) and specific subgroups of the general population (mean: 45%; 95% CI: 19–72) had higher prevalences than general practice (mean: 27%; 95% CI: 23–30) and general population (mean: 30%; 95% CI: 26–34) studies. Studies including only young adult patients (mean: 44%; 95% CI: 30–58) reported a higher prevalence of dentine hypersensitivity than those including other age groups, such as children and older adults (mean: 30%; 95% CI: 26–33). Finally, studies that recruited participants using a consecutive sample of patients (11%; 95% CI: 8–13) had lower prevalences than those using random (29%; 95% CI: 24–33) or convenience (33%; 95% CI: 29–37) sampling methods.

The systematic review of the studies[15–58] reporting the prevalence of dentine hypersensitivity has identified limitations in the published literature, particularly in the recruitment strategy, measurement process, and analyses. When reporting the recruitment strategy, reference to the target population to which the investigators intend to generalize the results should be provided, in addition to the number of potential participants who were approached to participate but declined. A comparison of demographic characteristics among all these groups should be routinely included. Most important, the sampling strategy should use a systematic (e.g., consecutive) or random sampling scheme, not one of convenience. Although the systematic review and meta-regression did not identify the method of diagnosis as an influential factor in the variation of prevalence estimates among studies, the reporting of these methods of data collection can be improved in future studies by stating the question used for the self-reports and describing in detail the clinical examination (e.g., number of teeth and surfaces tested, type, amount of time, distance and force used for the stimulus, protection of adjacent teeth, and assessment of the participant's response to the stimulus). When analyzing the results of prevalence studies and calculating sample sizes, the clustering of participants with dentine hypersensitivity within communities or practices should be taken into account through the use of appropriate statistical methods. Overall, research in this area would benefit from adherence to the STROBE statement on STrengthening the Reporting of Observational studies in Epidemiology. STROBE is part of the

EQUATOR network, an international collaboration of researchers who have provided checklists to facilitate the quality of the conduct and reporting of different study designs.[60]

In conclusion, despite many studies over several decades, we know relatively little about the prevalence of dentine hypersensitivity and its impact on daily life. Considering all available data, the best estimate of dentine hypersensitivity based on the published studies is approximately 10%, and the average from all studies is 33%. The extremely high degree of heterogeneity among studies can be only partially explained by the study characteristics; a new prevalence study could expect to find a prevalence of dentine hypersensitivity from 5% to 61%. To fully understand the burden of dentine hypersensitivity on patients and clinicians, methodological improvements are necessary. Most importantly, the impact of dentine hypersensitivity on daily life should be more often formally assessed because the mere response to stimulation during a clinical examination may not warrant a need for treatment if it is not a significant problem or concern for the patient.

Acknowledgment

The authors thank Rashmi Malhotra for helping with the study search and data extraction.

References

1. Holland GR, Narhi MN, Addy M, Gangarosa L, Orchardson R. Guidelines for the design and conduct of clinical trials on dentine hypersensitivity. *J Clin Periodontol* 1997;**24** (11):808–13.
2. Cunha-Cruz J, Wataha JC, Zhou L, Manning W, Trantow M, Bettendorf MM, et al. Treating dentin hypersensitivity: therapeutic choices made by dentists of the northwest PRECEDENT network. *J Am Dent Assoc* 2010;**141**(9):1097–105.
3. Gillam DG, Bulman JS, Newman HN. A pilot assessment of alternative methods of quantifying dental pain with particular reference to dentine hypersensitivity. *Community Dent Health* 1997;**14**(2):92–6.
4. Kakar A, Kakar K. Measurement of dentin hypersensitivity with the Jay Sensitivity Sensor Probe and the Yeaple probe to compare relief from dentin hypersensitivity by dentifrices. *Am J Dent* 2013;**26 Spec No B**:21B–8B.
5. Sowinski JA, Kakar A, Kakar K. Clinical evaluation of the Jay Sensitivity Sensor Probe: a new microprocessor-controlled instrument to evaluate dentin hypersensitivity. *Am J Dent* 2013;**26 Spec No B**:5B–12B.
6. Dowell P, Addy M. Dentin hypersensitivity—a review. 1. Etiology, symptoms and theories of pain production. *J Clin Periodontol* 1983;**10**(4):341–50.
7. Hjermstad MJ, Fayers PM, Haugen DF, Caraceni A, Hanks GW, Loge JH, et al. Studies comparing numerical rating scales, verbal rating scales, and visual analogue scales for assessment of pain intensity in adults: a systematic literature review. *J Pain Symptom Manage* 2011;**41**(6):1073–93.

8. Heaton LJ, Barlow AP, Coldwell SE. Development of labeled magnitude scales for the assessment of pain of dentin hypersensitivity. *J Orofac Pain* 2013;**27**(1):72–81.
9. Schiff T, Dotson M, Cohen S, De Vizio W, McCool J, Volpe A. Efficacy of a dentifrice containing potassium nitrate, soluble pyrophosphate, PVM/MA copolymer, and sodium fluoride on dentinal hypersensitivity: a twelve-week clinical study. *J Clin Dent* 1994;**5 Spec No**:87–92.
10. Olley RC, Wilson R, Moazzez R, Bartlett D. Validation of a cumulative hypersensitivity index (CHI) for dentine hypersensitivity severity. *J Clin Periodontol* 2013;**40** (10):942–7.
11. Splieth CH, Tachou A. Epidemiology of dentin hypersensitivity. *Clin Oral Investig* 2013;**17**(Suppl. 1):S3–8.
12. Slade GD. Derivation and validation of a short-form oral health impact profile. *Community Dent Oral Epidemiol* 1997;**25**(4):284–90.
13. Adulyanon S, Vourapukjaru J, Sheiham A. Oral impacts affecting daily performance in a low dental disease Thai population. *Community Dent Oral Epidemiol* 1996;**24**(6):385–9.
14. Boiko OV, Baker SR, Gibson BJ, Locker D, Sufi F, Barlow AP, et al. Construction and validation of the quality of life measure for dentine hypersensitivity (DHEQ). *J Clin Periodontol* 2010;**37**(11):973–80.
15. Jensen AL. Hypersensitivity controlled by iontophoresis: double blind clinical investigation. *J Am Dent Assoc* 1964;**68**:216–25.
16. Graf H, Galasse R. Morbidity, prevalence and intraoral distribution of hypersensitive teeth. *J Dent Res* 1977;**162**:479.
17. Flynn J, Galloway R, Orchardson R. The incidence of 'hypersensitive' teeth in the West of Scotland. *J Dent* 1985;**13**(3):230–6.
18. Locker D, Grushka M. Prevalence of oral and facial pain and discomfort: preliminary results of a mail survey. *Community Dent Oral Epidemiol* 1987;**15**(3):169–72.
19. Fischer C, Fischer RG, Wennberg A. Prevalence and distribution of cervical dentine hypersensitivity in a population in Rio de Janeiro, Brazil. *J Dent* 1992;**20**(5):272–6.
20. Murray LE, Roberts AJ. The prevalence of self-reported hypersensitive teeth. *Arch Oral Biol* 1994;**39**(Suppl.):129S.
21. Chabanski MB, Gillam DG, Bulman JS, Newman HN. Prevalence of cervical dentine sensitivity in a population of patients referred to a specialist Periodontology Department. *J Clin Periodontol* 1996;**23**(11):989–92.
22. Chabanski MB, Gillam DG, Bulman JS, Newman HN. Clinical evaluation of cervical dentine sensitivity in a population of patients referred to a specialist periodontology department: a pilot study. *J Oral Rehabil* 1997;**24**(9):666–72.
23. Irwin CR, McCusker P. Prevalence of dentine hypersensitivity in a general dental population. *J Ir Dent Assoc* 1997;**43**(1):7–9.
24. Liu HC, Lan WH, Hsieh CC. Prevalence and distribution of cervical dentin hypersensitivity in a population in Taipei, Taiwan. *J Endod* 1998;**24**(1):45–7.
25. Rees JS. The prevalence of dentine hypersensitivity in general dental practice in the UK. *J Clin Periodontol* 2000;**27**(11):860–5.
26. Gillam DG, Seo HS, Newman HN, Bulman JS. Comparison of dentine hypersensitivity in selected occidental and oriental populations. *J Oral Rehabil* 2001;**28**(1):20–5.
27. Taani DQ, Awartani F. Prevalence and distribution of dentin hypersensitivity and plaque in a dental hospital population. *Quintessence Int* 2001;**32**(5):372–6.
28. Clayton DR, McCarthy D, Gillam DG. A study of the prevalence and distribution of dentine sensitivity in a population of 17–58-year-old serving personnel on an RAF base in the Midlands. *J Oral Rehabil* 2002;**29**(1):14–23.

29. Rees JS, Addy M. A cross-sectional study of dentine hypersensitivity. *J Clin Periodontol* 2002;**29**(11):997–1003.
30. Taani SD, Awartani F. Clinical evaluation of cervical dentin sensitivity (CDS) in patients attending general dental clinics (GDC) and periodontal specialty clinics (PSC). *J Clin Periodontol* 2002;**29**(2):118–22.
31. Rees JS, Jin LJ, Lam S, Kudanowska I, Vowles R. The prevalence of dentine hypersensitivity in a hospital clinic population in Hong Kong. *J Dent* 2003;**31**(7):453–61.
32. Rees JS, Addy M. A cross-sectional study of buccal cervical sensitivity in UK general dental practice and a summary review of prevalence studies. *Int J Dent Hyg* 2004;**2**(2):64–9.
33. Udoye CI. Pattern and distribution of cervical dentine hypersensitivity in a Nigerian tertiary hospital. *Odontostomatol Trop* 2006;**29**(116):19–22.
34. Bamise CT, Olusile AO, Oginni AO, Dosumu OO. The prevalence of dentine hypersensitivity among adult patients attending a Nigerian teaching hospital. *Oral Health Prev Dent* 2007;**5**(1):49–53.
35. Ommerborn MA, Schneider C, Giraki M, Schafer R, Singh P, Franz M, et al. In vivo evaluation of noncarious cervical lesions in sleep bruxism subjects. *J Prosthet Dent* 2007;**98**(2):150–8.
36. Chi D, Milgrom P. The oral health of homeless adolescents and young adults and determinants of oral health: preliminary findings. *Spec Care Dentist* 2008;**28**(6):237–42.
37. Kehua Q, Yingying F, Hong S, Menghong W, Deyu H, Xu F. A cross-sectional study of dentine hypersensitivity in China. *Int Dent J* 2009;**59**(6):376–80.
38. Bamise CT, Kolawole KA, Oloyede EO, Esan TA. Tooth sensitivity experience among residential university students. *Int J Dent Hyg* 2010;**8**(2):95–100.
39. Que K, Ruan J, Fan X, Liang X, Hu D. A multi-centre and cross-sectional study of dentine hypersensitivity in China. *J Clin Periodontol* 2010;**37**(7):631–7.
40. Amarasena N, Spencer J, Ou Y, Brennan D. Dentine hypersensitivity in a private practice patient population in Australia. *J Oral Rehabil* 2011;**38**(1):52–60.
41. Azodo CC, Amayo AC. Dentinal sensitivity among a selected group of young adults in Nigeria. *Niger Med J* 2011;**52**(3):189–92.
42. Bartlett DW, Fares J, Shirodaria S, Chiu K, Ahmad N, Sherriff M. The association of tooth wear, diet and dietary habits in adults aged 18–30 years old. *J Dent* 2011;**39**(12):811–6.
43. Oderinu OH, Savage KO, Uti OG, Adegbulugbe IC. Prevalence of self-reported hypersensitive teeth among a group of Nigerian undergraduate students. *Niger Postgrad Med J* 2011;**18**(3):205–9.
44. Bahsi E, Dalli M, Uzgur R, Turkal M, Hamidi MM, Colak H. An analysis of the aetiology, prevalence and clinical features of dentine hypersensitivity in a general dental population. *Eur Rev Med Pharmacol Sci* 2012;**16**(8):1107–16.
45. Colak H, Aylikci BU, Hamidi MM, Uzgur R. Prevalence of dentine hypersensitivity among university students in Turkey. *Niger J Clin Pract* 2012;**15**(4):415–9.
46. Colak H, Demirer S, Hamidi M, Uzgur R, Koseoglu S. Prevalence of dentine hypersensitivity among adult patients attending a dental hospital clinic in Turkey. *West Indian Med J* 2012;**61**(2):174–9.
47. Dhaliwal JS, Palwankar P, Khinda PK, Sodhi SK. Prevalence of dentine hypersensitivity: a cross-sectional study in rural Punjabi Indians. *J Indian Soc Periodontol* 2012;**16**(3):426–9.
48. Tengrungsun T, Jamornnium Y, Tengrungsun S. Prevalence of dentine hypersensitivity among Thai dental patients at the Faculty of Dentistry, Mahidol University. *Southeast Asian J Trop Med Public Health* 2012;**43**(4):1059–64.

49. Wang Y, Que K, Lin L, Hu D, Li X. The prevalence of dentine hypersensitivity in the general population in China. *J Oral Rehabil* 2012;**39**(11):812–20.
50. Ye W, Feng XP, Li R. The prevalence of dentine hypersensitivity in Chinese adults. *J Oral Rehabil* 2012;**39**(3):182–7.
51. Cunha-Cruz J, Wataha JC, Heaton LJ, Rothen M, Sobieraj M, Scott J, et al. The prevalence of dentin hypersensitivity in general dental practices in the northwest United States. *J Am Dent Assoc* 2013;**144**(3):288–96.
52. Que K, Guo B, Jia Z, Chen Z, Yang J, Gao P. A cross-sectional study: non-carious cervical lesions, cervical dentine hypersensitivity and related risk factors. *J Oral Rehabil* 2013;**40**(1):24–32.
53. Rahiotis C, Polychronopoulou A, Tsiklakis K, Kakaboura A. Cervical dentin hypersensitivity: a cross-sectional investigation in Athens, Greece. *J Oral Rehabil* 2013;**40**(12):948–57.
54. Rane P, Pujari S, Patel P, Gandhewar M, Madria K, Dhume S. Epidemiological study to evaluate the prevalence of dentine hypersensitivity among patients. *J Int Oral Health* 2013;**5**(5):15–9.
55. Vijaya V, Sanjay V, Varghese RK, Ravuri R, Agarwal A. Association of dentine hypersensitivity with different risk factors—a cross sectional study. *J Int Oral Health* 2013;**5**(6):88–92.
56. West NX, Sanz M, Lussi A, Bartlett D, Bouchard P, Bourgeois D. Prevalence of dentine hypersensitivity and study of associated factors: a European population-based cross-sectional study. *J Dent* 2013;**41**(10):841–51.
57. Costa RS, Rios FS, Moura MS, Jardim JJ, Maltz M, Haas AN, et al. Prevalence and risk indicators of dentin hypersensitivity in adult and elderly populations from Porto Alegre, Brazil. *J Periodontol* 2014.
58. Scaramucci T, de Almeida Anfe TE, da Silva Ferreira S, Frias AC, Sobral MA. Investigation of the prevalence, clinical features, and risk factors of dentin hypersensitivity in a selected Brazilian population. *Clin Oral Investig* 2014;**18**(2):651–7.
59. Higgins JPT, Green S, editors. *Cochrane handbook for systematic reviews of interventions version 5.0.2 [updated September 2009]*. Available from: <http://www.cochrane-handbook.org>: The Cochrane Collaboration.
60. EQUATOR Network. *Enhancing the quality and transparency of health research*. Available from: <www.equator-network.org> [accessed 23.04.14].

The management of dentine hypersensitivity

David Gillam and Elena Talioti

Centre for Adult Oral Health, Institute of Dentistry, Barts and The London School of Medicine and Dentistry, QMUL, London

Introduction/overview

Dentine hypersensitivity (DH) is a worldwide clinical condition that impacts on the everyday life of numerous people.[1–3] The reported prevalence varies depending on the methods used to collect data (see Chapters 2 and 3). However, the condition may affect up to 74% of the population, and this figure may be higher in individuals with periodontal disease.[4,5] DH appears to have been previously underreported by lay people as well as underdiagnosed by clinicians, which may lead to the problem being either ignored or undertreated.[6–10] It is also important to recognize that the diagnosis of DH requires the exclusion of all other clinical conditions that have a similar pain history (e.g., fractured tooth syndrome, leaking restorations, caries). In a busy practice this may be a challenge.

To successfully manage DH, it is essential to eliminate any conflicting diagnoses and deploy simple and effective relief. A plethora of agents have been used, either as dentist-applied or over-the-counter (OTC) at-home products. These therapies have been evaluated either in the laboratory (*in vitro*) or in the clinical environment (*in situ* or *in vivo*) and have been reported to be effective either through their tubular-occluding properties or through their nerve-desensitizing properties, based on the principle of the hydrodynamic theory.[11–13]

One concern regarding the management of DH is whether the condition can be managed and monitored to the satisfaction of the clinician and all affected persons. For example, a number of treatments available to the clinician have been impractical to implement in the clinical environment.[7,10,14,15]

To resolve some of these issues, a recent UK Expert Forum on DH[16] produced simple guidelines that can be readily applied in general practice. The Forum acknowledged that a single approach to DH may not satisfy everyone (Figure 4.1). The proposed scheme links the diagnosis, prevention, and treatment

Dentine Hypersensitivity. DOI: http://dx.doi.org/10.1016/B978-0-12-801631-2.00004-X

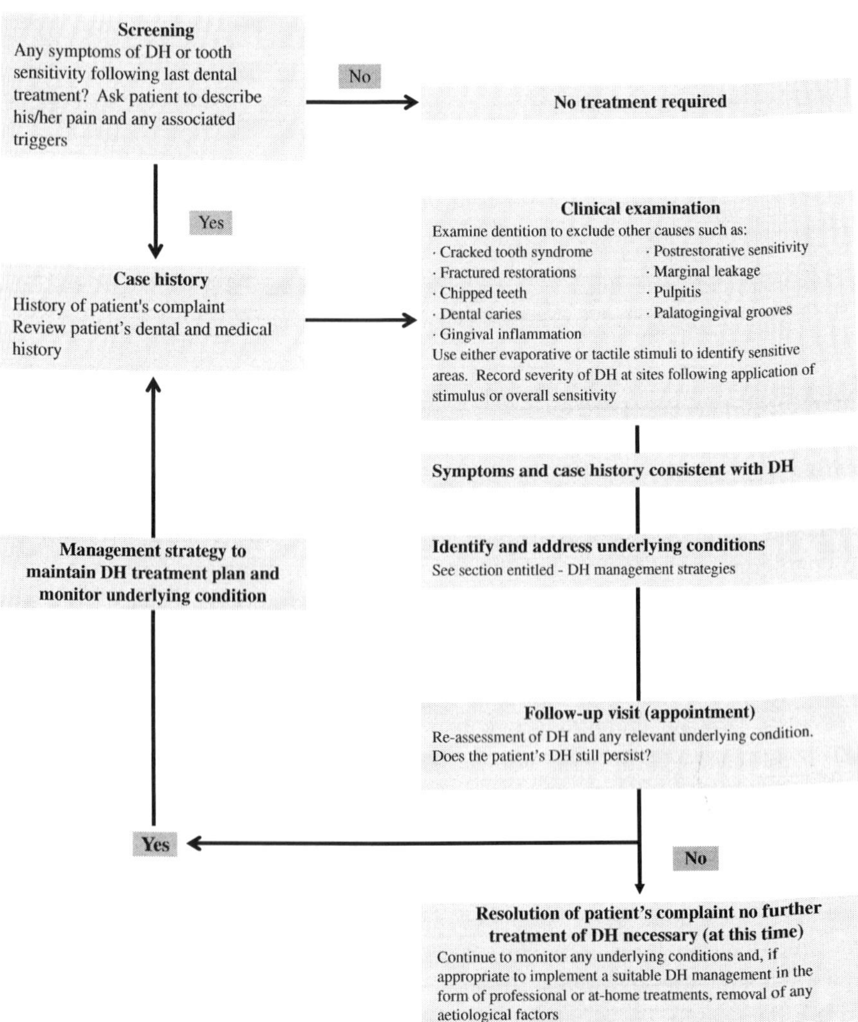

Figure 4.1 DH management guidelines.

strategies to three specific groups of people rather than recommending a blanket management for all patients with DH. These groups include (1) people with gingival recession caused by mechanical trauma, (2) those with tooth wear lesions, and (3) those with periodontal disease and receiving periodontal treatment (Figures 4.2–4.4).

Therefore, the aim of this chapter is to provide an overview of recent treatment modalities and product innovations for the clinical management of DH. It also emphasizes the need to tailor management to specific clinical situations to resolve the pain associated with DH.

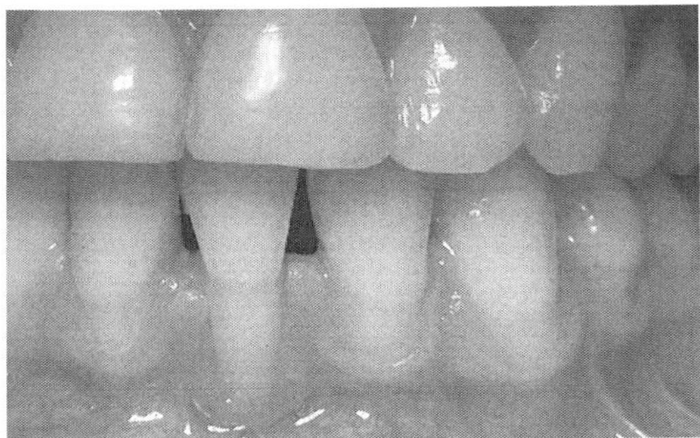

Figure 4.2 Gingival recession after orthodontic treatment associated with DH.
Source: George Belibasakis.

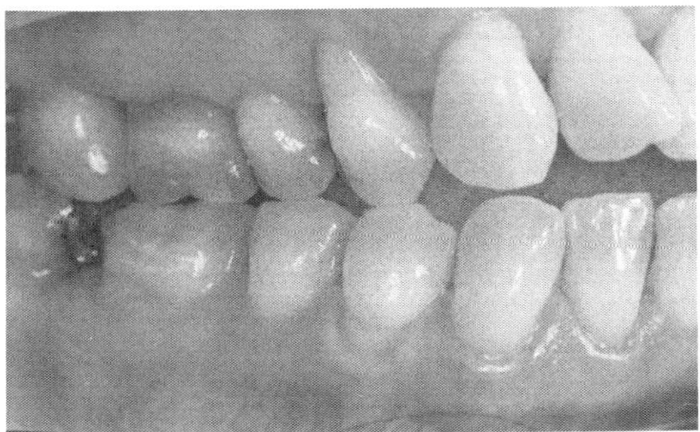

Figure 4.3 Clinical features of gingival recession with associated DH after periodontal surgery.
Source: P. Ower.

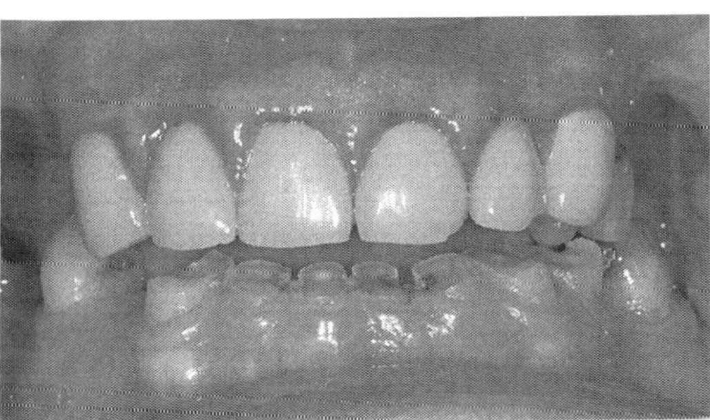

Figure 4.4 Clinical photograph illustrating the effect of severe attrition of the lower anterior teeth and gingival inflammation around the upper maxillary incisors.
Source: Dr. Wendy Turner.

Etiology, predisposing factors, and clinical features

The etiological factors and clinical features associated with DH are addressed in Chapter 2. It is important, however, to acknowledge that there may be different presenting features that should be taken into account when formulating management strategies. For example, there are people who have relatively healthy mouths and DH as a result of meticulous and perhaps overzealous oral hygiene, people who experience it as a result of tooth wear, and people who experience DH as a result of periodontal disease or its treatment and who may also have esthetic concerns related to the loss of gingival tissue (gingival recession)[16] (Figures 4.2–4.4). An essential step in the initial management of DH is to correctly diagnose the underlying condition by excluding conditions with a similar clinical presentation (Table 4.1). The second step is the selection of a suitable desensitizing product, depending on the extent and severity of the problem[10,15] (Figure 4.1).

Table 4.1 Overall management strategy options for treating DH

Gingival recession	Tooth wear	Periodontal treatment
Clinical evaluation		
• Clinical measurement of the gingival recession defect • Take study casts and clinical photographs to monitor condition over time • Check and monitor periodontal health • Identification and correction of predisposing or precipitating factors • Use of pain scores to assess and monitor DH (e.g., Visual Analogue Scales [VAS])	• Identify cause of tooth wear (enamel loss) • Record severity of lesions, if possible, using a recognized index[17,18] • Take study casts and clinical photographs to monitor condition over time • Check and monitor periodontal health • Use of pain scores to assess and monitor DH (e.g., VAS)	• Periodontal disease or periodontal treatment as the primary cause of exposure of dentine and associated DH • Check and monitor periodontal health. (6-point pocket charting) • Use of pain scores to assess and monitor DH (e.g., VAS)
Patient education (including preventive advice)		
– Show patient the affected site(s) – Explain probable cause for recession.	• Show patient the site(s) and explain probable cause of the tooth wear lesion(s)	• Reinforce the need for good oral hygiene • Show patient the site(s) affected by periodontal disease and explain

(Continued)

Table 4.1 **(Continued)**

Gingival recession	Tooth wear	Periodontal treatment
• Explain factors triggering sensitive teeth episodes • Encourage patients to modify their oral hygiene regimen in order to reduce damage to gingivae (e.g., reducing brushing force, correction of toothbrush technique) • Reduce excessive consumption of acid foods and drinks	• Recommend an oral hygiene regimen to minimize risk of further tooth wear. • Where appropriate recommend reducing frequency of consumption of acidic food and drink.	probable cause of the exposed dentine • Guide the patient to improve "at home" oral hygiene regimen. • Instruction on measures of reducing periodontal risk factors, for example, diabetes, smoking, and obesity.
Corrective clinical outcomes		
• Reduce excessive consumption of acid foods and drinks • Manufacture of silicone gingival veneers • Orthodontic treatment • Restorative correction of recession defect and subgingival margins of fillings and crowns • Polymers: sealants/ varnishes/resins/dentine bonding agents • Laser obturation of dentinal tubules • Use of desensitizing polishing pastes • Pulpal extirpation (root canal treatment)	• Provide high fluoride remineralizing treatment (pre-emptive phase) • Provide professional desensitizing treatment to relieve DH • Encourage patient to seek advice from medical practitioner, if tooth wear caused by working environment or reflux/excesssive vomiting (psychiatric evaluation may also be appropriate) • Restorative correction in the form of composite build up, crowns may also be appropiate	Initial phase Nonsurgical periodontal procedure(s). DH treatment (including desensitizing polishing pastes/fluoride varnishes) Reevaluation Follow-up assessment on periodontal status and DH Corrective phase • Surgical periodontal procedure(s) • DH treatment (including desensitizing polishing pastes/fluoride varnishes) Follow-up management maintenance phase • Supportive periodontal therapy • Ongoing monitoring of periodontal health

(*Continued*)

Table 4.1 **(Continued)**

Gingival recession	Tooth wear	Periodontal treatment
		• DH treatment (including desensitizing polishing pastes/fluoride varnishes) • Oral hygiene advice
Recommendations for home use (including toothpaste/mouth rinses)		
• Oral hygiene implementation as per recommendation • Strontium chloride/ strontium acetate • Potassium nitrate/ chloride/citrate/oxalate • Calcium compounds: • Calcium carbonate and arginine and caesin phosphopeptide + amorphous calcium phosphate • Bioactive glass • Nano/hydroxyapatite • Fluoride in higher concentration (2800/ 5000 ppm F [prescription]) • Amine/stannous fluoride	• Oral hygiene implementation as per recommendation • Toothpastes and mouth rinses (see recommendations for gingival recession)	• Oral hygiene implementation as per recommendation • Regular brushing with an antibacterial toothpaste to aid plaque control • Short period, the use of a 0.2% chlorhexidine solution for plaque control • Use of a desensitizing mouthrinse twice daily for DH control (when appropriate)

Source: Gillam et al.[16]

Methods for product evaluation

Several investigators have made recommendations for determining an ideal desensitizing product or technique,[19,20] but concern has also been raised regarding whether there can be a single modality that would satisfy every eventuality.[16] Currently, most, if not all, of the treatments are based on the principles implicit in the hydrodynamic theory as proposed by Brännström,[11] and as such they may be effective on the basis of dentine tubule occlusion or nerve excitation prevention.

Potential desensitizing products are therefore evaluated in the laboratory (*in vitro*) environment or using an animal model (e.g., the cat) for nerve desensitization[21,22] to test their ability within these two modes of action. It is important to avoid extrapolating laboratory results to the clinic because new products may have different effects in real people. When *in vitro* or animal studies have been successful, clinical studies are required to assess the efficacy of the products in participants with DH.

The conduct of DH studies and the methods involved in evaluating effectiveness in reducing pain have been the subject of a number of recommendations.[23,24] Data from studies before Holland and colleagues' recommendations in 1997[23] may not be comparable with those subsequently reported in terms of standardized methods including selection of participants and eliciting pain history.[25]

Published studies evaluating desensitizing toothpastes once demonstrated improvements in DH ranging from 30% to 80% reductions in sensitivity when compared with other toothpastes and placebo controls.[26] However, the results from these studies are somewhat conflicting and difficult to interpret, in part because of different methods, including patient selection criteria. An inherent problem, however, in clinical studies of the efficacy of desensitizing products is the interference of placebo or Hawthorne effects.[20,27,28] The extent to which the placebo effect complicates the interpretation of the results is difficult to predict, although these effects are similar to those reported in other medical and dental therapeutic studies.[29] The introduction of Holland and colleagues' guidelines for conducting DH studies provided a stronger framework to evaluate the efficacy of desensitizing treatment modalities and may have eliminated some of the problems seen before 1997.[23] An additional matter to consider when assessing the efficacy of these products, in particular the OTC toothpastes and mouth rinses, is that most of the evaluated product formulations are different than those currently available. However, one benefit of *in vitro* studies is that they provide a rapid screening mechanism for determining a particular mode of action of desensitizing products.

An important critique within the context of this book was the selection of outcomes in these early studies and whether the objective or subjective methods involved in the testing are relevant to the needs of affected people on a day-to-day basis. The relevance of such methods and patient-based outcomes is at the heart of this book. Pain is a subjective experience that is affected by participants' previous pain experiences, psychological profiles, levels of stress, and pain thresholds. It is therefore difficult to accurately assess the exact level of pain attributable to DH. The question may be asked about the best way to measure the pain response in DH pain studies, because there are a number of measures available, such as VAS and verbal rating scales (VRS).[30,31] Chapter 13 deals with this issue in detail. Added to this is the variation in stimuli to provoke pain, such as air blasts from a dental air syringe and force-controlled dental probes.

The use of oral health-related quality of life (OHQoL) appears to be promising, but the nature of the relationship between interventions and subjective outcomes warrants careful consideration.[1–3,32,33] Although an ideal result from DH studies would be the absence of any symptoms or impact, a more realistic target may be the reduction to an acceptable level so that affected persons can undertake normal day-to-day activities.[20]

In-office (professionally applied) treatment modalities

The clinician has a vast range of in-office treatment (professionally applied) modalities for treating DH, depending on the extent and severity of the problem, such as primers, sealants, and varnishes (including fluoride), conventional glass ionomer cements, resin-reinforced glass ionomers/compomers, adhesive resin bonding systems, and laser technology, all with varying degrees of reported efficacy.[15,34,35] Tables 4.2–4.5 summarize key work regarding these treatments. Laser technology has also been advocated, but the results are equivocal and their use appears to be limited in the clinical environment.[9,80–82,84–86] The evidence from the published literature therefore would suggest that there is no in-office (professionally applied) treatment modality that is universally accepted or used consistently in clinical practice.[9,16]

Table 4.2 Published studies supporting the efficacy of in-office professionally applied products (solutions and products)

Type, chemical/concentration	Product and clinical support
Fluorides	
sodium fluoride, stannous fluoride, hydrogen fluoride	Dentinbloc, Colgate Oral Pharmaceuticals, Canton MA[39,40]
Nerve desensitizers ***Potassium nitrate***	
1–15% solutions 5%, 10% in gel	41 42
Guanethidine monosulfate	
1% guanethidine monosulfate	Ismelin, Ciba-Geigy[43]
Oxalate	
3% Potassium oxalate 3% Potassium oxalate 6.8% Ferric oxalate	Protect, Sunstar Butler, Chicago, IL[44] Oxa-gel, Art-dent Ltd, Araraquara SP, Brazil[45] Sensodyne Sealant, GSK, Jersey City NJ[46]
Calcium phosphates	
1.5 M calcium chloride + 1.0 M potassium oxalate	ACP-CPP, GC corporation, Tokyo, Japan[47,48]

Source: Orchardson & Gillam,[15] Pashley et al.,[36] and modified. Gillam.[37,38]

Professionally applied treatments often involve patients with DH localized to one or two teeth with moderate to severe discomfort. As such, they would require immediate attention within the dental clinic. Generally speaking, the mode of action of these treatment modalities is by tubular occlusion, either through resin impregnation into the dentinal tubules or by the alteration of the dentine surface when using laser technology.[15,87]

Patients may also experience DH after restorative procedures (i.e., postoperative sensitivity), which may require the application of a desensitizer underneath a crown or restoration.[88,89] Several investigators have also reported patients experiencing postoperative sensitivity after periodontal therapy, including scaling and surgery.[15,90,91] DH after periodontal therapy may be alleviated initially by the application of desensitizing polishing pastes, such as Pro-Argin™ (Colgate Professional) or Novamin® products (Nupro® Sensodyne prophylaxis paste, Dentsply International), before necessitating a desensitizing toothpaste for home use.[37,38,49,50,53,54] Surgical intervention, including gingival grafting techniques used to treat marginal gingival defects or correct esthetic concerns, may reduce DH and have a positive impact on quality of life.[5,92] However, there is insufficient clinical evidence to conclude that these surgical root coverage procedures predictably reduce DH.[93] The use of desensitizing products in bleaching gels designed to whiten teeth have also been reported to alleviate bleaching sensitivity in patients with or without preexisting DH.[88]

Table 4.3 Published studies supporting the efficacy of in-office professionally applied products (desensitizing polishing pastes)

Type, chemical/concentration	Product and clinical support
Desensitizing pastes	
Colgate® Sensitive Pro-Relief™ Desensitizing paste with 8% arginine and calcium carbonate ($Arg/CaCO_3$) Elmex Sensitive Professional desensitizing paste. Pro-Argin™ Formulation, Gaba International, Switzerland	Pro-Argin™ (Colgate Professional)[49–52]
15% calcium sodium phosphosilicate (CSPS; NovaMin®)	Nupro® Sensodyne prophylaxis paste (Dentsply International).[53,54]
MI Paste Plus™ containing casein phosphopeptide-amorphous calcium phosphate (CCP-ACP) with 0.2% sodium fluoride	(MI Paste Plus™ GC Corporation, www.gcamerica.com). According to the manufacturer both MI Paste and *MI Paste Plus* have FDA 510 K clearance as a prophylaxis paste and for treating DH

Source: Orchardson & Gillam,[15] Pashley et al.,[36] and modified. Gillam.[37,38]

Table 4.4 Published studies supporting the efficacy of in-office professionally applied products (adhesives, resins, and cements)

Type	Product and clinical support
Fluoride varnish	Duraphat, Colgate Oral Pharmaceuticals, Canton, MA[55–57]
	Fluoline, PD Dental, Altenwalde, Germany[58]
	Fluor Protect (Ivoclar Vivadent)[59]
Oxalic acid + resin	MS Coat, Sun Medical Co, Shiga, Japan[60]
	Pain Free, Parkell Co, Farmingale, NY[40]
Sealants, primers	Seal & Protect, Dentsply Konstanz, Germany[61]
	Dentin Protector, Vivadent, Germany[62]
	Gluma Desensitizer, Heraeus Kulzer, Dormogen, Germany[58,63–67]
	Gluma Alternate, Heraeus Kulzer, Wehrheim, Germany[64]
	Health-Dent Desensitizer, Health-Dent Inc, Oswego, NY[58,64]
	HEMA + other, Hurri-Seal Dentin Desensitizer Beutlich Pharmaceuticals LLC, Waukegan, IL, HemaSeal & Cide (Germiphene Corporation, Canada)[68]
	Prime & Bond 2.1, Dentsply Caulk, Milford DE[69]
	Scotchbond (Single Bond), 3M Dental Products, St Paul, MN[58,60,70]
	One-Step Bisco Schaumburg IL[66]
Etch + primer	Scotchbond, 3M Dental Products, St Paul, MN[70]
	Systemp.desensitizer, Ivoclar Vivadent, Schaan, Liechtenstein[71]
Etch + primer + adhesive	Scotchbond Multi System, 3M Dental Products, St Paul, MN[64]
Primer + adhesive	SE Bond, Kuraray, Okayama, Japan[58]
Glass ionomer cements/resin-reinforced glass ionomer	Vitrabond-like (3M-ESPE, St. Paul, MN), Fuji VI1 (GC Asia Dental Pte Ltd, Singapore)[67,72,73]

Source: Orchardson & Gillam,[15] Pashley et al.,[36] and modified. Gillam.[37,38]

Table 4.5 Published studies supporting the efficacy of in-office professionally applied products (miscellaneous in-office professionally applied techniques)

Treatment modality and clinical support	
Iontophoresis	74–77
Hypnosis/hypnotherapy	78,79
Laser therapy	80–84

Source: Orchardson & Gillam,[15] Pashley et al.,[36] and modified. Gillam[37,38]

Toothpastes, mouth rinse formulations, and topically applied varnishes

From the published literature, a plethora of products have been recommended for the treatment of DH, such as fluoride, calcium phosphate compounds [e.g., amorphous calcium phosphate (ACP), beta-tricalcium phosphate (β-TCP), and tooth mousse (casein phosphopeptide-amorphous calcium phosphate; CPP-ACP)], strontium, potassium-containing products, and have been commercially available in both developed and developing countries over the past 40–50 years, and varying claims of clinical efficacy have been reported by several investigators.[13,15,25,37,38] Other treatment modalities have also been developed in the past 20 years, such as Novamin® (GSK-GlaxoSmithKline) and Pro-Argin® (Colgate) toothpastes, mouth rinses, and desensitizing polishing pastes. The main mode of action of these formulations is generally considered to be through tubular occlusion or potassium ion diffusion through the dentine tubules (Tables 4.6–4.8).

One important aspect of the *in vitro* studies of the effect of these products on the surface dentine is the ability to characterize the surface deposit and determine whether the active ingredient has the capability to bind or integrate into dentine through mechanical or chemical means. For example, Novamin® and Pro-Argin® may provide a source of available calcium and phosphate ions[164–166] that might remineralize and harden the dentine, protecting against subsequent acid challenge.

Evidence from *in vitro* studies,[167–170] however, suggests that these results are attributable to the abrasive component of a toothpaste that smears the surface dentine and, to some degree, occludes the tubule openings, other than being attributable to active ingredients *per se* (Figures 4.5 and 4.6).[171] The degree to which the abrasive influences the blocking of the dentine tubules is contentious, but it may contribute to the mechanical smearing effect and also may play a part in resistance to an acid challenge.[169,172,173] According to Addy and Mostafa, the retention of the surface deposit on the dentine may be influenced by washing.[174] For example, in their *in vitro* study, the silica-containing toothpastes were largely resistant to this process but the other abrasive systems were not resistant. Other ingredients within the toothpaste may also affect the dentine surface, for example, sodium lauryl sulfate (a surfactant) has been demonstrated to remove the smear layer.[175]

Most approaches appear to have a degree of success when treating the condition. As noted, a recurring problem when evaluating their efficacy is the diverse way that the clinical studies have been conducted. Consequently, there is limited agreement on the relative effectiveness of the treatments.[9,15,35,170] Furthermore, clinicians not only appear to be uncertain about the best way to manage DH but also express dissatisfaction with the various therapies available.[9] This suggests a need for clear guidance for treating DH.

Table 4.6 Characteristics of selected tubular-occluding toothpastes and mouth rinses

Product	Composition	Proposed mode of action
Colgate Pro-Argin™	Pro-Argin toothpastes containing 80,000 ppm (8%) arginine, bicarbonate, calcium carbonate, and 1450 ppm fluorine as NaMFP (also Elmex Sensitive)	Based on SensiStat® (Kleinberg[95]) The arginine complex binds to the tooth surface and allows the calcium carbonate to slowly dissolve and release calcium. *in vitro* and *in vivo* studies have been published in support of both laboratory and clinical claims for the product[25,96–104,107–109,166]. The mode of action is by tubule occlusion
	Pro-Argin mouthwash containing 0.8% arginine, pyrophosphate and PVM/MA copolymer	Recent *in vitro* and *in vivo* studies have been published in support of both laboratory and clinical claims for the product[105–106]. The mode of action is by tubule occlusion
SensiShield® (Novamin®)	Composed of calcium phosphorus, sodium and silica (CSPS (NovaMin®), 1450 ppm fluoride)	Novamin® in contact with saliva and water reacts and releases Ca and PO_4 ions. Sodium ions in the Novamin particles exchange with hydrogen cations which in turn allows the calcium and phosphate ions to be released. A calcium phosphate layer is formed then subsequently crystallizes into hydroxycarbonate apatite. The exposed dentine surface appears to act as a nucleation site for these ions to

(*Continued*)

Table 4.6 (Continued)

Product	Composition	Proposed mode of action
		form hydroxycarbonate apatite and bypasses the intermediate phase of ACP formation. Mainly *in vitro* support for Novamin® although more recently clinical data supporting the clinical efficacy of the product has been published.[110–124] The mode of action is by tubule occlusion
ACP	ACP is inorganic in nature and is made by combining soluble salts of calcium and phosphate through a two-phase system containing Ca in one part and PO_4 in another. When mixed together they react to form an amorphous phosphate material that precipitates on to the tooth surface	ACP is highly soluble and susceptible to acid attack and as the ACP is not protected and has no delivery system it has lower substantivity. ACP is not bioavailable after the product is rinsed away. Previously incorporated in Enamelon toothpaste no longer available) that relied on a dual chamber system in the toothpaste tube. Available in Enamel Care toothpaste (Church & Dwight). Limited and equivocal published data for effectiveness of ACP in the treatment of DH
Recaldent (CPP-ACP)	Casein phosphates (CPP) are peptides derived from milk protein casein that are complexed with calcium (Ca) and phosphate (PO_4). In this complex the CPP maintains the Ca and PO_4 ions in an	CPP-ACP uses peptides derived from the milk protein casein to maintain Ca and PO_4 in an ACP. The CPP binds to surfaces such as plaque, bacteria, and soft tissue providing a bioavailable Ca and PO_4

(*Continued*)

Table 4.6 **(Continued)**

Product	Composition	Proposed mode of action
	amorphous form (ACP). The milk-derived peptide containing amorphous Ca and PO_4 is the driving mechanism that binds to plaque, bacteria and the tooth surface	at the surface of the tooth without precipitation. The ACP is released during acidic challenges. Stabilization of ACP by the CPP ensures the delivery of Ca and PO_4 ions into the tooth structure before they crystallize. Most *in vitro* and *in vivo* studies support the anticaries benefit of the product. There appears to be limited clinical data on its effect on reducing DH[47,48,125–130]

Source: Mason et al.[94] modified.

Recent advances in the management of DH

As indicated, there does not appear to be an universally accepted effective treatment modality for DH. This has led to further research into new biomimetic substances and reformulations of existing products (toothpastes, mouth rinses, desensitizing polishing pastes) such as Bioactive glasses (Novamin®) and hydroxyapatite/nano-hydroxyapatite/nano-carbonate apatite crystal (HAP) strontium acetate (Colgate Pro-Argin®) toothpastes (Tables 4.6–4.8).

More recently, a 1.4% potassium oxalate mouth rinse (Listerine® Advanced Defence Sensitive) has been developed, although further long-term clinical studies are required to determine its effectiveness.[176–179] Other recent developments include a biomimetic self-assembling peptide (for treating the early caries lesion), a hydroxyapatite toothpaste, and fluoride varnishes (5% sodium fluoride and TCP, 3M-ESPE, St. Paul, MN), or 5% sodium fluoride and sodium trimetaphosphate (TMP) gel or varnish.[131,180–184] Although these novel products were initially designed to aid remineralization in chairside use, they may also be used in toothpastes, mouth rinses, gels, and varnishes for DH. The mode of action of these products appears to be by tubular occlusion.

Table 4.7 Characteristics of selected tubular-occluding toothpastes and mouth rinses

Product	Composition	Proposed mode of action
Blanx® Biorepair® Nanit®active (Henkel) UltraDEX® recalcifying and whitening toothpaste and oral rinse (Periproducts Ltd, UK) _	Hydroxyapatite, sodium monofluorophosphate (MFP)	According to product literature Nanit®active induces a process referred to as neomineralization. The Nanit®active nanoparticles react with the calcium and phosphate ions in saliva and a new protective layer is formed on the tooth surface (1–2 μm). Limited data available at present, UltraDEX® Recalcifying and Whitening Toothpaste and Oral Rinse is a relative recent product on the UK market.[131–138] The mode of action is by tubule occlusion
Strontium salts Sensodyne)	10% Strontium chloride (original), no fluoride 8% Strontium acetate and 1050 ppm fluoride strontium acetate, sodium monofluorophosphate	Hydrated technology. *In vitro* and *in vivo* studies support both the proposed mode of action and clinical effectiveness of both the acetate and chloride variants of the product.[94,139–148] The mode of action is by tubule occlusion. Jackson[149] however suggested that none of the studies on strontium toothpastes (pre-2000) demonstrated a consistent, improvements in participants' symptoms when compared with negative control toothpastes. Karim & Gillam[150] also support these conclusions

(Continued)

Table 4.7 (Continued)

Product	Composition	Proposed mode of action
Stannous fluoride Crest® ProHealth™	0.454% Stannous fluoride	Anhydrous technology. *In vitro* and *in vivo* studies support the proposed mode of action and clinical effectiveness of the stannous ion. Reformulated as a stabilized stannous fluoride and sodium hexametaphosphate toothpaste, uses hexametaphosphate to limit stains associated with the use of stannous ions.[151–157] The mode of action is by tubule occlusion
Colgate SnF_2	Stannous fluoride, potassium nitrate (5%)	Dual chamber delivery system. *In vitro* and *in vivo* studies support the proposed mode of action and clinical effectiveness of the product. The presence of potassium indicates its use as a nerve desensitizer; however, this product is no longer commercially available[94]
Amine fluoride (Elmex Sensitive)	Elmex® SENSITIVE PLUS toothpaste containing amine fluoride (1400 ppm F) Elmex Sensitive mouthrinse containing 250 ppm F (125 ppm F from amine fluoride, 125 ppm F from potassium fluoride	Amine fluoride leads to the formation of a protective layer on the dentine containing calcium fluoride, which promotes remineralization and tubular occlusion. Limited published data available.[158,159] The mode of action is by tubule occlusion

Source: Mason et al.[94] modified.

Table 4.8 Characteristics of selected nerve depolarizing toothpastes and mouth rinses

Product	Composition	Proposed mode of action
Sensodyne	Potassium nitrate, sodium fluoride (NaF)	Hydrated toothpaste technology. 2–5% Potassium-containing toothpastes and mouth rinses in the form of nitrate, chloride, or citrate have been formulated. Evidence of a desensitizing action based on historical animal studies.[21,22] No evidence of tubular occlusion was observed when potassium ions were tested *in vitro*.[13] Evidence suggests that potassium-containing toothpastes are effective in reducing DH[13,160] although there is no evidence to suggest this is by nerve depolarization.[13] A recent clinical study reported a transient depolarizing effect when potassium ions are applied on exposed dentine.[161] There is insufficient clinical data to state categorically whether potassium salts *per se* are effective in reducing DH[13,149,150,162,163]
Colgate	Potassium nitrate, sodium monofluorophosphate	
Crest	Potassium sodium fluoride (NaF)	
Sensodyne	Potassium chloride, sodium fluoride (NaF)	
Colgate	Potassium citrate, sodium monofluorophosphate	

Source: Mason et al.[94] modified.

Clinical management of DH

Naturally, the first stage of the management of DH involves appropriate diagnosis. Clinicians must recognize that the nature of DH as a diagnosis of exclusion requires them to identify, and rule out, conditions with a similar presentation to provide a definitive diagnosis (see Chapter 2). Those conditions include, but are not limited to, cracked (fractured) tooth syndrome, leaking restorations, and caries.[15,85,185] A thorough assessment of the symptoms including the history, the extent, and the severity and a clinical examination and special tests should lead to a differential diagnosis. This detailed assessment will enable the clinician to take into account all the possible etiological factors, therefore enabling successful management. The monitoring of DH is essential, yet even this may be problematic in dental practice. Gillam and colleagues have consequently recommended a simplified management algorithm taking into account the various presenting features associated with DH.[16]

Specific DH management strategies

By now, clinicians will have recognized that a simple "one size fits all" solution to the various presentations of DH may not be successful. Although a generic strategy

Figure 4.5 Coverage of the exposed dentine after a 2-min application of a fluoride toothpaste.[171]
Source: Gillam[37].

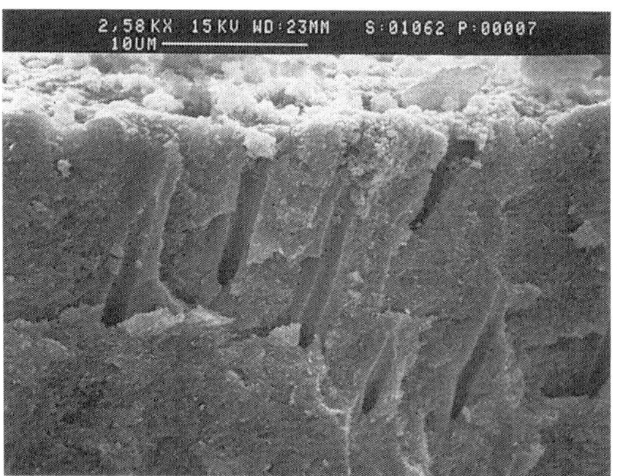

Figure 4.6 Evidence of tubular occlusion of dentine tubules by toothpaste ingredients (e.g., silica) after a 2-min application of a fluoride toothpaste.[171]
Source: Gillam[37].

(Figure 4.1) may appear more straightforward, specific strategies for each of the underlying causes may be of benefit for the successful management of DH (Figures 4.2–4.4).

Orchardson and Gillam recommended a stepwise approach to managing DH, depending on its extent, severity, and underlying cause.[15] Their steps begin with a noninvasive approach, supplemented with preventive measures, and escalated to more invasive treatments if the pain is unresponsive or increasing or if the initial diagnosis may have been incorrect.

An important aspect of care is to manage the patient's expectations throughout treatment. Patients in pain might expect treatment to resolve their problems within a very short time period. Although it may be appropriate for clinicians to recommend the use of an OTC product (Table 4.1), it would also be prudent to indicate that a toothpaste or mouthwash may take up to 4 weeks to be effective. If the pain is severe and localized to one or more teeth, then a professionally applied product may resolve the patient's symptoms.

Clinicians should recognize that no one desensitizing product can fully resolve all the various presenting features of DH. They should therefore deploy a range of products. The guidelines address this particular problem and advise on management for people with gingival recession, tooth wear, and periodontitis (Figure 4.1, Table 4.1).[16]

Gingival recession from mechanical trauma

Gingival recession is a multifactorial condition with predisposing and precipitating factors.[186,187] Patients with gingival recession from mechanical trauma may have meticulous oral hygiene associated with overzealous toothbrushing[185] (appearance similar to Figure 4.2). When mechanical trauma has been identified as an etiological factor, treatment should include encouraging patients to modify their oral hygiene regimen to reduce any further damage to the gingivae (e.g., reducing brushing force, correction of toothbrush technique) (Figure 4.1, Table 4.1).[16]

DH and tooth wear lesions

Tooth wear refers to loss of tooth substance caused by abrasion, attrition, erosion, and abfraction (Figure 4.4).[188,189] Many cases involve a combination of these factors. Acidic foods and drinks may lead to tooth wear when combined with mechanical cleaning (e.g., the frequency and timing of the intake of acidic components in the diet and the time of toothbrushing).[190,191] Continued tooth wear may lead to the exposure of the underlying dentine and the risk of pain from DH (Figure 4.4). There may be a strong, progressive relationship between DH and erosive tooth wear, which is important to recognize for prevention and treatment.[192] An overall management strategy for patients with tooth wear and associated DH must prevent further damage of the dental hard tissues (Table 4.1, Figure 4.1).[16]

DH and periodontal disease and treatment

Both periodontitis and its treatment can contribute to tissue damage and loss of periodontal tissue[5,15] (Figure 4.3). DH is therefore more common in people with periodontal disease and who have undergone nonsurgical and surgical treatments.[4,5,90,91]

The overall management strategy for patients with periodontal disease or as a result of periodontal treatment should also include patient education, correction, and maintenance of any underlying periodontal problems after the clinical evaluation of the primary cause of the exposure of the dentine and associated DH (Table 4.1).

Conclusion

Numerous treatment modalities have been reported to have a degree of efficacy against DH in clinical studies, but their real benefits in terms of clinical relevance to quality of life remain unclear. There is also uncertainty regarding the effectiveness of these treatments in clinical practice or everyday home use. Clinicians should acknowledge there is no "one treatment fits all" approach to fully resolve the problem for everyone with DH.

A UK Expert Forum on DH has provided practical, evidence-based guidance for the diagnosis, prevention, and treatment of specific presenting features of people with DH.[16] These simple guidelines can be readily applied when managing DH in general dental practice and may enable both clinicians and people affected with DH to be involved in the treatment process.

Acknowledgment

The authors have based this chapter on book chapters and publications by Gillam et al.[16] and Gillam.[37,38]

References

1. Bekes K, John MT, Schaller H-G, Hirsch C. Oral health-related quality of life in patients seeking care for dentin hypersensitivity. *J Oral Rehabil* 2008;**36**:45–51.
2. Boiko OV, Baker SR, Gibson BJ, Locker D, Sufi F, Barlow APS, et al. Construction and validation of the quality of life measure for dentine hypersensitivity (DHEQ). *J Clin Periodontol* 2010;**37**:973–80.
3. Bekes K, Hirsch C. What is known about the influence of dentine hypersensitivity on oral health-related quality of life. *Clin Oral Investig* 2013;**17**:S45–51.
4. Chabanski MB, Gillam DG, Bulman JS, Newman HN. The prevalence, distribution and severity of cervical dentine sensitivity (CDS) in a population of patients referred to a specialist periodontology department. *J Clin Periodontol* 1996;**23**:989–92.
5. Gillam D, Orchardson R. Advances in the treatment of root dentine sensitivity: mechanisms and treatment principles. *Endod Topics* 2006;**13**:13–33.

6. Schuurs AHB, Wesselink PR, Eijkman MAJ, Duivevnvoorden HJ. Dentists' views on cervical hypersensitivity and their knowledge of its treatment. *Endod Dent Traumatol* 1995;**11**:240–4.
7. Canadian Advisory Board on Dentin Hypersensitivity. Consensus-based recommendations for the diagnosis and management of dentin hypersensitivity. *J Can Dent Assoc* 2003;**69**:221–6.
8. Gillam DG, Bulman JS, Eijkman MAJ, Newman HN. Dentists' perceptions of dentine hypersensitivity and knowledge of its treatment. *J Oral Rehabil* 2002;**29**:219–25.
9. Cunha-Cruz J, Wataha JC, Zhou L, Manning W, Trantow M, Bettendorf MM, et al. Treating dentin hypersensitivity: therapeutic choices made by dentists of the Northwest PRECEDENT network. *J Am Dent Assoc* 2010;**141**:1097–105.
10. Gillam DG. Current diagnosis of dentin hypersensitivity in the dental office: an overview. *Clin Oral Investig* 2013;**17**:S21–9.
11. Brännström M. A hydrodynamic mechanism in the transmission of pain-produced stimuli through the dentine. In: Anderson DJ, editor. *Sensory mechanisms in dentine*. Oxford: Pergamon; 1963. p. 73–9.
12. Brännström M, Åström A. The hydrodynamics of the dentin; its possible relationship to dentinal pain. *Int Dent J* 1972;**22**:219–27.
13. Orchardson R, Gillam DG. The efficacy of potassium salts as agents for treating dentine hypersensitivity. *J Orofac Pain* 2000;**14**:9–19.
14. Addy M, Urquhart E. Dentine hypersensitivity: its prevalence, aetiology and clinical management. *Dent Update* 1992;**19**:407–8 410–2.
15. Orchardson R, Gillam DG. Managing dentin hypersensitivity. *J Am Dent Assoc* 2006;**137**:990–8.
16. Gillam DG, Chesters RK, Attrill DC, Brunton P, Slater M, Strand P, et al. Dentine hypersensitivity—guidelines for the management of a common oral health problem. *Dent Update* 2013;**40**:514–24.
17. Smith B, Knight J. An index for measuring the wear of teeth. *Br Dent J* 1984;**156**: 435–8.
18. Bartlett D, Ganss C, Lussi A. Basic Erosive Wear Examination (BEWE): a new scoring system for scientific and clinical needs. *Clin Oral Investig* 2008;**12**:65–8.
19. Grossman LI. A systematic method for the treatment of hypersensitive dentine. *J Am Dent Assoc* 1935;**22**:592–602.
20. Gillam DG. Clinical trial designs for testing of products for dentine hypersensitivity—a review. *Periodontal Abstr* 1997;**45**:37–46.
21. Markowitz K, Bilotto G, Kim S. Decreasing intradental nerve activity in the cat with potassium and divalent cations. *Arch Oral Biol* 1991;**36**:1–7.
22. Markowitz K, Kim S. The role of selected cations in the desensitization of intradental nerves. *Proc Finn Dent Soc* 1992;**88**:39–54.
23. Holland GR, Narhi MN, Addy M, Gangarosa L, Orchardson R. Guidelines for the design and conduct of clinical trials on dentine hypersensitivity. *J Clin Periodontol* 1997;**24**:808–13.
24. *ADA Acceptance Program Guidelines—products for the treatment of dentinal hypersensitivity.* <http://www.docstoc.com/docs/2976618/Acceptance-Program-Guidelines-Products-for-the-Treatment-of-Dentinal-Hypersensitivity>; 2009 [accessed 08.03.14].
25. Cummins D. Recent advances in dentin hypersensitivity: clinically proven treatments for instant and lasting sensitivity. *Am J Dent* 2010;**23**:1A–13A [special issue].

26. Clark GE, Troullos ES. Designing hypersensitivity clinical studies. *Dent Clin North Am* 1990;**34**:531–44.
27. Gillam DG. An overview on conducting clinical studies for the evaluation of desensitising products for the treatment of dentine hypersensitivity. *Oral Health Dialogue* 2011;**2**:6–14.
28. West NX. Dentine hypersensitivity and the placebo response: a comparison of the effect of strontium acetate, potassium nitrate and fluoride toothpastes. *J Clin Periodontol* 1997;**24**:209–15.
29. Curro FA, Friedman M, Leight RS. In: Addy M, Embery G, Edgar WM, Orchardson R, editors. *Design and conduct of clinical trials on dentine hypersensitivity*. London: Martin Dunitz Ltd; 2000. p. 299–314.
30. Gillam DG, Newman HN. Assessment of pain in cervical dentinal sensitivity studies. A review. *J Clin Periodontol* 1993;**20**:383–94.
31. Gillam DG, Orchardson R, Närhi MVO, Kontturi Närhi V. Present and future methods for the evaluation of pain associated with dentine hypersensitivity. In: Addy M, Embery G, Edgar WM, Orchardson R, editors. *Tooth wear and sensitivity*. London: Martin Dunitz Ltd; 2000. p. 283–97.
32. Baker SR, Gibson BJ, Sufi F, Barlow A, Robinson PG. The dentine hypersensitivity experience questionnaire (DHEQ): a longitudinal validation study. *J Clin Periodontol* 2014;**41**:52–9. Available from: http://dx.doi.org/doi:10.1111/jcpe.12181.
33. Machuca C, Baker SR, Sufi F, Mason S, Barlow A, Robinson PG. Derivation of a short form of the dentine hypersensitivity experience questionnaire. *J Clin Periodontol* 2014;**41**:46–51. Available from: http://dx.doi.org/doi:10.1111/jcpe.12175.
34. Pashley DH. Potential treatment modalities for dentine hypersensitivity: in office products. In: Addy M, Embery G, Edgar WM, Orchardson R, editors. *Tooth wear and sensitivity*. London: Martin Dunitz Ltd; 2000. p. 351–65.
35. Lin PY, Cheng YW, Chu CY, Chien KL, Lin CP, Tu YK. In-office treatment for dentin hypersensitivity: a systematic review and network meta-analysis. *J Clin Periodontol* 2013;**40**:53–64.
36. Pashley DH, Tay FR, Haywood VB, Collins MC, Drisko Cl. Dentin hypersensitivity: Consensus based recommendations for the diagnosis and management of dentin hypersensitivity. *Inside Dent* 2008;**4**(Sp Is):1–35.
37. Gillam DG. Treatment modalities for dentin hypersensitivity. In: Tara S, Clarkson BH, editors. *Clinician's guide to the diagnosis and management of tooth sensitivity*. Berlin, Heidelberg: Springer-Verlag; 2014 Available from: http://dx.doi.org/doi:10.1007/978-3-642-45164-5_6.
38. Gillam DG. Treatment approaches for dentin hypersensitivity. In: Tara S, Clarkson BH, editors. *Clinician's guide to the diagnosis and management of tooth sensitivity*. Berlin, Heidelberg: Springer-Verlag; 2014. p. 51–79.
39. Thrash WJ, Jones DL, Dodds WJ. Effect of a fluoride solution on dentinal hypersensitivity. *Am J Dent* 1992;**5**:299–302.
40. Morris MF, Davis RD, Richardson BW. Clinical efficacy of two dentin desensitizing agents. *Am J Dent* 1999;**12**:72–6.
41. Hodosh M. A superior desensitiser-potassium nitrate. *J Am Dent Assoc* 1974;**88**:831–2.
42. Frechoso SC, Menéndez M, Guisasola C, Arregui I, Tejerina JM, Sicilia A. Evaluation of the efficacy of two potassium nitrate bioadhesive gels (5% and 10%) in the treatment of dentine hypersensitivity. A randomised clinical trial. *J Clin Period ontol* 2003;**30**:315–20.

43. Dunne SM, Hannington-Kiff JG. The use of topical guanethidine in the relief of dentine hypersensitivity: a controlled study. *Pain* 1993;**54**:165–8.
44. Camps J, Pashley D. *In vivo* sensitivity of human root dentin to air blast and scratching. *J Periodontol* 2003;**74**:1589–94.
45. Pillon FL, Romani IG, Schmidt ER. Effect of a 3% potassium oxalate topical application on dentinal hypersensitivity after subgingival scaling and root planing. *J Periodontol* 2004;**75**:1461–4.
46. Gillam DG, Newman HN, Davies EH, Bulman JS, Troullos ES, Curro FA. Clinical evaluation of ferric oxalate in relieving dentine hypersensitivity. *J Oral Rehabil* 2004;**31**: 245–50.
47. Geiger S, Matalon S, Blasbalg J, Tung M, Eichmiller FC. The clinical effect of amorphous calcium phosphate (ACP) on root surface hypersensitivity. *Oper Dent* 2003;**28**: 496–500.
48. Rosaiah K, Aruna K. Clinical efficacy of amorphous calcium phosphate, GC tooth mousse and Gluma desensitizer in treating dentin hypersensitivity. *Int J Dent Clin* 2011;**3**:1–4.
49. Hamlin D, Williams KP, Delgado E, Zhang YP, DeVizio W, Mateo LR. Clinical evaluation of the efficacy of a desensitizing paste containing 8% arginine and calcium carbonate for the in-office relief of dentin hypersensitivity associated with dental prophylaxis. *Am J Dent* 2009;**22**:16A–20A [special issue].
50. Schiff T, Delgado E, Zhang YP, DeVizio W, Mateo LR. Clinical evaluation of the efficacy of an in-office desensitizing paste containing 8% arginine and calcium carbonate in providing instant and lasting relief of dentin hypersensitivity. *Am J Dent* 2009;**22**: 8A–15A [special issue].
51. Li Y, Lee S, Mateo LR, Delgado E, Zhang YP. Comparison of clinical efficacy of three professionally applied pastes on immediate and sustained reduction of dentin hypersensitivity. *Compend Contin Educ Dent* 2013;**34**:6–12 [accessed 10.05.13].
52. Hamlin D, Mateo LR, Dibart S, Delgado E, Zhang YP, DeVizio W. Comparative efficacy of two treatment regimens combining in-office and at-home programs for dentin hypersensitivity relief: a 24-week clinical study. *Am J Dent* 2012;**25**: 146–52.
53. Milleman JL, Milleman KR, Clark CE, Mongiello KA, Simonton TC, Proskin HM. NUPRO Sensodyne prophylaxis paste with Novamin for the treatment of dentin hypersensitivity: a 4-week clinical study. *Am J Dent* 2012;**25**:262–8.
54. Neuhaus KW, Milleman JL, Milleman KR, Mongiello KA, Simonton TC, Clark CE, et al. Effectiveness of a calcium sodium phosphosilicate containing prophylaxis paste in reducing dentine hypersensitivity immediately and 4 weeks after a single application: a double-blind randomized controlled trial. *J Clin Periodontol* 2013;**40**:349–57.
55. Gaffar A. Treating hypersensitivity with fluoride varnish. *Compend Contin Educ Dent* 1999;**20**:27–33.
56. Corona SA, Nascimento TN, Catirse AB, Lizarelli RF, Dinelli W, Palma-Dibb RG. Clinical evaluation of low-level laser therapy and fluoride varnish for treating cervical dentinal hypersensitivity. *J Oral Rehabil* 2003;**30**:1183–9.
57. Merika K, Heftit AF, Preshaw PM. Comparison of two topical treatments for dentine sensitivity. *Eur J Prosthodont Restor Dent* 2006;**14**:38–41.
58. Duran I, Sengun A. The long-term effectiveness of five current desensitizing products on cervical dentine sensitivity. *J Oral Rehabil* 2004;**31**:351–6.

59. Ritter AV, de L Dias W, Miguez P, Caplan DJ, et al. Treating cervical dentin hypersensitivity with fluoride varnish—a randomized clinical study. *J Am Dent Assoc* 2006;**137**:1013–20.
60. Prati C, Cervellati F, Sanasi V, Montebugnoli L. Treatment of cervical dentin hypersensitivity with resin adhesives: 4-week evaluation. *Am J Dent* 2001;**14**:378–82.
61. Baysan A, Lynch E. Treatment of cervical sensitivity with a root sealant. *Am J Dent* 2003;**16**:135–8.
62. Schwarz F, Arweiler N, Georg T, Reich E. Desensitizing effects of an Er:YAG laser on hypersensitive dentine. *J Clin Periodontol* 2002;**29**:211–5.
63. Dondi dall'Orologio G, Malferrari S. Desensitizing effects of Gluma and Gluma 2000 on hypersensitive dentin. *Am J Dent* 1993;**6**:283–6.
64. Dondi dall'Orologio G, Lorenzi R, Anselmi M, Opisso V. Dentin desensitizing effects of Gluma Alternate, Health-Dent Desensitizer and Scotchbond Multi-Purpose. *Am J Dent* 1999;**12**:103–6.
65. Dondi dall'Orologio G, Lone A, Finger WJ. c. *Am J Dent* 2002;**15**:330–4.
66. Kakaboura A, Rahiotis C, Thomaidis S, Doukoudakis S. Clinical effectiveness of two agents on the treatment of tooth cervical hypersensitivity. *Am J Dent* 2005;**18**:291–5.
67. Polderman RN, Frencken JE. Comparison between effectiveness of a low-viscosity glass ionomer and a resin-based glutaraldehyde containing primer in treating dentine hypersensitivity—a 25.2-month evaluation. *J Dent* 2007;**35**:144–9.
68. Kolker JL, Vargas MA, Armstrong SR, Dawson DV. Effect of desensitizing agents on dentin permeability and dentin tubule occlusion. *J Adhes Dent* 2002;**4**:211–21.
69. Swift Jr EJ, May Jr KN, Mitchell S. Clinical evaluation of Prime & Bond 2.1 for treating cervical dentin hypersensitivity. *Am J Dent* 2001;**14**:13–6.
70. Ferrari M, Cagidiaco MC, Kugel G, Davidson CL. Clinical evaluation of a one-bottle bonding system for desensitizing exposed roots. *Am J Dent* 1999;**12**:243–9.
71. Stewardson DA, Crisp RJ, McHugh S, Lendenmann U, Burke FJ. The effectiveness of Systemp.desensitizer in the treatment of dentine hypersensitivity. *Prim Dent Care* 2004;**11**:71–6.
72. Hansen EK. Dentin hypersensitivity treated with a fluoride-containing varnish or a light-cured glass-ionomer liner. *Eur J Oral Sci* 1992;**100**:305–9.
73. Tantbirojn D, Poolthong S, Leevailoj C, Srisawasdi S, Hodges JS, Randall RC. Clinical evaluation of a resin-modified glass-ionomer liner for cervical dentin hypersensitivity treatment. *Am J Dent* 2006;**19**:56–60.
74. Gangarosa LP, Park NH. Practical considerations in iontophoresis of fluoride for desensitizing dentin. *J Prosthet Dent* 1978;**39**:173–8.
75. Brough KM, Anderson DM, Love J, Overman PR. The effectiveness of iontophoresis in reducing dentin hypersensitivity. *J Am Dent Assoc* 1985;**111**:761–5.
76. Gupta M, Pandit IK, Srivastava N, Gugnani N. Comparative evaluation of 2% sodium fluoride iontophoresis and other cavity liners beneath silver amalgam restorations. *J Indian Soc Pedod Dent* 2010;**28**:68–72.
77. Aparna S, Setty S, Thakur S. Comparative efficacy of two treatment modalities for dentinal hypersensitivity: a clinical trial. *Indian J Dent Res* 2010;**21**:544–8. Available from: http://dx.doi.org/doi:10.4103/0970-9290.74213.
78. Starr CB, Mayhew RB, Pierson WP. The efficacy of hypnosis in the treatment of dentin hypersensitivity. *Gen Dent* 1989;**37**:13–5.
79. Eitner S, Bittner C, Wichmann M, Nickenig HJ, Sokol B. Comparison of conventional therapies for dentin hypersensitivity versus medical hypnosis. *Int J Clin Exp Hypn* 2010;**58**:457–75.

80. Renton-Harper P, Midda M. Nd: YAG laser treatment of dentinal hypersensitivity. *Br Dent J* 1992;**172**:13–6.
81. Kimura Y, Wilder-Smith P, Yonaga K, Matsumoto K. Treatment of dentine hypersensitivity by lasers: a review. *J Clin Periodontol* 2000;**27**:715–21.
82. Yilmaz HG, Kurtulmus-Yilmaz S, Cengiz E, Bayindir H, Aykac Y. Clinical evaluation of Er,Cr:YSGG and GaAlAs laser therapy for treating dentine hypersensitivity: a randomized controlled clinical trial. *J Dent* 2011;**39**:249–54.
83. Yilmaz HG, Cengiz E, Kurtulmus-Yilmaz S, Leblebicioglu B. Effectiveness of Er,Cr: YSGG laser on dentine hypersensitivity: a controlled clinical trial. *J Clin Periodontol* 2011;**38**:341–6.
84. Umberto R, Claudia R, Gaspare P, Tenore G, Alessandro DV. Treatment of dentine hypersensitivity by diode laser: a clinical study. *Int J Dent* 2012;**2012** Article ID 858950, 8 pages.
85. West NX. The dentine hypersensitivity patient—a total management package. *Int Dent J* 2007;**57**:411–9.
86. He S, Wang Y, Li X, Hu D. Effectiveness of laser therapy and topical desensitising agents in treating dentine hypersensitivity: a systematic review. *J Oral Rehabil* 2011;**38**:348–58.
87. Gillam DG, Mordan NJ, Newman HN. The Dentin Disc surface: a plausible model for dentin physiology and dentin sensitivity. *Adv Dent Res* 1997;**11**:487–501.
88. Haywood VB. Dentine hypersensitivity: bleaching and restorative considerations for successful management. *Int Dent J* 2002;**52**:366–96.
89. Jalandar SS, Pandharinath DS, Arun K, Smita V. Comparison of effect of desensitizing agents on the retention of crowns cemented with luting agents: an *in vitro* study. *J Adv Prosthodont* 2012;**4**:127–33.
90. Von Troil B, Needleman I, Sanz M. A systematic review of the prevalence of root sensitivity following periodontal therapy. *J Clin Periodontol* 2002;**29**:173–7.
91. Lin YH, Gillam DG. The prevalence of root sensitivity following periodontal therapy: a systematic review. *Int J Dent* 2012;**2012**:407023 Available from: http://dx.doi.org/doi:10.1155/2012/407023. Epub October 31, 2012.
92. de Oliveira DW, Marques DP, Aguiar-Cantuária IC, Flecha OD, Gonçalves PF. Effect of surgical defect coverage on cervical dentin hypersensitivity and quality of life. *J Periodontol* 2013;**84**:768–75 Available from: http://dx.doi.org/doi:10.1902/jop.2012.120479. Epub August 16, 2012.
93. de Oliveira DW, Oliveira-Ferreira F, Flecha OD, Gonçalves PF. Is surgical root coverage effective for the treatment of cervical dentin hypersensitivity? A systematic review. *J Periodontol* 2013;**84**:295–306 Available from: http://dx.doi.org/doi:10.1902/jop.2012.120143. Epub May 2012.
94. Mason S, Hughes N, Sufi F, Bannon L, Maggio B, North M, et al. A comparative clinical study investigating the efficacy of a dentifrice containing 8% strontium acetate and 1040 ppm fluoride in a silica base and a control dentifrice containing 1450 ppm fluoride in a silica base to provide immediate relief of dentin hypersensitivity. *J Clin Dent* 2010;**21**:42–8.
95. Kleinberg I. SensiStat, a new saliva-based composition for simple and effective treatment of dentinal sensitivity pain. *Dent Today* 2002;**21**:42–7.
96. Ayad F, Ayad N, Zhang YP, DeVizio W, Cummins D, Mateo LR. Comparing the efficacy in reducing dentin hypersensitivity of a new toothpaste containing 8.0% arginine, calcium carbonate, and 1450 ppm fluoride to a commercial sensitive toothpaste

containing 2% potassium ion: an eight week clinical study on Canadian adults. *J Clin Dent* 2009;**20**:10–6 [special issue].

97. Docimo R, Montesani L, Maturo P, Costacurta M, Bartolino M, DeVizio W, et al. Comparing the efficacy in reducing dentin hypersensitivity of a new toothpaste containing 8.0% arginine, calcium carbonate, and 1450 ppm fluoride to a commercial sensitive toothpaste containing 2% potassium ion: an eight-week clinical study in Rome, Italy. *J Clin Dent* 2009;**20**:17–22 [special issue].
98. Nathoo S, Delgado E, Zhang YP, DeVizio W, Cummins D, Mateo LR. Comparing the efficacy in providing instant relief of dentin hypersensitivity of a new toothpaste containing 8.0% arginine, calcium carbonate, and 1450 ppm fluoride relative to a benchmark desensitizing toothpaste containing 2% potassium ion and 1450 ppm fluoride, and to a control toothpaste with 1450 ppm fluoride: a three-day clinical study in New Jersey, USA. *J Clin Dent* 2009;**20**:123–30.
99. Schiff T, Delgado E, Zhang YP, DeVizio W, Cummins D, Mateo LR. The clinical effect of a single direct topical application of a dentifrice containing 8.0% arginine, calcium carbonate, and 1450 ppm fluoride on dentin hypersensitivity: the use of a cotton swab applicator versus the use of a fingertip. *J Clin Dent* 2009;**20**:131–6.
100. Schiff T, Mateo LR, Delgado E, Cummins D, Zhang YP, DeVizio W. Clinical efficacy in reducing dentin hypersensitivity of a dentifrice containing 8.0% arginine, calcium carbonate, and 1450 ppm fluoride compared to a dentifrice containing 8% strontium acetate and 1040 ppm fluoride under consumer usage conditions before and after switch-over. *J Clin Dent* 2011;**22**:128–38.
101. Petrou I, Heu R, Stranick M, Lavender S, Zaidel L, Cummins D, et al. A breakthrough therapy for dentin hypersensitivity: how dental products containing 8% arginine and calcium carbonate work to deliver effective relief of sensitive teeth. *J Clin Dent* 2009;**20**:23–31 [special issue].
102. Que K, Fu Y, Lin L, Hu D, Zhang YP, Panagakos FS, et al. Dentin hypersensitivity reduction of new toothpaste containing 8.0% arginine and 1450 ppm fluoride: an 8-week clinical study on Chinese adults. *Am J Dent* 2010;**23A**:28A–35A [special issue].
103. Kakar A, Kakar K, Sreenivasan PK, DeVizio W, Kohli R. Comparison of the clinical efficacy of a new dentifrice containing 8.0% arginine, calcium carbonate, and 1000 ppm fluoride to a commercially available sensitive toothpaste containing 2% potassium ion on dentin hypersensitivity: a randomized clinical trial. *J Clin Dent* 2012;**23**:40–7.
104. Fu Y, Li X, Que K, Wang M, Hu D, Mateo LR, et al. Instant dentin hypersensitivity relief of a new desensitizing dentifrice containing 8.0% arginine, a high cleaning calcium carbonate system and 1450 ppm fluoride: a 3-day clinical study in Chengdu, China. *Am J Dent* 2010;**23**:20A–7A [special issue].
105. Hu D, Stewart B, Mello S, Arvanitidou L, Panagakos F, De Vizio W, et al. Efficacy of a mouthwash containing 0.8% arginine, PVM/MA copolymer, pyrophosphates, and 0.05% sodium fluoride compared to a negative control mouthwash on dentin hypersensitivity reduction. A randomized clinical trial. *J Dent* 2013;**41**:S26–33 Available from: http://dx.doi.org/doi:10.1016/j.jdent.2012.10.001. Epub February 1, 2013.
106. Boneta ARE, Galán Salás RM, Mateo LR, Stewart B, Mello S, Arvanitidou LS, et al. Efficacy of a mouthwash containing 0.8% arginine, PVM/MA copolymer, pyrophosphates, and 0.05% sodium fluoride compared to a commercial mouthwash containing 2.4% potassium nitrate and 0.022% sodium fluoride and a control mouthwash

containing 0.05% sodium fluoride on dentine hypersensitivity: a six-week randomized clinical study. *J Dent* 2013;**41**:S34–41 Available from: http://dx.doi.org/doi:10.1016/j.jdent.2012.11.004. Epub February 1, 2013.

107. Boneta ARE, Ramirez K, Naboa J, Mateo LR, Stewart B, Panagokos F, et al. Efficacy in reducing dentine hypersensitivity of a regimen using a toothpaste containing 8% arginine and calcium carbonate, a mouthwash containing 0.8% arginine, pyrophosphate and PVM/MA copolymer and a toothbrush compared to potassium and negative control regimens: an eight-week randomized clinical trial. *J Dent* 2013;**41**:S42–9.
108. Yan B, Yi J, Li Y, Chen Y, Shi Z. Arginine-containing toothpastes for dentin hypersensitivity: systematic review and meta-analysis. *Quintessence Int* 2013;**44**:709–23.
109. Li R, Tang XJ, Li YH, Wang Y, Chen J, Wang QH, et al. Arginine-containing desensitizing toothpaste for dentine hypersensitivity: a meta-analysis. *Chinese J Evid Based Med* 2011;**11**:570–5.
110. Gillam DG, Tang JY, Mordan NJ, Newman HN. The effects of a novel Bioglass® dentifrice on dentine sensitivity: a scanning electron microscopy investigation. *J Oral Rehabil* 2002;**29**:305–13.
111. Burwell A. Tubule occlusion of a Novamin-containing dentifrice compared to Recaldent-containing dentifrice—a Remin/Demin study *in vitro*. *Novamin Res Rep* 2006.
112. Du Min Q, Bian Z, Jiang H, Greenspan DC, Burwell AK, Zhong J, et al. Clinical evaluation of a dentifrice containing calcium sodium phosphosilicate (Novamin) for the treatment of dentin hypersensitivity. *Am J Dent* 2008;**21**:210–4.
113. Salian S, Thakur S, Kulkarni S, LaTorre G. A randomized controlled clinical study evaluating the efficacy of two desensitizing dentifrices. *J Clin Dent* 2010;**21**:82–7.
114. Litkowski L, Greenspan DC. A clinical study of the effect of calcium sodium phosphosilicate on dentin hypersensitivity—proof of principle. *J Clin Dent* 2010;**21**:77–81.
115. Pradeep AR, Sharma A. Comparison of clinical efficacy of a dentifrice containing calcium sodium phosphosilicate to a dentifrice containing potassium nitrate and to a placebo on dentinal hypersensitivity: a randomized clinical trial. *J Periodontol* 2010;**81**:1167–73.
116. Narongdej T, Sakoolnamarka R, Boonroung T. The effectiveness of a calcium sodium phosphosilicate desensitizer in reducing cervical dentin hypersensitivity: a pilot study. *J Am Dent Assoc* 2010;**141**:995–9.
117. Sharma N, Roy S, Kakar A, Greenspan DC, Scott R. A clinical study comparing oral formulations containing 7.5% calcium sodium phosphosilicate (Novamin), 5% potassium nitrate, and 0.4% stannous fluoride for the management of dentin hypersensitivity. *J Clin Dent* 2010;**21**:88–92.
118. Earl JS, Topping N, Elle J, Langford RM, Greenspan DC. Physical and chemical characterization of the surface layers formed on dentin following treatment with a fluoridated toothpaste containing Novamin. *J Clin Dent* 2011;**22**:68–73.
119. Wang Z, Jiang T, Sauro S, Pashley DH, Toledano M, Osorio R, et al. The dentine remineralization activity of a desensitizing bioactive glass-containing toothpaste: an *in vitro* study. *Aust Dent J* 2011;**56**:372–81 Available from: http://dx.doi.org/doi:10.1111/j.1834-7819.2011.01361.x. Epub October 13, 2011.
120. West NX, Macdonald EL, Jones SB, Claydon NC, Hughes N, Jeffery P. Randomized *in situ* clinical study comparing the ability of two new desensitizing toothpaste technologies to occlude patent dentin tubules. *J Clin Dent* 2011;**22**:82–9.
121. Rajesh KS, Hedge S, Arun Kumar MS, Shetty DG. Evaluation of the efficacy of a 5% calcium sodium phosphosilicate (Novamin®) containing dentifrice for the relief of dentinal hypersensitivity: a clinical study. *Indian J Dent Res* 2012;**23**:363–7.

122. Ananthakrishna S, Raghu TN, Koshy S, Kumar N. Clinical evaluation of the efficacy of bioactive glass and strontium chloride for treatment of dentinal hypersensitivity. *J Interdiscipl Dent* 2012;**2**:92–7.
123. Acharya AB, Surve SM, Thakur SL. A clinical study of the effect of calcium sodium phosphosilicate on dentin hypersensitivity. *J Clin Exp Dent* 2013;**5**:e18–22.
124. Thanatvarakorn O, Nakashima SA, Prasansuttiporn T, Ikeda M, Tagami J. *In vitro* evaluation of dentinal hydraulic conductance and tubule sealing by a novel calcium-phosphate desensitizer. *J Biomed Mater Res B Appl Biomater* 2013;**101**:303–9.
125. Giniger M, Macdonald J, Ziemba S, et al. The clinical performance of professionally dispensed bleaching gel with added amorphous calcium phosphate. *J Am Dent Assoc* 2005;**136**:383–92.
126. Azarpazhooh A, Limeback H. Clinical efficacy of casein derivatives: a systematic review of the literature. *J Am Dent Assoc* 2008;**139**:915–24.
127. Walsh LJ. The effects of GC tooth mousse on cervical dentinal sensitivity: a controlled clinical trial. *Int Dent* 2010;**SA 12**:4–12.
128. Madhavan S, Nayak M, Shenoy A, Shetty R, Prasad K. Dentinal hypersensitivity: a comparative clinical evaluation of CPP-ACP F, sodium fluoride, propolis, and placebo. *J Conserv Dent* 2012;**15**:315–8.
129. Reynolds EC. Anticariogenic complexes of amorphous calcium phosphate stabilized by casein phosphopeptides: a review. *Spec Care Dent* 1998;**18**:8–16.
130. Yengopal V, Mickenautsch S. Caries preventive effect of casein phosphopeptide-amorphous calcium phosphate (CPP-ACP): a meta-analysis. *Acta Odontol Scand* 2009;**67**:321–32.
131. Hill R, Gillam DG, Karukhina N. The tubular occluding properties of a novel biomimetic hydroxyapatite toothpaste. *PER/IADR Cong* 2012; **September**:12–5 [Poster abstract no. 640].
132. Park JJ, Park JB, Kwon YH, Herr Y, Chung JH. The effect of microcrystalline hydroxyapatite containing toothpaste in the control of tooth hypersensitivity. *J Korean Acad Periodontol* 2005;**35**:577–90.
133. Rimondini L, Palazzo B, Iafisco M, Canegallo F, et al. The remineralizing effect of a carbonate-hydroxyapatite microparticles on dentine on dentine. *Mater Sci Forum* 2007;**539**:602–5.
134. Kang S-J, Kwon Y-H, Park J-B, Herr Y, Chung J-H. The effects of hydroxyapatite toothpaste on tooth hypersensitivity. *J Korean Acad Periodontol* 2009;**39**:9–16.
135. Kim SH, Park JB, Lee CW, Koo KT, Kim TI, Seol YJ, et al. The clinical effects of a hydroxyapatite containing toothpaste for dentine hypersensitivity. *J Korean Acad Periodontol* 2009;**39**:87–94.
136. Orsini G, Procaccini M, Manzoli L, Giuliodori F, Lorenzini A, Putignano A. A double-blind randomized-controlled trial comparing the desensitizing efficacy of a new dentifrice containing carbonate/hydroxyl-apatite nanocrystals and a sodium fluoride/potassium nitrate dentifrice. *J Clin Periodontol* 2010;**37**:510–7.
137. Shetty S, Kohad R, Yeltiwar R. Hydroxyapatite as an in-office agent for tooth hypersensitivity: a clinical and scanning electron microscopic study. *J Periodontol* 2010;**81**:1781–9.
138. Tschoppe P, Zandim DL, Martus P, Kielbassa AM. Enamel and dentine remineralization by nano-hydroxyapatite toothpastes. *J Dent* 2011;**39**:430–7.
139. Pawlowska J. Strontium chloride its importance in dentistry and prophylaxis. *Czas Stomatol* 1956;**9**:353–61.

140. Kun L. Biophysical study of dental tissues under the effect of a local strontium application. *Schweiz Monatsschr Zahnheilkd* 1976;**86**:661–76.
141. Gedalia I, Brayer L, Katter N, Richter M, Stabholz A. The effect of fluoride and strontium application on dentine. *In vivo* and *in vitro* studies. *J Periodontol* 1978;**49**:269–72.
142. Pearce N, Addy M, Newcombe RG. Dentine hypersensitivity: a clinical trial to compare 2 strontium desensitising toothpastes with a conventional fluoride toothpaste. *J Periodontol* 1994;**65**:113–9.
143. Ling TY, Gillam DG. The effectiveness of desensitizing agents for the treatment of cervical dentine sensitivity (CDS)—a review. *J West Soc Periodontol Periodontal Abstr* 1996;**44**:5–12.
144. Gillam DG, Jackson RJ, Bulman JS, Newman HN. Comparison of 2 desensitizing dentifrices with a commercially available fluoride dentifrice in alleviating cervical dentine sensitivity. *J Periodontol* 1996;**67**:737–42.
145. West NX, Addy M, Jackson RJ, Ridge DB. Dentine hypersensitivity and the placebo response. A comparison of the effect of strontium acetate, potassium nitrate and fluoride toothpastes. *J Clin Periodontol* 1997;**24**:209–15.
146. Kobler A, Kub O, Schaller H, Gernhardt CR. Clinical effectiveness of a strontium chloride-containing desensitizing agent over 6 months: a randomized, double-blind, placebo-controlled study. *Quintessence Int* 2008;**39**:321–5.
147. Hughes N, Mason S, Jeffery P, Welton H, Tobin M, O'Shea C, et al. A comparative clinical study investigating the efficacy of a test dentifrice containing 8% strontium acetate and 1040 ppm sodium fluoride versus a marketed control dentifrice containing 8% arginine, calcium carbonate, and 1450 ppm sodium monofluorophosphate in reducing dentinal hypersensitivity. *J Clin Dent* 2010;**21**:49–55.
148. Kumar RV, Shubhashini N, Sheshan H, Kranti K. A clinical trial comparing a stannous fluoride based dentifrice and a strontium chloride based dentifrice in alleviating dentinal hypersensitivity. *J Int Oral Heath* 2010;**2**:37–50.
149. Jackson RJ. Potential treatment modalities for dentine hypersensitivity: home use products. In: Addy M, Edgar M, Orchardson R, editors. *Tooth wear and sensitivity*. London: Martin Dunitz Ltd; 2000. . p. 328–38..
150. Karim BFA, Gillam DG. The efficacy of strontium and potassium toothpastes in treating dentine hypersensitivity: a systematic review. *Int J Dent* 2013. Article ID 573258, 13 pages <http://dx.doi.org/10.1155/2013/573258>.
151. Schiff T, Saletta L, Baker RA, Winston JL, He T. Desensitizing effect of a stabilized stannous fluoride/sodium hexametaphosphate dentifrice. *Compend Contin Educ Dent* 2005;**26**:35–40.
152. Schiff T, He T, Sagel L, Baker R. Efficacy and safety of a novel stabilized stannous fluoride and sodium hexametaphosphate dentifrice for dentinal hypersensitivity. *J Contemp Dent Pract* 2006;**2**:1–8.
153. White DJ, Lawless MA, Fatade A, Baig A, von Koppenfels R, Duschner H, et al. Stannous fluoride/sodium hexametaphosphate dentifrice increases dentin resistance to tubule exposure *in vitro*. *J Clin Dent* 2007;**18**:55–9.
154. Ni LX, He T, Chang A, Sun L. The desensitizing efficacy of a novel stannous-containing sodium fluoride dentifrice: an 8-week randomized and controlled clinical trial. *Am J Dent* 2010;**23**:17B–21B [special issue].

155. Einwag J, Hermann J, He T, Day T, Zhang Y, Anastasia MK, et al. A clinical assessment of the efficacy of a stannous-containing sodium fluoride dentifrice on dentinal hypersensitivity. *J Contemp Dent Pract* 2010;**11**:8–26.
156. He T, Chang J, Cheng R, Li X, Sun L, Biesbrock AR. Clinical evaluation of the fast onset and sustained sensitivity relief of a 0.454% stannous fluoride dentifrice compared to an 8.0% arginine-calcium carbonate-sodium monofluorophosphate dentifrice. *Am J Dent* 2011;**24**:336–40.
157. Ganss C, von Hinckeldey J, Tolle A, Schulze K, Klimek J, Schlueter N. Efficacy of the stannous ion and a biopolymer in toothpastes on enamel erosion/abrasion. *J Dent* 2012;**40**:1036–43 Available from: http://dx.doi.org/doi:10.1016/j.jdent.2012.08.005 Epub August 20, 2012.
158. Renggli HH. Effekt von Aminfluorid-Zahnpasten auf überempfindliche Zahnhälse. *Acta Med Dent Helv* 1997;1–5 <http://www.gaba-info.it/htm/833/it_IT/Renggli-H-H-1997. pdf> [accessed 08.03.14].
159. Zappa U. *The effect of Elmex SENSITIVE rinse on hypersensitive teeth. Universität Basel, Switzerland Report.* <http://www.gaba.com/data/docs/download/1211/en/Studie-Zappa-1999.pdf>. [accessed 08.03.14].
160. Markowitz K. The original desensitizers: strontium and potassium salts. *J Clin Dent* 2009;**20**:145–51.
161. Ajcharanukul O, Kraivaphan P, Wanachantararak S, Vongsavan N, Matthews B. Effects of potassium ions on dentine sensitivity in man. *Arch Oral Biol* 2007;**52**:632–9.
162. Poulsen S, Errboe M, Lescay Mevil Y, Glenny AM. Potassium containing toothpastes for dentine hypersensitivity. *Cochrane Database Syst Rev* 2006;**19**:CD001476 [Review]. doi:10.1002/14651858.CD001476.pub2.
163. Pol DG, Jonnala J, Chute M, Gunjikar T, Pol S. Potassium nitrate in the treatment of dentinal hypersensitivity—a mini analysis of studies. *J Indian Dent Assoc* 2010;**4**:399–403.
164. Burwell AK, Litkowski LJ, Greenspan DC. Calcium sodium phosphosilicate (Novamin): remineralization potential. *Adv Dent Res* 2009;**21**:35–9.
165. Greenspan DC. Novamin and tooth sensitivity—an overview. *J Clin Dent* 2010;**21**:61–5.
166. Lavender SA, Petrou I, Heu R, Stranick M, Cummins D, Kilpatrick-Liverman L, et al. Mode of action studies on a new desensitizing dentifrice containing 8.0% arginine, a high cleaning calcium carbonate system and 1450 ppm fluoride. *Am J Dent* 2010;**23**:14A–9A [special issue].
167. Greenhill JD, Pashley DH. The effects of desensitizing agents on the hydraulic conductance of human dentin *in vitro*. *J Dent Res* 1981;**60**:686–98.
168. Mostafa P, Addy M, Morgan T. Scanning electronmicroscopic, X-ray diffraction analysis, atomic absorption and fluoride probe measurements of the uptake of toothpaste ingredients onto dentine. *J Dent Res* 1983;**62**:433 [Abstract no. 165].
169. Pashley DH, O'Meara JA, Kepler EE, Galloway SE, Thompson SM, Stewart SP. Dentin permeability. Effects of desensitizing toothpastes *in vitro*. *J Periodontol* 1984;**55**:522–5.
170. Gillam DG. *The assessment and treatment of cervical dentinal sensitivity* [DDS thesis]. Edinburgh, Scotland: University of Edinburgh; 1992.

171. Mordan NJ, Gillam DG, Critchell J, Curro FA, Ley F. Effects of abrasive components on dentine: an SEM study. *J Dent Res* 2002;**81**:A-374 [special issue, Abstract no. 3010].
172. Absi EG, Addy M, Adams D. Dentine hypersensitivity: uptake of various sensitizing toothpastes onto dentine *in vitro* SEM investigation. *J Dent Res* 1989;**68**:573 [special issue, Abstract no. 117].
173. Parkinson CR, Willson RJ. A comparative *in vitro* study investigating the occlusion and mineralization properties of commercial toothpastes in a four-day dentin disc model. *J Clin Dent* 2011;**22**:74–81.
174. Addy M, Mostafa P. Dentine hypersensitivity. II. Effects produced by the uptake *in vitro* of toothpastes onto dentine. *J Oral Rehabil* 1989;**16**:35–48.
175. Absi EG, Addy M, Adams D. Dentine hypersensitivity—the effect of toothbrushing and dietary compounds on dentine *in vitro*: an SEM study. *J Oral Rehabil* 1992;**19**:101–10.
176. Pashley DH. How can sensitive dentine become hypersensitive and can it be reversed? *J Dent* 2013;**41**:S49–55.
177. Sharma D, McGuire JA, Amini P. Randomized trial of the clinical efficacy of a potassium oxalate-containing mouthrinse in rapid relief of dentin sensitivity. *J Clin Dent* 2013;**24**:62–7.
178. Sharma D, McGuire JA, Gallob JT, Amini P. Randomised clinical efficacy trial of potassium oxalate mouthrinse in relieving dentinal sensitivity. *J Dent* 2013;**41**:S40–8.
179. Sharma D, Hong CX, Heipp PS. A novel potassium oxalate-containing tooth-desensitising mouthrinse: a comparative *in vitro* study. *J Dent* 2013;**41**:S18–27.
180. Brunton PA, Davies RP, Burke JL, Smith A, Aggeli A, Brookes SJ, et al. Treatment of early caries lesions using biomimetic self-assembling peptides—a clinical safety trial. *Br Dent J* 2013;**215**:E6. Available from: http://dx.doi.org/doi:10.1038/sj.bdj.2013.741
181. Karlinsey RL, Mackey AC, Schwandt CS. Effects on dentin treated with eluted multi-mineral varnish *in vitro*. *Open Dent J* 2012;**6**:157–63 Available from: http://dx.doi.org/doi:10.2174/1874210601206010157. Epub October 5, 2012.
182. Danelon M, Takeshita EM, Sassaki KT, Delbem AC. *In situ* evaluation of a low fluoride concentration gel with sodium trimetaphosphate in enamel remineralization. *Am J Dent* 2013;**26**:15–20.
183. Moretto MJ, Delbem AC, Manarelli MM, Pessan JP, Martinhon CC. Effect of fluoride varnish supplemented with sodium trimetaphosphate on enamel erosion and abrasion: an *in situ/ex vivo* study. *J Dent* 2013;**41**:1302–6 Available from: http://dx.doi.org/doi:10.1016/j.jdent.2013.09.008. Epub October 3, 2013.
184. Manarelli MM, Delbem AC, Lima TM, Castilho FC, Pessan JP. *In vitro* remineralizing effect of fluoride varnishes containing sodium trimetaphosphate. *Caries Res* 2014;**48**:299–305. [Epub ahead of print].
185. Dababneh RH, Khour AT, Addy M. Dentine hypersensitivity—an enigma? A review of terminology, mechanisms, aetiology and management. *Br Dent J* 1999;**187**:606–11 [Discussion 603].
186. Kassab MM, Cohen RE. The etiology and prevalence of gingival recession. *J Am Dent Assoc* 2003;**134**:220–5.
187. Marini MG, Greghi SLA, Passanezi E, Sant'ana ACP. Gingival recession: prevalence, extension and severity in adults. *J Appl Oral Sci* 2004;**12**:250–5.

188. Addy M. Dentine hypersensitivity; definition, prevalence, distribution and aetiology. In: Addy M, Embery G, Edgar WM, Orchardson R, editors. *Tooth wear and sensitivity: clinical advances in restorative dentistry*. London: Martin Dunitz Ltd; 2000. p. 239–48.
189. Addy M. Dentine hypersensitivity: new perspectives on an old problem. *Int Dent J* 2002;**52**:367–75.
190. Addy M, Hunter ML. Can tooth brushing damage your health? Effect on oral and dental tissues. *Int Dent J* 2003;**53**:177–86.
191. Addy M. Toothbrushing, tooth wear and dentine hypersensitivity—are they associated? *Int Dent J* 2005;**65**:261–7.
192. West NX, Sanz M, Lussi A, Bartlett D, Bouchard P, Bourgeois D. Prevalence of dentine hypersensitivity and study of associated factors: a European population-based cross-sectional study. *J Dent* 2013;**41**:841–51.

The importance of subjective assessments of dentine hypersensitivity

Finbarr Allen
Cork Dental School & Hospital, Wilton, Cork, Ireland

Introduction

Over the past century, there has been a seismic shift in health status in developed countries. Life expectancy is a commonly used statistic to indicate that improved health status has arisen from improvements in socioeconomic status and health care delivery systems. In the early part of the twentieth century, acute illnesses were associated with high mortality rates, which, in turn, influenced life expectancy across the population. Infectious diseases were often fatal, and the prognosis for these diseases was hampered by poor sanitation and lack of antimicrobial medications. As our understanding of population health has improved, it is now recognized that life expectancy can be strongly influenced by easy-to-access care delivery systems and wider social factors such as good housing and sanitation.

At the present time, life expectancy in industrialized countries is increasing rapidly in line with improvements in economic status at the national level and better access to health care, which is underpinned by advances in scientific knowledge.[1] This is by no means equitable across all societies, and there is evidence of disparity and social inequality. For example, residents of deprived areas continue to have a poorer life expectancy than residents of affluent areas, and this continues to be a burden on health care systems across the western world.[2] Furthermore, there are between-country inequalities. Life expectancy in the so-called third-world countries is well below that of their more affluent neighbors.

While recognizing that there are disparities, there has been a dramatic shift in population profile from one characterized by moderate life expectancy and high prevalence of acute illness to long life expectancy and high prevalence of chronic illness. In many cases, there are no cures for chronic disease and health care involves palliative treatment to ameliorate symptoms. This care is ongoing, frequently requires input from a range of health care professionals in different care settings, and is in demand by an increasing proportion of the population, in line with improved life expectancy.

This scenario has created a major problem for health care systems across the globe because amelioration of the effects of chronic disease is very costly in

Dentine Hypersensitivity. DOI: http://dx.doi.org/10.1016/B978-0-12-801631-2.00005-1

economic terms. Health policymakers now face challenges of how to control the rapidly expanding costs of health care at the same time as ensuring access to quality health care for the widest possible spectrum of the population. At the same time as this economic and political challenge, the thinking of how to assess the impact of disease on health status is changing. The purpose of this chapter is to address health assessment methodology in general and, specifically, how this methodology may be used in the assessment of dental hypersensitivity.

Assessment of dental disease and health

Trends in oral health status indicate a dramatically reduced prevalence of edentulousness, with larger numbers of adults keeping more of their teeth into old age. The nature of dental disease is such that it is cumulative over time, and many of the more severe manifestations of dental disease present late in life. The burden of disease management is lifelong, and it is chronic in nature. Dental pain is most often associated with primary and secondary caries, failing restorations, and exposed root surfaces. Pain in the head and neck region is also associated with mucosal conditions, neuropathic facial pain, and myofacial pain. Each of these conditions can affect any age group; thus, the prevalence of self-reported dental pain is widespread. The nature of pain is such that it is highly subjective, and subjective assessment of pain is influenced by factors such as personality and coping skills. Objective measures of pain do not adequately capture the patient experience of pain or the impact of treatment to moderate pain symptoms. They do not predict patient behaviors in response to painful conditions and, in fact, there is often quite a difference between objectively measured disease and subjective symptom status. Accordingly, some form of subjective assessment is needed to fully elucidate the impact of pain on daily living.

Over the past few decades, there has been a move away from sole use of objective measures to assess disease. The limitations of the "biomedical" paradigm of health have been principally recognized, and this model only deals with disease. However, health and disease, in sociological terms, are not part of a continuum, but rather are independent dimensions of human experience.[3] This was recognized by the World Health Organization (WHO) as early as 1948, when health was defined as "a complete state of physical, mental, and social well-being, and not merely the absence of illness." More recently, distinctions have been made between disease ("abnormalities in anatomical structures, physiological or biochemical processes") and illness ("subjective perception of changes in his/her physical, mental, and social well-being").[4] Consequently, any measure of health needs to assess social and emotional aspects of health and to assess presence or absence of disease.

In the socioenvironmental model of health, each of these separate conceptual domains is recognized. In this model, the complex multidimensional nature of health is encompassed, including cultural, environmental, and psychosocial influences. The biopsychosocial model of health recognizes the need to include both objective and subjective methods to assess the impact of disease on health status.

Objective measures indicate stages in pathological processes and provide some guidance regarding clinical care requirements. For instance, when dental caries has reached the pulp chamber, then an endodontic procedure or extraction would be the clinical care choice. However, it becomes more complicated when, for instance, widespread gingival recession is evident. This can be objectively measured, but the measure of attachment loss does not indicate the subjective impact of this condition on the patient. Loss of periodontal attachment may have a negative impact on appearance or precipitate sensitivity, or even both. These consequences are not captured by the objective measure alone.

In recent medical and dental literature, "quality of life" has become a ubiquitous term. However, it is clear in various reports that this term means different things to different authors. As Locker[5] suggests, measurement of quality of life could involve measurement of practically anything of interest to anybody. The lack of a precise definition of "quality of life" suggests that the term only has meaning on an individual level. The reason for this is that certain attributes such as adequate income and secure employment, which are widely believed to improve quality of life, are not rated with equal importance by everybody. A more recent definition of "quality of life" has been suggested by Raphael et al., who believe that "quality of life is concerned with the degree to which a person enjoys the important possibilities of life."[6] In a commentary on this issue, Fayers and Machin refer to the assumption of causal links between physical functioning and quality of life.[7] It seems logical to assume that inability to function within "normal" limits is likely to impact negatively on quality of life. However, this assumption takes no account of an individual's ability to adapt to altered circumstances or the possibility that external factors may moderate the consequences of changed physical functioning status. However, the literature contains many examples of patients with serious, debilitating illnesses who have good life satisfaction and quality of life scores. The so-called disability paradox is a term used to describe this phenomenon.

In an effort to focus on the assessment of health and quality of life issues, the term "health-related quality of life" is now widely used. Other terms used include "subjective health status" and "patient-rated health status indicators."[8]

How do we measure "health-related quality of life?"

There has been a relatively large amount of development and evaluation work in this area over the past 20 years, but it is recognized that no single measure is appropriate for oral health. Different conceptual approaches have been taken in the development of the content of measures and scoring systems, and most describe themselves as "oral health-related quality of life" (OHQoL) measures. Before reviewing some of the existing measures, there are some fundamental measurement issues that a researcher should consider when planning to measure OHQoL.

1. Is the measure *valid*? A key property of any measure is its validity and, in essence, this is whether the intended measure actually measures what it is you are trying to measure. There are a number of aspects to this. To begin, it is important that the measure content is

appropriate and relevant to the topic undergoing study. This is face validity. In this regard, widely used generic measures of health such as the SF-36 or Sickness Impact Profile are of little use when assessing oral health outcomes. These measures do not contain items of specific relevance to oral health; thus, they have poor face validity for oral health. It is important that the questions in the measure are relevant to the condition in question, that there are enough questions relevant to the condition of interest, and that all levels of the condition are recorded. This is referred to as "content validity." To illustrate this concept, if one wanted to assess the impact of tooth loss, then there is a need to include specific questions related to chewing and appearance in the measure. The more relevant questions in the measure, the better its content validity.

Further aspects of validity are "construct," "concurrent," and "discriminant" validity. Construct validity refers either to the scores of a measure or to some underlying construct. It has two components of concurrent and discriminant validity. Intuitively, one might expect that OHQoL is likely to be poorer in patients with significant disease. In theory, there should be a positive correlation between these two domains. When testing the concurrent validity of the measure, demonstration of an association between disease status and OHQoL would be an indicator of good concurrent validity. It should be recognized that such correlation is not always very strong because objective and subjective measures do not measure precisely the same thing. Discriminant validity is the negative of concurrent validity. It refers to the lack of association between scores from the measure and unrelated constructs; therefore, we would not expect OHQoL to be related to eye color!

2. Is the measure *sensitive to change*? When a disease state has had significant impact, then the goal of intervention is to either eliminate the problem or reduce its impact. As mentioned, some of the consequences of disease are not fully reversible and the goal of treatment is to limit the impact. For example, complete loss of teeth is irreversible and complete replacement prostheses are provided to reduce the effects of tooth loss. Implant-retained prostheses offer the possibility of improved function and OHQoL when compared with conventional prostheses but at a much higher treatment cost. When evaluating the subjective impact of these treatments on OHQoL, it is important that a measure is sufficiently responsive to pre- and posttreatment changes. This "responsiveness" is critical in evaluating treatment interventions. It should be noted that some measures are more appropriate for use in cross-sectional studies than in longitudinal studies. These measures are likely to have poor to moderate responsiveness properties and are potentially unable to detect changes that occur after treatment intervention. This may be related to content, and it may also be a problem with short versions of measures. It is a known problem that removing items from a measure compromises its responsiveness. Short versions of measures (e.g., the Oral Health Impact Profile, OHIP-14) facilitate use in clinical settings, but there is a trade-off against compromised responsiveness properties. When planning to use an OHQoL questionnaire as an outcome measure, it is important to ensure that its responsiveness properties are understood. Shortened versions of measures are prone to "ceiling" and "floor" effects, whereby real change in either improvement or deterioration is masked by an insufficient number of items in the measure.

3. Is the measure *reliable*? Quality of life measures are subject to random error effects, as with any other form of measurement. It is important to establish that observations made from the measurement are as close to true measurement effects as possible. A variety of statistical methods are used to determine the reliability of a measure, with the most common being intraclass correlation coefficients (ICCs). This statistic is presented on a 0–1 scale, and scores close to 1 indicate strength of reliability. Scores less than 0.7 indicate moderate to poor reliability, and these items or scales are prone to random errors that compromise the accuracy of observations made. Another measure of reliability is test/retest reliability. Health status is subject to change, and patient-rated health status can change if a disease state progresses or if the perception of the problem is altered by change in circumstances. Ideally, responses given to an OHQoL measure should remain consistent in the absence of any change in the health state in the short term. This means that there should be no major difference in the responses given to a measure completed on two occasions with a short interval period in the absence of any intervention. Test/retest reliability is assessed by calculating ICCs, Cohen's Kappa scores, or Pearson's correlation coefficients for two OHQoL measures administered over a short time interval, for example, 1 month. Once again, the higher the score, the more reliable the measure.

4. Is the measure *practical to use*? This is of importance in descriptive population studies and in clinical trials. First, it is important to understand that the characteristics of quality of life measures required for each of these contexts is different. In a descriptive population study, the goal is to determine the prevalence of impacts of dental status and disease states in OHQoL. This information might determine the allocation of resources or service planning. Accordingly, it is important to use a measure that has good construct validity. This usually requires the measure to have large numbers of items likely to have high, medium, and low prevalence of impact. This facilitates the identification of subgroups of a population who are particularly prone to the effects of illness, with a high impact on OHQoL. On the contrary, the purpose of a clinical trial is to detect the impact of an intervention on OHQoL, and perhaps to compare different interventions on a particular disease state or profile. Therefore, it is vital to have sufficient items in the measure that are responsive to change after an intervention. In this context, it is undesirable to have a large number of items likely to have a low prevalence of impact, because these items will not be influenced by a therapeutic intervention in the majority of the sample. A measure with a high proportion of low prevalence impacts will be prone to ceiling or floor effects, or both, and this will potentially mask real change and underestimate the impact of an intervention.

Practical utility is an issue for both forms of study. In a population study, it may be impractical to include a large number of items because of the examiner and response burden and because of financial reasons. It is also important to avoid random removal of items, because the most widely used measures are empirically based and are multidimensional in nature. Two items per subscale are recommended, and the overall scores are based on summary of the scales that comprise the measure. Ideally, any measure used in a population study should allow for the possibility of comparison with other similar studies. There is a relative paucity of international data that make between-country comparisons, and most available data use the OHIP-14.

In clinical trials, responsiveness is reduced by eliminating items. The resultant loss of responsiveness to change may hamper measurement of actual change that has occurred. There is a balance between response burden and adequate responsiveness properties, and this should be considered before any clinical trial.

Finally, the measure you intend to use needs to be appropriate for the population of interest. It is not appropriate to use a measure derived from adult populations when the population includes children. A variety of child-specific measures have been developed (e.g., the Child Perceptions Questionnaire), and these are appropriate for use with children up to the age of 16 years. Some measures are described as generic, i.e., they are not specific to a particular condition and are intended for use as a general measure of health status. Most of the widely used OHQoL measures fall into this category, namely the OHIP, the General Oral Health Assessment Index (GOHAI), the Oral Impacts on Daily Performance (OIDP), and the Oral Health Quality of Life measure (OHQoL). The generic nature of these measures makes them useful for measuring oral health, but there is a loss of condition-specific questions. Researchers need to consider this and they must ensure that the measure they intend to use has appropriate content for the condition they wish to measure. This is particularly relevant when using shortened versions of generic measures.[9]

Interpretation of OHQoL data and measurement of pain symptoms

Interpretation of OHQoL data is not as intuitive as, for example, measurement of body temperature or blood pressure. The potential consequences of significantly elevated body temperature and blood pressure are well described. However, health-related quality of life data are intrinsically meaningless and are only of value when comparisons are made. This includes within-group and between-group comparisons in clinical trials, and comparison of reference groups in descriptive population studies. The latter is somewhat hampered by the lack of international data sets and inconsistency in collecting data. A simple example of this is a lack of consistency in the literature regarding the framing of the reference period for questions. When the OHIP-49 was first developed, the reference period for questions was 12 months. Since then, a variety of clinical trial studies have used reference periods of 1-, 3-, and 6-month. The potential impact of this has been assessed by Saila et al.,[10] and it does not appear to be problematic in descriptive population studies. However, in clinical trials, one might be interested in how quickly a treatment intervention is effective; therefore, the reference period should be framed accordingly. This is particularly relevant to pain assessment.

A further consideration is the relationship between frequency and severity. A commonly used tool for pain assessment is the visual analogue scale (VAS). This approach is not based on multidimensional scales but has been widely used as a measure of pain symptoms. This takes the form of a line 100 mm in length, with extreme response possibilities at either end. In a pain measurement scenario, this could be presented as, "How would you rate pain in your jaw at the present time?", with the response at one end of the line being "absolutely as bad as it could be," and at the other end the response would be "absolutely no problem whatsoever." The patient is asked to put a mark perpendicular to the response line somewhere between the two extremes. The distance to the mark is measured in millimeters. This gives some indication regarding how an individual patient is feeling about the pain at a given time. Patients can be asked to repeat this process over time as treatment progresses, and the information can be used to indicate whether they perceive improvement with treatment. A potential shortcoming of VAS is that these measures give an indication of the direction of change but do not explain why it has happened.

A further approach is to create measurement scales that allow patients to describe how severe the problem is and how frequently they experience the problem. This approach has been used in the construction of the OIDP. The dimension score is either the sum of responses or the product of severity and frequency scores. This approach allows the identification of patients with regular experience of severe symptoms, and it offers a more comprehensive analysis. In public health terms, these patients may be given priority for care programs over patients with problems of low severity and frequency. Finally, there is some concern that so-called OHQoL measures are actually capturing symptom status (functional, psychosocial, or both) rather than impact on quality of life. This has been discussed by Prutkin and Feinstein, who recommended using global ratings.[11] In a recent study by Locker and Quinonez,[12] they addressed this phenomenon in a population survey by asking respondents, who reported one or more impacts on the OHIP-14 measure to rate their quality of life on a 6-point scale ranging from "very poor" to "excellent." They showed some discordance between reported impacts and how participants rated their quality of life. The main discordance was that a proportion who reported impacts indicated that they were not bothered by these impacts to any great extent. Of those who reported one or more impacts either "very often" or "fairly often," only 36% indicated that their quality of life was affected. This would suggest that incorporation of both global ratings and an OHQoL measure would improve understanding of the consequences of functional and psychosocial impacts.

How to capture clinically relevant change

It is a standard practice to use statistical significance as a measure of importance when comparing data. However, it may be argued that this, although important, should not be the sole measure of interest when measuring health-related quality of life. One of our goals in health care is to provide treatments that improve the quality of life of our patients. Therefore, it is important to show that interventions are perceived by those patients to have resulted in clinically meaningful change. This can be change on a scale from small but important change to very large change. In health status measurement, the term "minimally important difference" (MID) is used to describe a change that patients perceive to be important. Jaescke et al.[13] have defined MID as "The smallest difference in score in the domain of interest which patients perceive as beneficial and which would mandate, in the absence of troublesome side effects and excessive cost, a change in the patient's management." If this minimal difference is not reported by patients, then even if the difference is shown to be statistically significant, the intervention is not effective. The MID can be calculated in two ways, either using a distribution- or an anchor-based approach. The distribution-based approach involves calculation of either a standardized response mean (SRM) or an effect size (ES). The SRM is the ratio of the mean change (mean $T1$ − mean $T2$) to the standard deviation of that change. An ES is the ratio of the mean change (mean $T1$ − mean $T2$) to the standard deviation of the $T1$ measurement. These scores are indicators of the clinical significance of change, whether observed changes are small, moderate, or large. Kazis et al.[14] showed how pre- and posttreatment changes for asthma management could be measured as statistically significant but not clinically meaningful. Studies such as this highlight the pitfalls in using statistical significance to support the use of one intervention over another.

Cohen[15] has provided benchmarks for interpretation of ESs and describes an ES of 0.2 as a small change, 0.3–0.7 as a moderate change, and 0.8 or more as a large change. When evaluating the Dentine Hypersensitivity Experience Questionnaire (DHEQ), Baker et al.[16] used ES statistics to describe the responsiveness of that measure.

In the anchor-based approach, pre- and posttreatment differences in quality of life scores are compared to self-reported global change scores ranging from a great deal worse to a great deal better. Locker et al.[17] used this approach for determining the MID for the OHIP-14 when used with elderly patients undergoing routine dental treatment. The MID was 5 scale points or approximately 10% of the scale range of 56 points. This benchmark would appear to be a very useful aid in evaluating the impact of treatment rather than reliance on the traditional method of statistical significance.

Relevance for measurement of dentine hypersensitivity

Pain and a patient's ability to cope with pain determine treatment-seeking behaviors. In an analysis of data from the 1998 UK Adult Dental Health survey, Nuttal et al.[18] analyzed the frequency of impacts reported "very often" or "fairly often" on the OHIP-14 scale. The most frequently reported of a possible 15 response combinations were those indicating the experience of *pain/discomfort* with no other impact (23.8% of respondents) and *pain/discomfort* with a form of *disability* (12.1% of respondents). This gives some indication of the relatively high prevalence of impact of pain symptoms in the population. Dentine hypersensitivity is most commonly associated with erosive toothwear and gingival recession, both of which have a relatively high prevalence in the population.[19,20] By its very nature, dentine hypersensitivity cannot be measured by objective assessment alone. There is a focus on developing noninvasive methods of treating this condition, and subjective assessment is vital to determining the most effective treatment strategies for managing dentine hypersensitivity. Impacts of pain caused by dentine hypersensitivity can be detected by generic measures such as the OHIP, but a condition-specific measure can more fully elucidate the true impact of the condition. Newly developed measures of dentine hypersensitivity need to be fully evaluated in clinical trials, but ideally they should have good validity, reliability, and responsiveness properties.

Conclusion

A significant body of work has been undertaken over the past 20 years to develop measures that have been used to capture subjective assessment of health status. These have been used in population surveys and clinical trials, and it is now possible to determine the impact of disease and how it affects patients' daily lives in a systematic way. There has been a focus on issues of terminology, interpretation of health-related quality of life data, and measurement issues. It is important to use these measures appropriately and to recognize that no single measure will be suitable in cross-sectional and longitudinal studies. Generic measures are widely used and can facilitate external comparisons, but they may not fully capture specific disease states. The use of disease- or condition-specific measures can overcome this problem. However, a balance must be struck between sufficient numbers of questions and practical utility. When contemplating which measure to use, it is recommended that the measure is valid for the disease state of interest and has good reliability and responsiveness properties. Finally, when reporting data from clinical intervention studies, it is recommended that either ESs or SRMs are reported to describe the magnitude of change that has occurred after treatment.

References

1. Lutz W, Sanderson W, Scherbov S. The coming acceleration of global population ageing. *Nature* 2008;**451**:716–9.
2. Wilkinson R, Marmot M, editors. Social determinants of health: the solid facts. Copenhagen: World Health Organisation; 1998.
3. Locker D. Measuring oral health: a conceptual framework. *Community Dent Health* 1988;**5**:3–18.
4. Culyer AJ. *Health indicators*. Oxford: Martin Robertson; 1983.
5. Locker D. Concepts of oral health, disease and the quality of life. In: Slade GD, editor. *Measuring oral health and quality of life*. Chapel Hill, NC: University of North Carolina: Dental Ecolog; 1997. p. 11–24.
6. Raphael D, Brown I, Renwick R, Rootman I. *Quality of life theory and assessment: what are the implications for health promotion*. University of Toronto, Centre for Health Promotion; 1994.
7. Fayers P, Machin D. *Quality of life: assessment, analysis and interpretation*. Chichester: Wiley Publications Ltd.; 2000.
8. Fitzpatrick R, Fletcher A, Gore D, Spiegelhalter D, Cox D. Quality of life measures in health care. I: application and issues in assessment. *Br Med J* 1992;**305**:1074–7.
9. Locker D, Allen F. Developing short-form measures of oral health-related quality of life. *J Public Health Dent* 2002;**62**:13–20.

10. Saila S, Lahti S, Nuttall N, Sanders A, Steele J, Allen PF, et al. Effect of 1-month versus 12-month reference period on responses to the 14-item Oral Health Impact Profile. *Eur J Oral Sci* 2007;**115**:246–9.
11. Prutkin JM, Feinstein AR. Quality-of-life measurements: origin and pathogenesis. *Yale J Biol Med* 2002;**75**:79–93.
12. Locker D, Quinonez C. To what extent do oral disorders compromise the quality of life? *Community Dent Oral Epidemiol* 2011;**39**:3–11.
13. Jaescke R, Singer J, Guyatt G. Measurement of health status: ascertaining the minimum clinically important difference. *Control Clin Trials* 1989;**10**:407–15.
14. Kazis LE, Anderson JJ, Meenan RF. Effect sizes for interpreting changes in health status. *Med Care* 1989;**27**:S178–89.
15. Cohen J. *Statistical power analysis for the behavioural sciences*. New York, NY: Academic Press; 1977.
16. Baker S, Gibson BJ, Sufi F, Barlow A, Robinson PG. The Dentine Hypersensitivity Experience Questionnaire: a longitudinal validation study. *J Clin Periodontol* 2014;**41**:52–9.
17. Locker D, Jokovic A, Clarke M. Assessing the responsiveness of measures of oral health-related quality of life. *Community Dent Oral Epidemiol* 2004;**32**:10–8.
18. Nuttall NM, Slade GD, Sanders AE, Steele JG, Allen PF, Lahti S. An empirically derived population response model of the short form of the oral health impact profile. *Community Dent Oral Epidemiol* 2006;**34**:18–24.
19. Amarasena N, Spencer J, Ou Y, Brennan D. Dentine hypersensitivity in a private practice patient population in Australia. *J Oral Rehabil* 2011;**38**:52–60.
20. West NX, Sanz M, Lussi A, Bartlett D, Bouchard P, Bourgeois D. Prevalence of dentine hypersensitivity and study of associated factors: a European population-based cross-sectional study. *J Dent* 2013;**41**:841–51.

Part Two

The Subjective Experience of Dentine Hypersensitivity

The everyday impact of dentine sensitivity: personal and functional aspects*

Barry J. Gibson[1], Olga V. Boiko[1], Sarah R. Baker[1], Peter G. Robinson[1], Ashley P.S. Barlow[2], Tess Player[2] and David Locker[3]
[1]School of Clinical Dentistry, Claremont Crescent, University of Sheffield, Sheffield, UK, [2]GlaxoSmithKline Consumer Healthcare, Weybridge, UK, [3]Faculty of Dentistry, University of Toronto, Canada

Introduction

Patient-centered research in dentistry specifically concerned with the measurement of oral conditions and their impact on everyday life is relatively recent. Researchers have been exploring the patient-centered perspective since the 1970s.[1] Since then, a series of sociodental indicators have been developed.[2–5] Many of these indicators were discussed at a conference in North Carolina in the 1990s. The proceedings of this conference were then published in a report.[6] There are now numerous so-called oral health-related quality of life measures, although the term is somewhat of a misnomer for oral health status.[7] This chapter focuses on oral health status and explores qualitative aspects of measurement of the impact of dentine sensitivity on everyday life.

The most widely used measure of oral health status is the Oral Health Impact Profile (OHIP).[6,8–10] The OHIP has been shown to have good discriminant and construct validity and, because it focuses on problems specific to oral health, it is said to be of more use for measuring the outcomes of oral disorders than generic health status measures such as the SF-36.[11] One key advantage of the OHIP over other oral health status measures is that it is based on a conceptual model of oral health.[5] Although the claim that it represents this conceptual model has recently been challenged, it remains the most significant measure of general oral health status.[12]

Data regarding the appropriateness of the OHIP for the measurement of the impact of dentine sensitivity have been published.[13] A clinical population in Germany receiving treatment for the condition were experiencing more impacts and had poorer oral health than a sample of the general population. The study used an adapted version of the OHIP-49; although there was a difference in mean scores of 22.3 between the two samples, this difference was small on the overall scale of

*Previously published as: Gibson B, Boiko OV, Baker S, Robinson PG, Barlow A, Player T, Locker D. The everyday impact of dentine sensitivity: personal and functional aspects. *Social Science and Dentistry* 2010;**1**:11–20.

Dentine Hypersensitivity. DOI: http://dx.doi.org/10.1016/B978-0-12-801631-2.00006-3

245. The extent to which OHIP is appropriate for the specific problems associated with the impact of dentine sensitivity remains unclear because, although it was based on interviews with participants about the impact of their oral health, these basic qualitative data have never been published. Therefore we cannot evaluate the extent to which it can address the specific problems and impacts associated with dentine sensitivity.[14]

From even a brief review of the literature on dentine sensitivity, it is clear that there is an urgent need to develop a person-centered approach to the condition. Existing reviews adopt a clinical perspective and focus very little on patients' perspectives.[15–18] A consensus statement published in 1997 defined as dentine sensitivity as a "short, sharp pain arising from exposed dentine in response to stimuli typically thermal, evaporative, tactile, osmotic, or chemical and which cannot be ascribed to any other form of dental defect or pathology" (p. 809).[19] There is little mention of lay perspectives of the condition and, further still, little reference to general models of health conditions and their impacts.

The literature focuses on several aspects of the condition such as its etiology, incidence, and measures of pain. In each case there is little reference to lay perspectives. Etiologically speaking, several theories have been proposed to explain the underlying mechanism for dentine sensitivity, with the evidence appearing to support some form of hydrodynamic mechanism (See chapter 2).[15–19] The literature also describes the numerous etiological factors associated with dentine sensitivity, and it has been suggested that it takes a combination of these factors, most notably erosion and abrasion, to result in sufficient tissue loss for sensitivity to occur.[16,18,20] There is little reflection on the wider social and political implications of discovering that it is the discipline of toothbrushing in combination with an erosive diet that can act as one of the biggest predisposing factors for dentine sensitivity. The internalization of the dental discipline in populations through various toothbrushing regimes may well have had some unintended consequences.[21–23]

Difficulties associated with the diagnosis of the condition have led to widely varying estimates of its prevalence (Chapter 3). It should not be surprising because of the differential diagnosis of the condition, and overestimates of its incidence thrive. In a study of regular attendees at a general practice population in London, 52% of subjects reported having dentine sensitivity, with females having a higher prevalence than males. Data collected from a Korean sample at the same time reported an incidence of 55.4%.[24] Likewise, 57.2% of an Irish general practice population using questionnaires reported the condition.[25] Population studies contrast sharply with studies that use clinical diagnoses when the prevalence of the condition declines. For example, Rees and Addy reported that only 201 of 4841 patients (4.1%) had clinically diagnosed dentine sensitivity.[26] This figure cannot be considered a valid estimate because data were collected from 19 different general practices using 19 different examiners. There are no data indicating the level of agreement between examiners, and there was a wide variation in estimates between examiners and populations.[26]

When patient perspectives have been sought, this has been largely restricted to ratings of pain, usually in response to a stimulus within a clinical setting.[26–29]

Unfortunately, pain experiences are heavily modified and dependent on personal and environmental factors.[30,31] However, the extensive literature on pain remains to be integrated into ideas of dentine sensitivity. For example, some pain scales confuse the sensory, affective, and timeline aspects of pain in one scale, and others almost exclusively measure sensory as opposed to affective pain.[27,31–33] Other research makes little or no reference to the experience of pain in an everyday context.

Patient-centered research on dentine sensitivity has focused on the main triggers for the condition, e.g., cold drinks.[24,28,29,34] What is clear from the existing literature is that when patients' experiences and perspectives are sought, these views are only sought minimally. Although the everyday impact of sensitivity was investigated in a limited manner, this research has made little reference to the burgeoning literature on the impact of oral conditions on everyday life.[5,8,24] According to this research, dentine sensitivity impacted on toothbrushing in 8.7% of cases, 28.2% of participants could not drink cold water without some discomfort, and 26% could not eat ice cream without discomfort. Likewise, 46.6% of participants reported not avoiding the area of discomfort in relation to eating, drinking, and teeth-cleaning, whereas 10.5% would avoid the area of discomfort. Patients were unable to complete most day-to-day activities without undue discomfort. The condition was also described as a low-grade issue that often persisted for more than 5 years.[24] Therefore, we aimed to explore the impact of dentine sensitivity on everyday life.

Materials and methods

Participants were purposively recruited from a general population with the specific aim of securing a range of experiences and views about the everyday impact of dentine sensitivity.[35] It was intended that sufficient participants would be needed to achieve saturation of information, but not so many as to prohibit detailed analysis. We expected to interview 20–30 participants on this premise. Participants were currently experiencing sensitivity in their teeth and were adults (older than age 18 years). Participants were initially identified through the research team's contacts and through snowball sampling.

During data collection and analysis, the research team identified some people who described themselves as having quite severe sensitivity. It is very important to note that these participants were recruited from a snowball sample from the immediate social circle of the researchers. After analyzing data from 13 people who described themselves as having "sensitive teeth," the research team considered where to sample next. The team was aware, from previous research on dentine sensitivity, that some people do not use the term sensitivity and are, in fact, reluctant to do so, but nonetheless they experience the condition. In the interest of getting the full range of experience and impact of sensitivity on everyday life, we decided to purposively sample another 10 participants who did not describe themselves as having "sensitivity." This group was recruited from a recruitment agency on the basis that they experienced "twinges and discomfort" in their teeth. During the interviews, we only used the term "sensitivity" if the patients used it at any point in time.

The study was granted ethical and research governance approval from the University of Sheffield research ethics committee. Participants were phoned or e-mailed, or both, and invited to participate. They were asked about their sensitive teeth, age, and availability for the study. The goal of the study was described in general terms. After this initial approach, participants were sent a written information sheet and consent form. They were then called and an interview date was arranged at a suitable time. Written consent was obtained on the day of the interview. Interviews lasted from 20 to 40 min; on completion, participants were given a small honorarium to thank them for their time.

Interviews began through the use of standard questions that explored the general impact of dentine sensitivity on the lives of patients to cover all aspects of their experience with the condition. The interview also included the use of a visual analogue scale rating the intensity of the pain and elicited stories about the context of pain experiences. Interviewers attempted to be as open as possible to participants' narratives and flexible in switching between interview topics. At times, the interviewer would repeat questions to elicit more detailed responses about various aspects of the experience of dentine sensitivity. Interviews were then transcribed as soon as possible after the interviews and the recordings were deleted. To preserve anonymity, all identifying information was avoided during the interview and any emerging identifying information was removed from the transcripts.

Data analysis

Data were analyzed from a framework obtained from the data and informed by the literature on chronic illness, coping, and illness beliefs, along with the general literature on the biopsychosocial impact of oral health.[36–41] It is important to note that these findings were to be used to develop items for a questionnaire to measure the impact of dentine sensitivity on everyday life. Data analysis also focussed on detailing the range of impacts associated with the condition.

Results

A total of 23 interviews were conducted. Fifteen participants were female and eight were male. The principal impacts on everyday life were pain and impacts on functional status and everyday activities, such as eating, drinking, talking, toothbrushing, and social interaction in general. Impacts appeared to be related to a range of individual and environmental influences. Data suggested that these broad variables interacted in complex ways, leading to a characteristically subtle but complex condition that can have significant impacts on everyday life. What follows is a presentation of the results starting with the everyday impact of dentine sensitivity.

The impact of dentine sensitivity on everyday life

Dentine sensitivity primarily manifests through pain for those who are affected. However, accounts from participants encompassed experiences of pain and its duration, frequency, intensity, and localization. The following is a typical account:

> *I think so, yeah, fortunately it's not such a strong sensation and it doesn't happen so often that I'm aware of it, but yeah, it's in the molars, both sides (S1.4.p.1).*

According to many participants the sensitivity was a very unusual pain, often described as "sensations" rather than pain. Therefore, there was a wide variety of descriptions of the sensitivity, pain, or sensations. In total, there were 22 nominalizations for this pain, including terms like "*brain freeze*," "*nails on a blackboard*," and "*needles*." Although the ways participants described their sensations differed, they were commonly characterized as a sharp, stabbing, or shooting pain or as a mild twinge. Less frequent terms included tingles or a shivery feeling. For some the word "pain" as a descriptor sounded a bit too "*harsh*" (S1.4.p.1).

The nature of pain interacted with various triggers (Figure 6.1) for the condition. For example, sensations were described as "lingering" or "itchy," sometimes lasting for 20 min as a reaction to consuming sweet things and chocolate. In contrast, with cold foods (ice cream, popsicles) and cold air, the sensation tended to be more intense but lasted only a few seconds. Often the sensation lasted while the stimulus was in the mouth, for example, soon after drinking hot coffee. In some instances, the duration of the sensation remained during the activity, for example, until toothbrushing had stopped. On some occasions, the sensations were said to have lasted for full days, especially after recent dental treatment. Interestingly, cold foods and drinks caused, more often than not, the corresponding feeling of freezing: "*you feel it's very*

(1) Foods: ice cream, ice popsicles, honey, sugar, chocolate, chips, cereals, pineapple, cold fruit, apple, melon, grapes, and sorbet.

(2) Drinks: cold lemonade, soda, hot tea, and coffee.

(3) Physical pressures: toothbrushing, flossing, scale and polish, tongue touches, and metal touches.

(4) Cold air when breathing and at dentist, cold water when rinsing

(5) Whitening toothpastes.

Figure 6.1 The triggers associated with the everyday manifestation of dentine sensitivity.

cold, it's frozen and then it's gone" (S1.22.p.6). Physical pressures including those self-inflicted, like running the tongue over the teeth to cause pain or vigorous toothbrushing, were also reported to initiate the sensations. In most cases, however, the sensation was sudden and instant, difficult to predict, or difficult to prepare for. In a few cases, participants reported sensitivity to two or three triggers.

Despite some cautiousness to label the sensation "pain," all interviewees registered a mark on a visual analogue pain scale. The intensity was most commonly rated as 3 or 4 out of 10 on the scale. It appeared that the intensity varied depending on the trigger. In the case of thermal triggers, the intensity was dependent on the temperature; correspondingly, colder (frozen) and hotter things caused more intense sensations. Triggers of different quality were also rated differently. A respondent articulated that:

> *If it is cold while I'm eating melon or warm tea that's probably about six on a scale of one to ten. If it's after grinding you'd probably get close to a ten depending on a severity of what has happened (S1.5.p.3).*

Therefore, caution should be taken when obtaining pain ratings in clinical settings. The triggers interact to give a complex and subtle experience of pain in everyday life. This is in keeping with the psychological literature where anticipation of high-intensity pain will result in high-intensity ratings at stimulus impact compared with those anticipating low-intensity pain, even if the actual intensity of the stimulus is constant. So, in the previous quotation, it is not difficult to see how pain ratings for sensitivity can vary dramatically according to context.

The common approach to measuring pain intensity in the dental literature measures only one component of the overall experience. Ignoring the sensory and affective components of pain will exclude a full understanding of the condition.[42] This is important because these components are related to the everyday experience of sensitivity. For example, recent research using a momentary within-person daily design found that although sensory and affective components together with activity limitations were related to the intensity of the pain, this relationship was nonlinear. It appeared that sensory characteristics of pain were related to higher levels of intensity, whereas affective qualities of pain were associated with low-intensity pain. At this stage, it is clear that we need to know more about each of these aspects of pain and how they relate to varied experiences in relation to dentine sensitivity.

The frequency of the pain appeared to be an important indicator of the condition but did not appear to predict its severity. There were differences in the frequency of the episodes, ranging from having sensations over some periods in their lives to fluctuations within a particular (most recent) period. Those who had experienced sensitivity for more than 5–10 years discussed the frequency of episodes with respect to different sensitivity treatments and also life circumstances (living in another country, family changes, etc.). The condition was also described as periodic and cyclical:

> *It comes in cycles so at the moment I've not had it in some time but if they are that way out then it might happen two–three times a week or more (S1.11.p.2).*

Participants could not provide a clear explanation for this periodicity.

Biological, personal, and environmental factors could explain the fluctuating nature of dentine hypersensitivity. The condition appears to have many similar characteristics associated with chronic headaches. Yet many questions remain. For example, what determines the severity of impact of dentine sensitivity? Is the frequency, severity, or unpredictability with which it is experienced most predictive of its impact?

Predictability

Last night I went for dinner at Phil and Christine's and she has defrosted a cheesecake and it was still cold, it was defrosted but still very cold and I remember just looking out the corner of my eye to make sure no one sees me because I get very self-conscious (S1.9.p.9).

The predictability of the sensitivity episodes was a central feature of the everyday impact of the condition. Participants' awareness of the *potential* pain in their teeth constituted an impact in itself. Accounts provided a range of experiences from indifference to strong anticipation of the sensations. One factor regarding anticipation and predictability appeared to be the length of time the individual had been experiencing the condition. In social science, the term used to describe this feature of illness is called the illness career. Those who had experienced problems for years had often become conditioned to it:

I am always aware of the potential problem (S1.3.p.7).
I do have to think twice before having something (S1.12.p.6).

In contrasting accounts, people did not anticipate possible pain while eating or drinking. In most of these cases, participants' teeth had development of sensitivity fairly recently and the sensations were not as intense: "*No, no, I don't think that my teeth will be sensitive, no*" (S 1.23.p.5). The feelings could take them by surprise:

It was like 'oh'! I did not expect it and that's why he [boyfriend] was like 'what's wrong?' and I was like 'my teeth just hurt a bit really' and that was all really (S1.16.p.7).

I get cross with myself because I am like 'you know that's going to set your teeth off why did you do that?...Why did you leave that in the fridge?' (S1.7.p.7).

In comparison with those with long illness careers, these participants did not expect the pain or could not predict when the sensation would occur. The narratives of predictability were more involved, complex, and accounted for delicate nuances such as the knowledge of how the sensations in their teeth were likely to initiate and escalate during exposure to triggers: *"I just flinch a bit, I know it is coming"* (S1.7.p.7). Such a need to predict pain could become a mental strain over a longer period than just when encountering triggers. The overall impact of the condition therefore may be more profound for the chronic sufferers than for the recently affected:

> *You never forget that you have it because the decisions you make and the way you do things is affected by it (S1.6.p.8).*

Constant awareness and an illness career suggest that sensitivity shares characteristics with some chronic illnesses. Living with sensitive teeth requires development of adjustment mechanisms and integrating sensitivity into the context of everyday life.

In chronic illness control (actual or perceived) over the management of pain is linked to lower pain ratings, less disability, better well-being, and better coping with pain.[43,44] There are two parts and knowing what that event is going to be like.[43] If the event is predictable, then there tends to be less surprise and anxiety, which in turn leads to less pain reactivity.[45] In contrast, unpredictable events can lead to anxiety and increased vigilance, which can, in turn, increase sensory receptivity.[46] Unpredictable pain has been related to increased ratings of anxiety and negative valence.[47] One interesting point worth considering is if "generalized hypervigilance," which involves increased sensitivity to a range of stimuli usually resulting in increased monitoring of the environment, can become established in dentine sensitivity.[48] What we can see in these data is that for some people the pain is unpredictable and for others it is predictable. The main factor that appeared to be related to predictability was the length of the illness career. For some this was a long-term illness with predictable consequences, whereas for others it was a health condition that caught them by surprise. These different frameworks are the subject of further exploration in our next work.

Emotional impact

It was common to find annoyance as a reaction to the inconvenience and discomfort caused by the sensitivity. However, several patterns were found in participants' accounts. An example of the first kind of annoyance related to the unexpected nature of the pain: *"I wince, like oww, oww, oww and it sort of lasts for quite a few seconds so it is really annoying"* (S1.9.p.1). An annoyance of a second kind arose from a reduced ability to enjoy food:

> *It is annoying, it is annoying, and it feels like I am not getting the full benefit, I am not enjoying food like probably other people who haven't got sensitive teeth (S1.19.p.5).*

Likewise, participants reported being frustrated because the sensations were "spoiling" their pleasure. In other words, they were frustrated at the restriction that sensitivity posed to them in their everyday life. An example of the third kind of annoyance apparent in a few accounts connected to the necessity to deploy coping strategies, changing the way they ate and drank was: "*I try not to chew in that area, which is a bit annoying*" (S1.10.p.1). Moreover, there was an understanding among the participants that some adjustments could be damaging and unhealthy:

> *I just sort of swallow it, it sort of feels, it is going down but as I say it is just frustrating as I feel I should not be eating all this and it is not helping digestion (S1.12.p.8).*

The fourth kind of annoyance was associated with guilt of eating things they believe they should not have eaten: *"Oh God, I have eaten ice cream and I shouldn't have done"* (S1.5.p.4). Sometimes participants reported feelings of anger for doing this. Finally, some participants described feeling full of self-reproach and guilt for not being able to take care of their teeth properly. During episodes of sensitivity, participants tended to recall having forgotten to use fluoride or sensitivity toothpaste recently or having failed to floss appropriately. There were also accounts of omissions in their diet and smoking as contributors to an increasing sensitivity in their teeth. Other times participants referred to the times, usually when they were teenagers, when they overlooked their oral health and somehow contributed to their current problems.

> *I wish I'd looked after my teeth better and I think 'why don't I go to the dentist,' so I do get annoyed but it is mainly with myself (S1.9.p.4).*

Functional impact

The functional impact reported by participants included restrictions in performing everyday tasks such as eating, drinking, taking care of their teeth, being outside on a cold day, and some sporting activities. Importantly, these restrictions were described in conjunction with adjustment mechanisms, such as coping. Separating these two in the data was very difficult:

> *I was making this honey and apples for my daughters and it was clear runny honey and I was really looking forward to eating it and I couldn't at that point, you know, the pain was I wouldn't call it unbearable but it was a put off, it put me off (S1.3.p.7).*

In that case, the sensitivity impacted on the pleasure of eating and forced the participants to decline the meal. Others reported sensitive teeth taking a lot of enjoyment out of food. For example, they were not happy always having to eat food at room temperature because some things tasted better cool; they also disliked having to eat foods that were *"easier to eat"* and modifying the ways they ate:

> *It is not as enjoyable, because I find myself chewing on one side and avoiding chewing on the left side so that's not particularly a nice thing (S1.10.p.4).*

Maintaining oral hygiene could be troublesome if toothbrushing and rinsing with cold water caused unpleasant sensations. Although there were no accounts of stopping oral hygiene practices, some participants developed ways of dealing with it by buying an electric toothbrush, moving water around their mouth, and other adjustments.

Finally, physical activities such as exercising outside or just being out on a cold and windy day, were restricted: *"I have to keep my mouth shut; the weather can affect my teeth as well"* (S1.19.p.2). Occasionally, there were responses like the following: *"I struggle when I am skiing"* (S1.16.p.5) or *"when I am swimming I try and keep my mouth shut"* (S1.5.p.10). Indirectly, this affected social functioning because occasionally it necessitated keeping the mouth closed and prevented people from talking.

Social impact

Sensitivity impacted on the social activities of some participants, although others did not notice any differences in the way they socialized. This impact was indirect and mostly concerned situations in which participants ate socially. Going out for a meal with others presented problems for some.

> *I am just aware it's painful, occasionally, if I am in company I will shut off from the conversation so I will miss sometimes what people are talking about… But during that minute you could miss something quite vital like that people are talking about and then you are peddling, back-peddling trying to catch up. Especially if it is a quick conversation or there is a debate going on and you just miss (S1.5.p.6,12).*

This exclusion from conversation is almost as painful as the sensitivity itself. This respondent also had another difficulty:

> *If it comes [food] and it is too warm or too cold I have to wait, which invariably means I delay everybody else at the table as I finish last. Or I end up only eating half the meal because everybody has finished and I am conscious of holding everybody else up (S1.5.p.9).*

Thus, the sensitivity led to a violation of etiquette. Moreover, such impacts could cause further embarrassment when the problem was exposed:

> *I am polite and cover it up or I take ages to eat it till it is warmed up a bit or do the funny thing with my teeth while nobody's looking (S1.6.p.8).*

Such modifications of food had to be hidden from others. This observation was supported by another account of the participant who drank with friends:

> *They bring me a drink from the bar, they always put ice in and I hate ice. If nobody's looking I scoop it out (S1.14.p.7).*

The last two accounts indicate that there can be a significant taboo regarding sharing feelings of dentine sensitivity with others, although this varied, with some people not sharing it because it was a problem not worth worrying about.

Coping with dentine sensitivity

The adaption strategies of people with sensitive teeth varied dramatically. For participants experiencing the condition over years, techniques of adjustment to sensitivity were often complex. Coping strategies could be divided into three major categories: avoidance coping, approach coping, and tolerating. Avoidance coping appeared to be quite common, with participants reporting avoiding cold drinks, frozen fruit, ice cubes, and ice cream and cold/hot/sweet food:

> *I just try to avoid having ice, too much ice in my drinks and you just sort of learn not to have things that set it off really and a bit like ice cream, I will have the odd one (S1.7.p.2).*

Occasionally, avoidance of hot drinks was mentioned: *"I would rather have a cup of cool coffee, so it is such an unpleasant pain, I am consciously avoiding it"* (S1.2.p.3). The boundary between avoidance and approach coping was flexible. Participants might go outside to eat some foods and use other active techniques of coping. In this respect, there was some approach coping. Two strategies were apparent: modifications of food and modification of the ways the food/drink was consumed.

Changes in the ways food was consumed involved techniques of avoiding contact with certain teeth and developing ways to melt, chew, bite, suck, lick and drink. Most participants emphasized minimizing contact with teeth affected by moving food and drinks to another side of the mouth: *"I try not to eat on that side of the mouth"* (S1.3.p.4) and *"I take it to the back of my mouth and eat it"* (S1.6.p.3). Similar habits were developed in relation to drinking: *"I am conscious and don't let the drink go into that area"* (S1.10.p.1). Other participants admitted the use of straws for the same reasons. Special techniques were invented by participants for eating ice cream, like biting in small pieces, avoiding chewing, melting ice cream in the mouth, sucking and licking, and using a spoon. Sensitivity to sticky food sometimes forced participants to lick the area affected or to even use fingers to clear it up:

> *With sticky foods like say I have currants on my cereal things like that, I am making sure all the time that I haven't got any food stuck on that level of my tooth because like I said I move some food with my finger (S1.4.p.2).*

Such strategies (i.e., guarding body parts) have been described in the literature as initially adaptive, but may become maladaptive over time.[48a]

Food modification as an approach coping strategy included warming cold foods/drinks and cooling hot foods/drinks. In most instances, bringing food to room temperature was enough to cease negative effects on sensitive teeth; however, at times participants resorted to a more drastic measures like putting ice cream in a microwave *"to take a chill out of it,"* holding the offending area with a finger, or using other agents to neutralize the unpleasant sensation, such as *"a bit of chewing gum"* (S1.12.p.6). Typically, participants allowed cold things to warm:

> *To drink water is quite difficult, but I find if I leave it on the side for a while and let it warm up a little then it is all right (S1.12.p.3).*

Meanwhile, some accounts described "accepting" foods and things that affected the teeth. For participants with recent experiences of sensitivity, being indifferent to the episodes was a relatively common strategy. Some believed that they did not have to do anything to avoid or protect their teeth from contact with triggers because the sensation was fairly insignificant and canceled by other feelings of pleasure or habitual comfort of eating. One respondent admitted:

> *I just get used to it by then, my teeth kind of like get used to it, after a few mouthfuls I am ok. That's why I just persevere with it really (S1.16.p.9).*

These accounts suggest some parallels with reactions to pain episodes in tolerating the sensitivity. This indicated that the impact of dentine sensitivity on everyday life can be restricted. Such restrictions are important and need to be considered in relation to the longevity of the condition and pain characteristics. Acceptance can also determine functional status and impairment in chronic pain, for which control beliefs and active coping are related to more positive mood.[48b] Acceptance is an important factor in avoiding disability and maintaining function.

Although some coping strategies prevented impacts from specific episodes of pain, others appeared to be aimed at the overall management of sensitivity. In this respect, there was an underlying continuum between active or approach coping and passive stoicism or avoidance coping strategies. Finally, like many other chronic conditions, a stoic approach was also apparent in some of the responses. Typical responses were a refrain of "just put up with it": "*Just try and get through it really*" (S1.10.p.3) or *"we have been taught to get on with things...don't make a fuss"* (S1.5.p.10). Stoicism as a strategy of control and this choice to tolerate the pain connects with other explanatory ideas in the model of sensitivity. The next two sections on illness beliefs and identity add some detail to this issue.

Illness beliefs

A range of social and cognitive variables have been linked to functional limitations in other conditions.[49] In addition, strong beliefs in the chronicity and negative consequences of the condition, and avoidance coping styles are associated with poorer physical and psychological outcomes.[50] Such beliefs form an important part of the explanation of variations in the response to different health conditions and their impact.

Illness beliefs summarize peoples' ideas about their condition, its causes and manifestations, and the evaluation of its impact. In this respect, there was a contrast between those whose sensitivity was a relatively new experience that presented mild discomfort compared with those for whom it was a more serious problem that had significant impact over a longer period of time. This is in keeping with the literature on illness beliefs. People who believe they are disabled by their pain also believe that they have become damaged in some way and should avoid certain activities. Such conditions increase impacts of pain on daily living. This difference was reflected in narratives of the onset of the condition. Some had accounts of when they began feeling sensitivity, whereas others had no specific memories. There were no specific accounts of a single event that caused sensitivity other than a reference to a specific time period.

> *I can remember being at umm secondary school and not being able to eat ice cream umm having the most terrific headaches if I ate ice cream, shooting pains at the front of my face. Err... so probably as far back as then... Many years yeah and then during pregnancy things got worse umm and then obviously as I have got older things have got worse because my gums have receded (S1.5.p.2).*

This story suggests sensitivity can be placed into the context of the respondent's life in conjunction with other life events and health status. Accordingly, certain life events were linked with an increase in sensitivity. Other participants with recent onset were less certain about when and why it had started. This indicated either that their sensitivity had a low intensity or that they as yet had a relatively unelaborated view of the condition. It was a characteristic of many accounts that only approximations could be given, such as "for a few months" or a "few years." One participant with a recent history of sensitivity admitted:

> *I would say within the last few months I've been aware of it but I would say that actually it's something, it's something that I've had for long time but I've not thought about it as much and therefore it's not been as much of an issue (S1.4.p.3).*

Knowledge of the condition varied between participants. Some sought detailed information from different sources, whereas others retained a relatively superficial knowledge and were largely unaware of the causes and prognosis of the condition. Again, this difference appeared to be governed by the degree to which people saw the condition either as an established fact of life or as something transitory. Participants who saw it as a condition worthy of some investigation appeared to utilize three major sources of information: "*what I've heard from the dentist and what I've seen on tv and on the back of the toothpastes packet*" (S1.9.p.4). In some instances the Internet was an obvious source of information.

The expertise of such accounts contrasted with accounts of little knowledge and lack of interest in getting more information. Why this was the case is not easily explained because the relative indifference to the causes of the condition was present in observations of chronic sufferers as well as participants new to the problem.

> *I don't really know anything about where it comes from or why it happens. You think that I would have probably questioned that but I suppose it is just something I've got used to (S1.19.p.5).*

Observations such as these demonstrated how a lay understanding of sensitivity emerged through a complex mixture of expert knowledge and sources such as the Internet and advertising. It also demonstrated how lay beliefs could become relatively undifferentiated. For example, contrary to clinical diagnosis, the condition was often considered by participants as the result of oral health and aging.

> *Perhaps, just as your teeth get older, your gums perhaps aren't as good as when you were younger (S1.14.p.1).*

For those who tended to build the bridges between the condition and the state of the gum, teeth, and oral health in general, sensitivity became just one of the many relatively small but irritating worries about their health. These small things appeared to add up and cause some considerable worry for participants: *"I'm consciously worried about why; the reasons behind my sensitivity cause me anxiety"* (S1.4.p.7).

> *I think the long term it worries me if they get worse and I suppose with the gum problem as well that worries me so it's the two together (S1.7.p.7).*

Illness beliefs encapsulated ideas about how participants evaluated the severity of the condition. Overall, observations about the experience of sensitivity ranged from seeing it as not a problem at all to being concerned and worried. Those who recently experienced sensations were less likely to be concerned about it:

> *It doesn't bother me. I know I can put up with it (S1.14.p.9).*
> *Just an inconvenience really or a nuisance (S1.20.p.7).*

When the extent of the pain and trouble over the years were more significant, these people tended to rate the impact as negative but relatively nonproblematic. Some anxiety, however, was expressed in relation to the long-term prognosis. The potential psychosocial impact was clearly marked in one interview:

> *It's not bothering me too much at this stage, if it was a case of there's nothing we can do you're going to have to live with this for the rest of your life then if it might present more of a psychological problem for me (S1.10.p.5).*

Sometimes the situation of the interview provoked reflections on the meaning of the sensitivity for their general health and well-being. Research of the sociology of health and illness has suggested that narratives could be understood as a story-telling activity that inspires reflexivity in making sense of health-related events. As the interview proceeded, some participants reported changing the way they observed the condition. An account of the respondent who initially declined the idea of sensitivity impacting on his life exemplified such changes:

> *It changed my eating habits as I told and about my teeth, the way I am thinking about them maybe it changed, maybe I have a pain and I think about the future and think something will happen to them. It made me think this way (S1.22.p.10).*

Conclusion

Dentine sensitivity was experienced in complex ways in everyday life. Although the professional definition of dentine sensitivity is that it is a "short, sharp pain...in response to stimuli typically thermal, evaporative, tactile, osmotic, or chemical...",[18,19] our data suggest those affected have a complex experience with a wide variety of triggers and responses. The sensations were not readily described as "pain." Although participants did rate the level of pain they were experiencing, this was more a result of being asked directly to do so, not because this is how they described the sensations. This links with previous works in the pain literature by Melzack,[42,51] which describe that pain experiences have different qualities with three major dimensions: sensory (e.g., needles); affective (e.g., vicious); and evaluative (e.g., annoying). These sensations are said to differ according to different types of pain, whether acute or chronic and, of course, across people because of a number of social, cultural, and individual characteristics.

In addition to the physical pressures during toothbrushing, flossing, and scaling, stimuli such as cold air while breathing could trigger sensations. Many of these stimuli are already recognized in the dentine sensitivity literature. What is not recognized is that these stimuli along with the situations in which they are experienced can have a significant affective impact on everyday life. Therefore, although the level of the pain associated with the sensations was often described to us as minor (mostly 3 or 4 on a scale of 1–10), they were nonetheless associated with significant impacts.

A common way of measuring dentine sensitivity has been through the use of visual analogue scales in clinical situations.[52] Although this form of measurement is standardized, it belies the fact that responses are heavily modified by the context in which measurement occurs. Although there is some recognition that the descriptors being used can affect ratings, there have been few attempts to explore the affective impact of dentine-sensitive pain from the perspective of a science of everyday life.[53] The findings of this study confirm that further research into the everyday nature of dentine-sensitive pain would be beneficial. It is apparent that these data connect to the psychological literature on pain experience. According to our findings there are several points that require further explanation and research. First, it is apparent that there are sensory and affective components of pain and that these interact in nonlinear ways; as yet, we do not have enough detailed data regarding dentine sensitivity to tell us more about these aspects of the condition. Second, there is an important element of predictability and control associated with the pain of dentine sensitivity. In this respect, the length of a person's illness career appears to be directly related to the predictability of the condition, and whether this translates into lower pain ratings remains the subject of further work. Finally, it appears that acceptance of a chronic condition is an important predictor of outcome in relation to functional status. We do not know how this affects the outcome of dentine sensitivity. It seems that there is a distinction between those who experience the condition very much within the framework as a chronic illness and those who see it as a set of problems associated with a normal healthy life.

The psychological literature makes a careful distinction between health and illness cognitions. Health is seen as not being ill, as a reserve of mental and physical strength, and as being in equilibrium.[54,55] Illness, however, involves not feeling normal, having specific symptoms of a specific condition, seeking to identify consequences about what can and cannot be done, a timeline for illness, and finally experiencing an absence of health.[40,41] From the data in this study, it is difficult to separate responses exclusively into a health or an illness framework because the experience of dentine sensitivity appears to have both frameworks associated with it. Dentine sensitivity does not appear to have any form of major physical crisis associated with it.[56] Dental disease and its effects are often ubiquitous features of everyday life in most developed populations. It seems that participants had some difficulty in establishing whether they were suffering from an illness. When dentine sensitivity was experienced within what seemed to be an illness framework, there seemed to be some support for Leventhal's[40,41] self-regulatory model of illness cognitions.[57]

Finally, these data suggest that what is required is a biopsychosocial understanding of the pain of dentine sensitivity that recognizes the centrality of the biological but that the actual experience of pain is dependent on psychological (illness beliefs, coping) and social components. In this chapter, we have focused on the psychological and personal factors. Further work is being undertaken to elaborate the social aspects of the condition. Clearly, an increased understanding of the contribution of these psychological and social variables to a person's daily experience of dentine sensitivity will help broaden the research agenda and improve our theoretical understanding of the treatment of the condition.

Acknowledgment

We acknowledge the very generous support of GlaxoSmithKline. We have found their continued collaboration on this subject both challenging and interesting.

References

1. Cohen L, Jago J. Toward the formulation of sociodental indicators. *Int J Health Serv* 1976;**6**:681–98.
2. Nikias M, Bailit H, Beck J, Cohen L, Conrad D, Giddon D, et al. Progress report of the committee on Sociodental Indicators of the Behavioural Science Group of the International Association of Dental Research. Erfurt, German Democratic Republic, September 7–11, 1980.
3. Reisine S, Fertig J, Cipes M, Lawler S, Miozza J. Impact of oral health on the quality-of-life. *J Dent Res* 1987;**66**:215.
4. Reisine ST. The impact of dental conditions on social functioning and the quality of life. *Annu Rev Public Health* 1988;**9**:1–19.
5. Locker D. Measuring oral health: a conceptual framework. *Community Dent Health* 1988;**5**:3–18.
6. Slade GD. Derivation and validation of a short-form oral health impact profile. *Community Dent Oral Epidemiol* 1997;**25**(4):284–90.
7. Locker D, Allen F. What do measures of 'oral health-related quality of life' measure?. *Community Dent Oral Epidemiol* 2007;**35**:401–11.
8. Slade G, Spencer A. Development and evaluation of the oral health impact profile. *Community Dent Health* 1994;**11**(1):3–11.
9. Slade G, editor. *Measuring oral health and quality of life*. Chapel Hill, NC: Department of Dental Ecology, University of North Carolina; 1997.
10. Slade GD. Quality of life outcomes from dental care among older adults. *J Dent Res* 1998;**77**:661.
11. Allen PF, McMillan AS, Walshaw D, Locker D. A comparison of the validity of generic- and disease-specific measures in the assessment of oral health-related quality of life. *Community Dent Oral Epidemiol* 1999;**27**(5):344–52.
12. Baker S, Gibson B, Locker D. Is the oral health impact profile measuring up? Investigating the scale's construct validity using structural equation modelling. *Community Dent Oral Epidemiol* 2008;**36**:532–41.
13. Bekes K, John M, Schaller H, Hirsch C. Oral health-related quality of life in patients seeking care for dentin hypersensitivity. *J Oral Rehabil* 2009;**36**(1):45–51.
14. McGrath C, Adu-Ababiof F, Zaki AS, Bedi R. An evaluation of an oral health related quality of life measure—OHQoL-UK(c) in Ghana. *J Dent Res* 1999;**78**(5):1059.
15. Dowell P, Addy M. Dentine hypersensitivity—a review: aetiology, symptoms and theories of pain production. *J Clin Periodontol* 1983;**10**:341–50.
16. Dababneh RH, Khouri AT, Addy M. Dentine hypersensitivity—an enigma? A review of terminology, epidemiology, mechanisms, aetiology and management. *Br Dent J* 1999;**187**(11):606–11.
17. Hypersensitivity CABoD. Consensus-based recommendations for the diagnosis and management of dentin hypersensitivity. *J Can Dent Assoc* 2003;**69**(4):221–6.
18. Orchardson R, Gillam DG. Managing dentin hypersensitivity. *J Am Dent Assoc* 2006;**137**(7):990–8.
19. Holland G, Narhi M, Addy M, Gangarosa L, Orchardson R. Guidelines for the design and conduct of clinical trials on dentine hypersensitivity. *J Clin Periodontol* 1997;**24**(11):808–13.
20. Absi E, Addy M, Adams D. Dentine hypersensitivity—the effect of toothbrushing and dietary compounds on dentine in vitro: an SEM study. *J Oral Rehabil* 1992;**19**:101–10.

21. Nettleton S. Protecting a vulnerable margin—towards an analysis of how the mouth came to be separated from the body. *Sociol Health Illn* 1988;**10**(2):156–69.
22. Nettleton S. Wisdom, diligence and teeth: discursive practices and the creation of mothers. *Sociol Health Illn* 1991;**13**(1):98–111.
23. Lupton D. *The imperative of health: public health and the regulated body*. London: Sage; 1995.
24. Gillam D, Seo H, Bulman J, Newman H. Perceptions of dentine hypersensitivity in a general practice population. *J Oral Rehabil* 1999;**26**:710–4.
25. Irwin C, McCusker P. Prevalence of dentine hypersensitivity in a general dental population. *J Ir Dent Assoc* 1997;**43**(1):7–9.
26. Rees JS, Addy M. A cross-sectional study of dentine hypersensitivity. *J Clin Periodontol* 2002;**29**(11):997–1003.
27. Al-Wahadni A, Linden G. Dentine hypersensitivity in Jordanian dental attenders: a case control study. *J Clin Periodontol* 2002;**29**:688–93.
28. Flynn J, Galloway R, Orchardson R. The incidence of 'hypersensitive' teeth in the West of Scotland. *J Dent* 1985;**13**:230–6.
29. Fischer C, Fischer R, Wennberg A. Prevalence and distribution of cervical dentine hypersensitivity in a population in Rio de Janerio. *Brazil J Dent* 1992;**20**:272–6.
30. Gracely R, Dubner R, McGrath P, Heft M. New methods of pain measurement and their application to pain control. *Int Dent J* 1978;**28**:52–65.
31. Tarbet W, Silverman G, Stolman J, Fratarcangelo P. Clinical evaluation of a new treatment of dentinal hypersensitivity. *J Periodontol* 1980;**51**:535–40.
32. Nagata T, Ishida H, Shinohara H, Nishikawa S, Kasahara S, Wakano Y, et al. Clinical evaluation of a potassium nitrate dentifrice for the treatment of dentinal hypersensitivity. *J Clin Periodontol* 1994;**21**(3):217–21.
33. Wara-aswapati N, Krongnawakul D, Jiraviboon D, Adulyanon S, Karimbus N, Pitiphat W. The effect of a new toothpaste containing potassium nitrate and triclosan on gingival health, plaque formation and dentine hypersensitivity. *J Clin Periodontol* 2005;**32**:53–8.
34. Chabanski MB, Gillam DG. Aetiology, prevalence and clinical features of cervical dentine sensitivity. *J Oral Rehabil* 1997;**24**(1):15–9.
35. Sandelowski M. Sample size in qualitative research. *Res Nurs Health* 1995;**18**(2):179–83.
36. Ritchie J, Spencer L. Qualitative data analysis for applied policy research. In: Bryman A, Burgess R, editors. *Analyzing qualitative data*. New York, NY: Routledge; 1994.
37. Bury M. Chronic illness as biographical disruption. *Sociol Health Illn* 1982;**4**(2):167–82.
38. Williams SJ. Chronic illness as biographical disruption or biographical disruption as chronic illness? Reflections on a core concept. *Sociol Health Illn* 2000;**22**(1):40–67.
39. Lazarus R, Folkman S. *Stress, appraisal and coping*. New York, NY: Springer; 1984.
40. Leventhal H, Meyer D, Nerenz D. The common sense representation of illness danger. In: Rachman S, editor. *Medical psychology*. New York, NY: Pergamon Press; 1980. p. 7–30.
41. Leventhal H, Benyamini Y, Brownlee S. Illness representations: theoretical foundations. In: Petrie K, Weinman J, editors. *Perceptions of health and illness*. Amsterdam: Harwood; 1997. p. 1–18.
42. Melzack R. *The puzzle of pain*. Middlesex: Penguin Books; 1973.
43. Miller SM. Why having control reduces stress: if I can stop the roller coaster, I don't want to get off. In: Garber J, Mep S, editors. *Human helplessness: theory and applications*. New York, NY: Academic Press; 1980.

44. Harkapaa K. Relationships of psychological distress and health locus of control beliefs with the use of cognitive and behavioural coping strategies in low back pain patients. *Clin J Pain* 1991;**7**:275–82.
45. Bolles RC, Fanselow MS. A perceptual-defensive-recuperative model of fear and pain. *Behav Brain Sci* 1980;**3**:291–301.
46. Rhudy JL. Fear and anxiety: divergent effects on human pain thresholds. *Pain* 2000;**84**:65–75.
47. Carlsson K, Andersson J, Petrovic P, Petersson KM, Ohman A, Ingvar M. Predictability modulates the affective and sensory-discriminative neural processing of pain. *NeuroImage* 2006;**32**:1804–14.
48. Lautenbacher S, Rollman GB. Somatisation, hypochrondriasis, and related conditions. In: Block AR, Kremer EF, Fernandez E, editors. *Handbook of pain syndromes: biopsychosocial perspectives*. Mahwah, NJ: Lawrence Erlbaum Associates, Inc; 1998. p. 613–32.
48a. Tan G, Jensen MP, Robinson-Whelen S, Thomby JI, Monga TN. Coping with chronic pain: a comparison of two measures. *Pain* 2001;**90**:127–33.
48b. Esteve R, Ramirez-Maestre C, Lopez-Martinez AE. Adjustment to chronic pain: The role of pain acceptance, coping strategies, and pain-related cognitions. *Annals of Behavioural Medicine* 2007;**33**:179–88.
49. Jensen MP, Turner JA, Romano JM. Changes in beliefs, catastrophising and coping are associated with improvements in multidisciplinary pain treatment. *J Consult Clin Psychol* 2001;**69**:655–62.
50. Hagger M, Orbell S. A meta-analytic review of the common-sense model of illness representations. *Psychol Health* 2003;**18**:141–84.
51. Melzack R. Pain: past, present and future. *Can J Exp Psychol* 1993;**47**:615–29.
52. Coleman T, Kinderknecht K. Cervical dentin hypersensitivity. Part 1: the air indexing method. *Quintessence Int* 2000;**31**(1):461–5.
53. Tammaro S, Berggren U, Bergenholtz G. Representation of verbal pain descriptors on a visual analogue scale by dental patients and dental students. *Eur J Oral Sci* 1997;**105**:207–12.
54. Herzlich C. *Health and illness*. London: Academic Press; 1973.
55. Blaxter M. *Health and lifestyles*. London: Routledge; 1990.
56. Moos R, Schaefer J. The crisis of physical illness: an overview and conceptual approach. In: Moos R, editor. *Coping with physcial illness: new perspectives*. New York, NY: Plenum; 1984.
57. Ogden J. *Health psychology: a textbook*. Berkshire, England: Open University Press; 2004.

Construction and validation of the quality of life measure for dentine hypersensitivity (DHEQ)*

Olga V. Boiko[1], *Sarah R. Baker*[1], *Barry J. Gibson*[1], *David Locker*[2], *Farzana Sufi*[3], *Ashley P.S. Barlow*[3] *and Peter G. Robinson*[1]

[1]School of Clinical Dentistry, Claremont Crescent, University of Sheffield, Sheffield, UK, [2]Faculty of Dentistry, University of Toronto, Toronto, Canada, [3]GlaxoSmithKline, Consumer Health Care, Weybridge, UK

Introduction

Self-reported assessments are increasingly used in dentistry to capture the psychosocial experiences of pain, discomfort, and malfunctioning, thereby supplementing clinical indicators.[1] Such research has been important in recognizing the long-term complex effects of oral conditions and can be used to evaluate clinical interventions and measurement of change.[2–4]

Research on oral health-related quality of life (OHQoL) has commonly used instruments such as the Oral Health Impact Profile (OHIP) that are generic for a number of oral health conditions and inquire about a broad spectrum of limitations and dysfunctions.[5,6] However, this breadth can be a disadvantage because generic measures may not detect the nuances of a specific condition or distinguish them from other impacts. Wong et al.[7] showed that many OHIP items are irrelevant to specific oral health states, which prompted their work on a new instrument, the OHIP-aesthetic. Elsewhere, OHIP-49 was found to be only partially responsive to changes after tooth whitening.[8] In relation to dentine hypersensitivity (DH), Bekes et al.[9] found that the generic OHIP-49 was insensitive to the particular impacts of DH. Although patients attending for treatment of hypersensitivity experienced more impacts and had poorer oral health than the general population, the difference in mean scores was less than 10% of the overall scale. All of these factors suggest that the impacts of specific oral conditions and, in particular, DH are not captured by generic measures.

DH is a "short, sharp pain arising from exposed dentine in response to stimuli, typically thermal, evaporative, tactile, osmotic, or chemical and which cannot be ascribed to any other dental defect or pathology."[10] Theories to explain the

*This chapter was previously published as: Boiko OV, Baker SR, Gibson BJ, Locker D, Sufi F, Barlow, APS, Robinson PG. Construction and validation of the quality of life measure for dentine hypersensitivity (DHEQ). *Journal of Clinical Periodontology* 2010; **37**: 973–980. doi:10.1111/j.1600-051X.2010.01618.x.

Dentine Hypersensitivity. DOI: http://dx.doi.org/10.1016/B978-0-12-801631-2.00007-5

underlying mechanism focus on a hydrodynamic mechanism, exacerbated by tissue lost to erosion and abrasion.[10–12] The condition is increasingly common. However, population studies contrast sharply with studies that use clinical diagnoses. Some clinical research estimates that the prevalence is as low as 3.8–4.1% in a UK general dental practice.[13] Other studies report significantly higher prevalence, often more than 50%.[14–16] These differences suggest that DH may be underreported and unrecognized by clinicians and patients.

Pain is the major symptom of the condition. Studies of patients' experiences have been restricted to ratings of pain, usually in response to a stimulus within a clinical setting.[13,17] There has been little consideration of the impact on everyday life. In one study, DH hindered toothbrushing in 8.7% of participants, 28.2% of participants could not drink cold water, and 26% could not eat ice cream without discomfort, and therefore 10% avoided the area of discomfort.[14] In light of these data, a qualitative study explored the daily experiences of people with DH.[18] The findings showed the depth and complexity of pain experiences associated with sensitivity, impacts on functional status and everyday activities such as eating, drinking, talking, toothbrushing, and social interaction, and also more subtle impacts on emotions and identity. The current chapter draws on those data to develop a condition-specific questionnaire for DH.

The first reason for constructing a condition-specific measure was the need to address the particular impacts of DH. We could further expect that such a measure could be more responsive to changes in the condition. Hence, the aim of the study was to develop and validate a specific measure of OHQoL in relation to DH (the Dentine Hypersensitivity Experience Questionnaire, DHEQ) based on an improved biopsychosocial understanding of the condition.

Materials and methods

The study was designed in seven stages based on a multistage impact approach.[19] The following sections report on the material and methods of each stage. Ethical approval was obtained from the University of Sheffield Ethics Committee.

Stage 1: Theoretical model

Initially, a theoretical model was chosen as the framework for the study to guide the interviews and questionnaire development. Three models were considered: Locker's model of oral health; the World Health Organization's International Classification of Functioning, Disability, and Health (ICF); and the Wilson and Cleary model linking clinical variables and quality of life.[20–22] The Wilson and Cleary model was selected based on its compatibility with the functional and coping impacts of DH.[22] This model is also compatible with the Locker model but provides a broader framework for understanding the relationship between clinical status, symptoms, functioning, perceived health, and overall quality of life (Figure 7.1).[20] The ICF was difficult to operationalize because it classifies conditions and impacts but is less clear on how these may be related.[21]

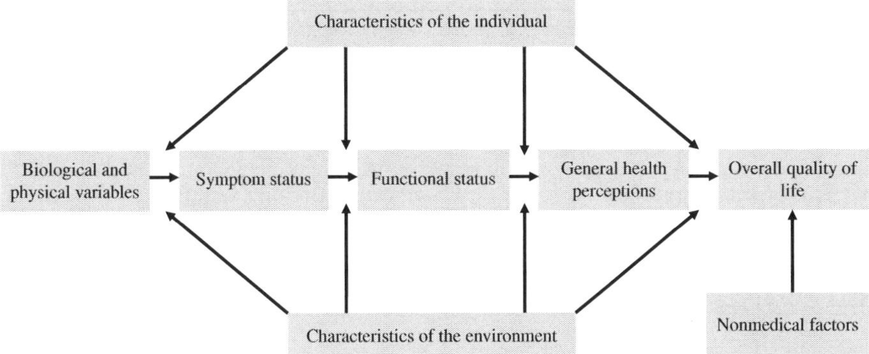

Figure 7.1 The Wilson and Cleary model.[22] Linking clinical variables with health-related quality of life: a conceptual model of patient outcomes.

Stage 2: Qualitative interviews

To identify the everyday impacts of sensitivity, 23 in-depth interviews (15 females and 8 males) were conducted until saturation was achieved.[18] Participants were recruited purposively from the general population using the criteria of adults with sensitive teeth and a range of ages, gender, and longevity of the condition. The number of female participants prevailed over male participants, which accords with the gender balance affected by DH.[13,23] Participants from young (18–40) and older groups (40–65) were evenly interviewed. Half of the participants recruited for the study characterized themselves as having DH, whereas half described twinges in their teeth in response to thermal (cold or hot) or physical stimuli (toothbrushing). This second group was recruited to reflect people who experienced symptoms consistent with DH but who may not identify themselves as having the condition.

On gaining consent, an interview was arranged at a mutually suitable time and place. Interviews lasted 30–40 min. The interviews were transcribed and then analyzed using framework analysis.[24] The preliminary interview guide was based on the theoretical model and previous data, including self-reported experiences in focus groups conducted by a consumer health care company.

Stage 3: Questionnaire development

The qualitative findings were used to generate the questionnaire items. The data were used to populate the theoretical model within the following domains:

- Pain (symptoms in the model)
- Functional restrictions, adaptation, avoidance, social impact, emotional impact, and identity (all regarded as functional limitation)
- A global oral health rating used to represent general health perceptions
- Effect on life overall (quality of life)

The scale component of the measure comprised only the items in the functional restrictions section (adaptation, avoidance, social impact, emotional impact, and identity subscales). Only the scores for these items are aggregated to measure OHQoL.

Response formats were chosen that were relevant to the domain. Particularly, the frequencies of some impacts were not recorded because of their intermittent nature and participants' strategies for avoiding pain stimuli. These strategies for coping formed part of the impact of the condition.

Stage 4: Focus groups

Face validation of DHEQ was undertaken by three focus groups. As in the qualitative stage, 20 participants were recruited on the basis of age, gender, educational background, and disease longevity. The gender split was 1:2 (male to female). One focus group (6 participants) involved people with long-standing DH and, therefore, participants tended to be older. The second group (7 participants) included participants of different ages, those with long-term symptoms, and those with "twinges and discomfort." The third group (7 participants) consisted of young people with "twinges and discomfort" who were new to the symptoms of DH. After completing DHEQ, the participants were asked about each item so that problems with wording and meaning could be identified and resolved.

Stage 5: Cross-sectional validation

To examine validity and reliability, the DHEQ was tested cross-sectionally in a quota sample of 163 adults recruited from the general population via online advertisements at the University of Sheffield (75%) and across the United Kingdom (25%). The questionnaire was distributed via postal mail. Three participants (1.8%) provided questionnaires missing more than 10% of the answers for the items and were excluded from further analysis. As intended, 64% of participants had self-reported DH and 34% described themselves as experiencing "twinges" but did not describe this as sensitive teeth. Demographic quotas included an equal split of genders and age. Participants broadly matched the socioeconomic status of the UK general population.

For test–retest reliability assessment, a proportion of the participants (25% of the sample) were sent a second copy of the questionnaire after 2 weeks, as in reliability tests of other OHQoL measures.[1] Assessment was based on data from 34 participants (21% of the sample) whose global rating of oral health was the same as with the first administration.

Analytical procedures

To assess the reliability and validity of the DHEQ in this general population sample, data were analyzed iteratively in four stages. First, the data were described using the numbers of missing responses, proportions and appropriate measures of central tendency, and spread for each of the items and scales scores. Item-impact values for the impact scales were calculated as the product of the mean score and percentage of people broadly agreeing they had that impact ("strongly agree," "agree," and "agree a little" responses on the item).[a]

Second, preliminary psychometric analysis of the scales was performed. Internal consistency and test–retest reliability were assessed using item and subscale total correlations, Cronbach's alpha, and intraclass correlation coefficients. Construct validity was assessed by correlating the impact scale and its subscales with the global oral health rating and the summary measure of impact on life overall.

Third, confirmatory factor analysis (CFA) was used to provide a further test of the within-construct validity of the scale. CFA is the first in the two-stage process of structural equation modeling (SEM; the measurement model).[25] CFA provides information on how scale items (e.g., "Having the sensations in my teeth takes a lot of the pleasure out of eating and drinking") measure underlying (latent) constructs (e.g., functional restrictions); therefore, it is a test of the validity of selected items (i.e., do items selected to measure a construct actually do so?) and the number of constructs that "fit" the data (e.g., 1 for "DH impacts" or 8 for "pain," "restrictions").

Fourth, the results of the CFA and item-impact analysis were used to revise the questionnaire. The revised version had one item removed from each of the restrictions and identity subscales. The approach and avoidance coping subscales were merged into a single adaptation subscale.

Fifth, the psychometric analyses were repeated on the revised structure. Data were analyzed using SPSS 16.0 and Amos 6.0. A P value of 0.05 was selected as the level of significance in hypothesis tests.

Stage 6: Follow-up interviews

A follow-up validation was stimulated by the relatively high proportion of neutral responses ("neither agree nor disagree") received during the validation study that provoked a question about the clarity of questions. Follow-up face-to-face interviews were conducted with 11 participants who used neutral responses in the cross-sectional validation. The participants were asked to verbally complete the questionnaire and comment on its clarity as they progressed.

Stage 7: Validation in a clinical population

A second cross-sectional validation of DHEQ was performed in a population with clinically diagnosed DH recruited during the preintervention stage of a sponsored randomized controlled trial before the participants received any interventions or other study procedures. Data were collected from 108 participants and analyzed using statistical measures similar to those in stage 5.

[a]Detailed description of items and the statistics on item impact are available on request.

Results

An initial pool of 50 items was generated from the qualitative data within the domains of pain (corresponding to a symptom in the original model), impact (corresponding to functional limitation), and effect on life overall (corresponding to quality of life) (Table 7.1). The impact scale had six subscales based on the initial domains of the Wilson and Cleary[22] model (see Appendix). For the purposes of construct validation, a standard global rating of oral health was added.

Two summary measures were created for the impact scale and its subscales. The "total score" was calculated as the sum of the item scores (1–7 Likert scale) per participant (possible range of 0–257 and 0–243 in the original and revised DHEQ, respectively). "Subscale scores" for each of the subscales were created in the same way. The "extent" of impacts was calculated as the number of impacts per participant with which each participant broadly agreed ("strongly agree," "agree," and "agree a little" responses; possible range of 0–36 and 0–34 for original and refined versions, respectively).

The three focus groups suggested minor modifications to item wording but generally understood the scales and reported no consistent problems in the use of DHEQ.

Validation in the general population sample

Descriptive results

The number of missing values was low. Six participants (3.7%) did not provide an answer to one item and three (1.8%) did not respond to another item of the original DHEQ. Nine other items received only one (0.6%) nonresponse. The visual analogue scales placed the pain of sensitivity at the middle range (Table 7.2).

The means for all summary measures of impact were close to the center of the possible range and there was substantial spread in the data, indicating no floor or ceiling effects (Table 7.3). The data are summarized in subscale scores (Table 7.3) for the original and revised versions of DHEQ in the left and right columns, respectively. Total score and extent data were approximately normally distributed.

Individual item weights are presented in Table 7.4 along with the frequencies, means (standard deviations), and item impacts for the original scale items (Q10-45). Item impact, calculated as mean score multiplied by the proportion with that impact, demonstrated a wide range (11.88–529.52).

Reliability and validity

Nearly all item-total correlations were more than 0.4. All were statistically significant (Table 7.4).

The questionnaire demonstrated high levels of internal consistency for all impact scores and subscales (Table 7.5). All correlations between the subscales and total scores were significant and consistent. The highest correlations were seen between "total score" and "emotional impact" ($r = 0.89$) and between "total score" and

Table 7.1 Format of the DHEQ

	Number and type of items	Purpose	Summary measure
Introductory descriptors	6 closed questions	Describe pain	Each item treated separately
Pain scales			
Intensity Bothersomeness Tolerability	3 visual analogue scales	Measure pain	Each item treated separately and scaled 0–10
Impact subscales			
Restrictions	5/4[a]	Measure restrictions in daily activity	7-point Likert scales coded: 1 = "strongly disagree" to 7 = "strongly disagree"
Approach coping	6[b]/12	Measure activities to cope and prevent sensitivity episodes	
Avoidance coping	6[b]	Measure impacts due to avoiding potential pain stimuli	
Social impact	5	Measure handicap	
Emotional impact	8	Measure emotional impact	
Identity	6/5[a]	Measure impact on personal identity	
Global oral health rating	1	Measure health perception	5-point Likert scale coded 1 = "excellent," 2 = "very good," 3 = "good," 4 = "fair," 5 = "poor," 6 = "very poor"
Overall effect	4	Measure effect on overall quality of life	5-point Likert scale coded 0 = "not at all" to 4 = "very much"

The gray shaded area forms the impact subscales of the DHEQ.
[a]One item was removed from each of the restrictions and identity subscales in the revised questionnaire.
[b]These two scales were merged in the revised version of the questionnaire.

Table 7.2 Mean scores for the pain scale among 160 participants in the general population sample

Pain scale (VAS)	Mean (SD)	Range
Intensity	5.5 (1.73)	1–9
Bothersomeness	5.3 (2.17)	1–10
Tolerability	4.4 (2.00)	1–9

Table 7.3 Total score, extent and subscales scores among 160 participants in the general population sample

	Original DHEQ			Revised DHEQ		
	No. items	Mean (SD)	Range	No. items	Mean (SD)	Range
Total score	36	138.6 (36.36)	40–228	34	130.96 (35.06)	34–219
Extent	36	16.52 (7.52)	1–35	34	14.43 (5.97)	1–33
Subscales						
Restrictions[a]	5	21.82 (5.44)	9–35	4	17.01 (4.98)	4–28
Approach coping	6	22.81 (7.60)	6–38	12	47.57 (14.01)	12–73
Avoidance coping[b]	6	24.76 (7.48)	6–41			
Social impact[b]	5	15.29 (6.53)	5–35	5	15.29 (6.53)	5–35
Emotional impact	8	32.67 (9.55)	8–53	8	32.67 (9.55)	8–53
Identity[a]	6	21.24 (6.78)	6–41	5	18.42 (5.68)	5–34
Global oral health rating	1	3.36 (1.14)	1–6	4	3.36 (1.14)	1–6
Effect on life overall	4	4.38 (3.06)	1–16	4	4.38(3.06)	1–16

The gray shaded area forms the impact subscales of the DHEQ.
[a]One item was removed from each of these two subscales in the revised questionnaire.
[b]These two scales were merged in the revised version of the questionnaire.

Table 7.4 Mean scores, item impacts, and item-total correlations among 160 participants in the general population sample

	Item	Mean	SD	% of people who had impact	Item impact	Item-total correlation (r_s)
10	Restrictions: pleasure out of eating	4.39	1.72	63	276.57	0.704
11	Restrictions: cannot finish meal	2.76	1.66	21	57.96	0.690
12	Restrictions: longer to finish meal	4.00	1.76	51	204.0	0.719
13	Restrictions: uncertainty when[a]	4.82	1.80	68	327.76	0.256
14	Restrictions: problems with eating ice cream	5.86	1.35	89	521.54	0.347
15	Adaptation: modification of eating	4.90	1.66	77	377.3	0.590
16	Adaptation: careful when breathing	4.29	2.04	58	248.82	0.558
17	Adaptation: warming food/drinks	4.24	1.76	54	228.96	0.550
18	Adaptation: cooling food/drink	3.24	1.92	33	106.92	0.623
19	Adaptation: cutting fruit	3.24	1.93	32	103.68	0.550
20	Adaptation: putting a scarf over mouth	2.91	1.77	24	69.84	0.478
21	Adaptation: avoiding cold drinks/foods	4.47	1.86	61	272.57	0.624
22	Adaptation: avoiding hot drinks/foods	3.07	1.80	35	107.45	0.650
23	Adaptation: avoiding contact with certain teeth	5.18	1.68	80	414.4	0.483
24	Adaptation: change toothbrushing	4.23	1.95	52	219.96	0.448
25	Adaptation: biting in small pieces	4.17	1.90	54	225.18	0.588
26	Adaptation: avoiding other food	3.64	1.81	36	131.04	0.665
27	Social: longer than others to finish	3.35	1.87	33	110.55	0.715
28	Social: choose food with others	3.32	1.79	32	106.24	0.730

(*Continued*)

Table 7.4 (Continued)

	Item	Mean	SD	% of people who had impact	Item impact	Item-total correlation (r_s)
29	Social: hide the way of eating	2.78	1.70	21	58.38	0.623
30	Social: unable to take part in conversations	1.98	1.19	6	11.88	0.483
31	Social: painful at the dentist	3.86	2.08	42	162.12	0.661
32	Emotions: frustrated not finding a cure	4.05	1.83	47	190.35	0.725
33	Emotions: anxious of eating contributes	4.25	1.77	55	233.75	0.590
34	Emotions: irritating sensations	5.36	1.30	88	471.68	0.663
35	Emotions: annoyed with myself for contributing	3.86	1.88	44	169.84	0.451
36	Emotions: guilty for contributing	3.44	1.88	34	116.96	0.668
37	Emotions: annoying sensations	5.42	1.38	85	460.7	0.510
38	Emotions: embarrassing sensations	2.64	1.42	12	31.68	0.441
39	Emotions: anxious because of sensations	3.65	1.79	40	146	0.715
40	Identity: difficult to accept	2.82	1.81	19	53.58	0.510
41	Identity: part of my life[a]	5.72	1.17	91	529.52	0.209
42	Identity: different from others	2.61	1.59	16	41.76	0.651
43	Identity: makes me feel old	3.38	1.95	34	114.92	0.565
44	Identity: makes me feel damaged	3.08	1.75	28	86.24	0.566
45	Identity: makes me feel unhealthy	3.64	1.81	44	160.16	0.527

All $P < 0.05$, Pearson correlation.
[a]Two items deleted from the impact scale after CFA (revised DHEQ).

Table 7.5 **Impact scale reliability in the general population sample**

		Original DHEQ		Revised DHEQ	
	No. items	**Cronbach's alpha ($n = 160$)**	**ICC ($n = 34$)**	**Cronbach's alpha ($n = 160$)**	**ICC ($n = 34$)**
Total score	36	0.91	0.93	0.86	0.92
Subscales					
Restrictions	5	0.50	0.76	0.76	0.76
Approach coping	6	0.74	0.82	0.86	0.88
Avoidance	6	0.78	0.81		
Social impact	5	0.76	0.83	0.76	0.83
Emotional impact	8	0.87	1.00	0.87	1.00
Identity	6	0.59	1.00	0.70	0.75

The "approach coping" and "avoidance" subscales were merged to an "adaptation" subscale in the revised DHEQ.

"avoidance" ($r = 0.85$) in the original version, and between "total scores" and the merged "adaptation" subscale ($r = 0.90$) in the revised DHEQ.

Cronbach's alpha for the total impact score in the original DHEQ was 0.91 (Table 7.5). Alphas for the subscales ranged from 0.50 for functional restrictions to 0.87 for emotional impact, indicating fair to good internal consistency reliability. Alphas improved for all subscales in the revised scale. In particular, the scores for "restrictions" and "identity" increased to 0.76 and 0.70, respectively. The merging of the "avoidance" and "approach" subscales into an "adaptation" subscale also improved internal reliability, as measured by Cronbach's alpha.

The test–retest reliability was calculated for 34 participants whose global rating was the same at baseline and 2 weeks later. The intraclass correlation coefficients were 0.93 and 0.92 for the original and revised versions of the impact scale, respectively, indicating very high agreement (Table 7.5). Test–retest reliability for both versions was lowest for the functional restrictions subscale (0.76) but was still very acceptable.

Total and subscale scores of the original and revised impact scales all correlated significantly with global oral health ratings, indicating good construct validity (Table 7.6). Both versions of DHEQ total scores were strongly and significantly correlated with the scores for the subscale of effect on everyday life.

Six items received a high proportion of neutral responses ("neither agree nor disagree"). Follow-up interviews suggested that participants chose neutral responses when items were not relevant to their experiences. Therefore, we identified no confusion in the questions or their meaning to people. Thus, the "neutral response" option was maintained.

The initial step of the CFA was to test the DHEQ using a first-order model with pain, restrictions, approach coping, avoidance coping, social impact, emotional

Table 7.6 Correlations between impact total and subscales scores with global oral health status

Global oral health rating	Original DHEQ		Revised DHEQ	
	r_s	*P*	r_s	*P*
Total score	0.23	<0.01	0.23	<0.01
Subscales				
Restrictions	0.15	0.06	0.16	0.04
Approach coping[a]	0.16	0.04	0.18	0.03
Avoidance coping[a]	0.17	0.03		
Social impact	0.24	<0.01	0.24	<0.01
Emotional impact	0.18	0.02	0.18	0.02
Identity	0.24	<0.01	0.25	<0.01

[a]The "approach coping" and "avoidance" subscales were merged into an "adaptation" subscale in the revised DHEQ.

impact, identity, and impact on life overall as the eight latent constructs. Items were not allowed to load on more than one construct, nor were their error terms allowed to correlate.

The model was examined using Amos 7.0 with maximum likelihood estimation and bootstrapping[26] as recommended for samples of <200.[27] We evaluated model fit using indices from the three fit classes: absolute fit; parsimony-adjusted; and comparative.[28] A χ^2/df ratio <3.0, RMSEA values 0.08 or less, CFI and TLI of 0.90 or more, and an SRMR <0.08 were taken to indicate an acceptable model fit.[28]

The eight-factor measurement model was an acceptable fit to the data in three of the *a priori* model fitting indices: χ^2/df = 1.947; RMSEA = 0.077 (90% confidence interval [CI]: 0.07–0.083); SRMR = 0.075; CFI = 0.782; and TLI = 0.763. Inspection of the standardized regression weights indicated that all items were significant measures of their respective constructs ($P < 0.01$), with the exception of "I am uncertain when I am going to have these sensations in my teeth" ($P = 0.06$) and "having these sensations in my teeth is now just a part of my life" ($P = 0.87$). The two items were therefore deleted from the revised DHEQ.

Because the correlation between the approach and avoidance coping factors was high (0.94), they were collapsed into a single factor (relabeled "adaptation") in the CFA rerun. The modification significantly improved the fit of the model ($\Delta\chi^2 = 138.56$; Δdf = 74; $P < 0.01$; revised model fit criteria: $\chi^2 = 1.955$; RMSEA = 0.077 [90% CI: 0.072–0.083]; SRMR = 0.069; CFI = 0.795; TLI = 0.778). The standardized beta weights for the three pain items were high (β range = 0.67–0.88). Similarly, for restrictions, the range was 0.74–0.80, with the lowest loading being for "problems eating ice cream" ($\beta = 0.38$). For the relabeled adaptation factor, the loadings were lower, ranging between 0.43 ("changed the way I brush my teeth") and 0.68 ("avoid very cold drinks or food"). In relation to social impacts, beta weights ranged between 0.67 and 0.81, with the exception of the item "going to the dentist is hard for me" (0.46). The range in beta weights for the emotional impact factor was between 0.44 ("I felt guilty because I might have contributed to the sensations I am having with my teeth") and 0.80 ("I've been anxious that something I eat or drink might cause sensations in my teeth"). Finally, the item loadings for the identity ($\beta = 0.63-0.78$) factors and the impact on life overall ($\beta = 0.64-0.86$) factors were consistently high. All items were highly significant indicators of their respective constructs (all $P < 0.01$) (full CFA data available on request).

Clinical sample validation

For validation among the clinical sample, there were few missing data and scores showed neither floor nor ceiling effects. Reliability analyses were restricted to internal consistency. Cronbach's alphas were high for the total score (0.82) and subscales (0.79–0.89). Total and subscale scores were significantly correlated with global ratings ($r = 0.26$). The mean scores and 95% CIs for the impact scale and its subscales in the clinical sample were: "total score," 147.6 ± 5.98; "restrictions" subscale, 18.91 ± 0.98; "adaptation," 55.83 ± 2.52; "social impact," 18.08 ± 1.23; "emotional impact," 35.89 ± 1.51; and "identity," 18.89 ± 1.11. In all cases except for identity, these CIs did not include the scores for the general population as presented in Table 7.3, indicating greater impact in the clinical sample.

Discussion

The purpose of this study was to develop and validate the DHEQ as an evaluative condition-specific measure of quality of life in people with DH. The study was one of the first to develop a disease-specific OHQoL measure and aimed to measure particular everyday impacts related to DH. The validation of DHEQ supports the feasibility of condition-specific instruments for measuring biopsychosocial impacts of other oral conditions. The results also offer the possibility of a much deeper understanding of a condition that has been described as an enigma.[29]

DHEQ was developed through meticulously following a series of stages and adopting a robust theoretical framework.[19,22] The interview guide was informed by the model and by previous data. Rich qualitative data were used to populate the model and instrument with items. Focus group data supported the face validity of the original DHEQ. The questionnaire was administered to a general population sample and refined via well-known analytical techniques. Follow-up interviews focusing on neutral responses further supported the face validity. Content validity is indicated by a wide range of responses. DHEQ has excellent internal reliability as measured by item-total correlations and Cronbach's alpha (0.86 and 0.82 for the revised DHEQ in population and clinical samples, respectively) that meets standard thresholds for measurements of this kind.[30,31] Test–retest reliability was also excellent. As a result of CFA, internal consistency in subscales was improved by deletion of two items and merging two subscales. Construct validity was indicated by significant correlations between total and subscales scores for global ratings of oral health and the subscale for impacts on everyday life ($r = 0.23$ and $r = 0.25$). The revised DHEQ demonstrated high validity and reliability scores in samples from general and clinical populations.

The findings support the value of condition- and disease-specific impacts and their measurement. Generic measures can be too vague in assessing the links between oral conditions and OHQoL. Locker and Allen[32] raised important critical points associated with the instrumental qualities of these measures. Weak links between conditions and impacts on quality of life are a common problem with generic measures. As demonstrated previously, OHIP did not distinguish the impact of DH from those of other conditions.[9] DHEQ provides a strong alternative to generic OHQoL instruments because of its direct reference to the problems associated with sensitive teeth. Moreover, measures such as OHIP are designed to detect handicaps and are therefore less relevant for the experiences of DH. This is not to say that DH does not cause impacts, because it is associated with tangible everyday discomfort. DHEQ retains the authenticity of the experiences in measuring the whole range of impacts and adaptation strategies, such as changes in eating practices, food withdrawal, mouth and teeth awareness, modified toothbrushing, associated emotional coping, identity changes, and others. Therefore, the discriminative capacity of this new measure is much higher than previous instruments could show.

The study also indicates possible implications for the development of new quality of life measures. Initially with reference to general health, the poor utility of generic instruments was contrasted with disease-specific (oral health) measures.[33] Ten years later, this project marks a shift toward measures that are not simply disease-specific in capturing combinations of impacts but that are theoretically based, in practical and linguistic terms, on the impacts of specific oral conditions. The approach adopted in the current study emphasizes a need for greater care in exploring the nuances of impacts and discriminating between different oral conditions. In this way, the methodologically difficult concept of quality of life can be operationalized through particular impacts of oral conditions, especially if a robust theoretical framework is used and supported by qualitative data.

The applications of condition-specific measures in randomized controlled trials are relatively new, but these may detect changes in functional and personal experiences of the condition.[34] Our research is highly suggestive of the ability of DHEQ to capture improvements in pain and other impacts. A forthcoming longitudinal study will test the evaluative properties of DHEQ.

The final version of DHEQ is presented in Appendix One of this book.

Acknowledgment

The authors thank people who participated in this study. They also acknowledge the valuable contributions of Dr. Stephen Mason and Mrs. Vicky Murysinowski of GlaxoSmithKline Consumer Healthcare to this research.

References

1. Jokovic A, Locker D, Stephens M, Kenny D, Tompson B, Guyatt G. Validity and reliability of a questionnaire for measuring child oral-health-related quality of life. *J Dent Res* 2002;**81**:459–63.
2. Awad MA, Feine JS. Measuring patient satisfaction with mandibular prostheses. *Community Dent Oral Epidemiol* 1998;**26**:400–5.
3. Baker SR, Pankhurst CL, Robinson PG. Utility of two oral health-related quality of life measures in patients with xerostomia. *Community Dent Oral Epidemiol* 2006;**34**:351–62.
4. Pearson NK, Gibson BJ, Davis DM, Geilier S, Robinson PG. The effect of a domiciliary denture service on oral health related quality of life: a randomised controlled trial. *Br Dent J* 2007;**203**:E3 568.
5. Slade G, Spencer A. Development and evaluation of the Oral Health Impact Profile. *Community Dent Health* 1994;**11**:3–11.
6. Slade GD. Derivation and validation of a short-form oral health impact profile. *Community Dent Oral Epidemiol* 1997;**25**:284–90.
7. Wong AHH, Cheung CS, McGraph C. Developing a short form of Oral Health Impact Profile (OHIP) for dental aesthetics: OHIP-aesthetic. *Community Dent Oral Epidemiol* 2007;**35**:64–72.
8. McGrath C, Wong AHH, Lo ECM, Cheung CS. The sensitivity and responsiveness of an oral health related quality of life measure to tooth whitening. *J Dent* 2005;**33** (8):697–702.
9. Bekes K, John MT, Schaller H-G, Hirsch C. Oral health-related quality of life in patients seeking care for dentin hypersensitivity. *J Oral Rehabil* 2009;**36**:45–51.
10. Dababneh RH, Khouri AT, Addy M. Dentine hypersensitivity: an enigma? A review of terminology, epidemiology, mechanisms, aetiology and management. *Br Dent J* 1999;**187**:606–11.
11. Holland GR, Narhi MN, Addy M, Gangarosa L, Orchadson R. Guidelines for the design and conduct of clinical trials on dentine hypersensitivity. *J Clin Periodontol* 1997;**24**:808–13.
12. Orchardson R, Gillam DG. Managing dentin hypersensitivity. *J Am Dent Assoc* 2006;**137**:990–8.
13. Rees JS, Addy M. A cross-sectional study of dentine hypersensitivity. *J Clin Periodontol* 2002;**29**:997–1003.
14. Gillam D, Seo H, Bulman J, Newman H. Perceptions of dentine hypersensitivity in a general practice population. *J Oral Rehabil* 1999;**26**:710–4.
15. Irwin C, McCusker P. Prevalence of dentine hypersensitivity in a general dental population. *J Ir Dent Assoc* 1997;**43**:7–9.
16. Rees JS, Jin LJ, Lam S, Kudanowska I, Vowles R. The prevalence of dentine hypersensitivity in a hospital clinic population in Hong Kong. *J Dent* 2003;**31**:453–61.
17. Al-Wahadni A, Linden GJ. Dentine hypersensitivity in Jordanian dental attenders: a case control study. *J Clin Periodontol* 2002;**29**:688–93.
18. Gibson B, Boiko O, Baker S, Robinson PG, Barlow A, Player T, et al. The everyday impact of dentine sensitivity: personal and functional aspects. *Soc Sci Dent* 2010;**1**:11–20.
19. Juniper E, Guyatt G, Jaeschke R. How to develop and validate a new health-related quality of life instrument. In: Spikler B, editor. *Quality of life and pharmacoeconomics in clinical trials*. Philadelphia, PA: Lippincott-Raven Publishers; 1996. p. 49–56.

20. Locker D. Measuring oral health: a conceptual framework. *Community Dent Health* 1988;**5**:3–18.
21. World Health Organisation. *International classification of functioning, disability and health*. Geneva: WHO; 2001.
22. Wilson IB, Cleary PD. Linking clinical variables with health-related quality of life: conceptual model of patient outcomes. *J Am Med Assoc* 1995;**273**:59–65.
23. Rees JS. The prevalence of dentine hypersensitivity in general dental practice in the UK. *J Clin Periodontol* 2000;**27**:860–5.
24. Ritchie J, Spencer L. Qualitative data analysis for applied policy research. In: Bryman A, Burgess R, editors. *Analyzing qualitative data*. New York, NY: Routledge; 1994. p. 173–93.
25. Kline RB. *Principles and practice of structural equation modelling*. 2nd ed. New York, NY: Guilford Press; 2005.
26. Arbuckle JL. *Amos 6.0 user's guide*. Philadelphia, PA: Amos Development Corporation; 2005.
27. Efron B, Tibshirani R. *An introduction to the bootstrap*. New York, NY: Chapman & Hall; CRC; 1993.
28. Brown TA. *Confirmatory factor analysis for applied research*. New York, NY: The Guilford Press; 2006.
29. Johnson RH, Zulgar-Nairn BJ, Koval JJ. The effectiveness of an electroniosing toothbrush in the control of dentinal hypersensitivity. *J Periodontol* 1982;**53**:353–9.
30. Nunnally J. *Psychometric theory*. New York, NY: McGraw-Hill; 1978.
31. Streiner DL, Norman GR. *Health measurement scales*. 3rd ed. Oxford: Oxford University Press; 2003.
32. Locker D, Allen F. What do measures of 'oral health-related quality of life' measure?. *Community Dent Oral Epidemiol* 2007;**35**:401–11.
33. Allen PF, McMillan AS, Walshaw D, Locker D. A comparison of the validity of generic- and disease-specific measures in the assessment of oral-health related quality of life. *Community Dent Oral Epidemiol* 1999;**27**:334–52.
34. Ozcelik O, Cenk Haytac M, Seydaoglu G. Immediate post-operative effects of different periodontal treatment modalities on oral health-related quality of life: a randomized clinical trial. *J Clin Periodontol* 2007;**34**:788–96.

Ice cream-related quality of life: constructing a questionnaire to capture changes in the impacts of dentine hypersensitivity

Peter G. Robinson, Sarah R. Baker and Barry J. Gibson
School of Clinical Dentistry, Claremont Crescent, University of Sheffield, Sheffield, UK

Although not authors of this chapter, two colleagues were absolutely integral to the work described in it. Our friend, the late Dr. David Locker of the University of Toronto, contributed a huge amount to the world's understanding of the effects of oral health. He brought his vast experience, deep understanding, and playfulness to this project, and we miss him dearly. Dr. Olga Boiko, now working at the University of Exeter, was tireless in recruiting all the participants, conducting all of the interviews, administering the questionnaires, and acting as the principal data analyst for the qualitative and quantitative aspects of the work. Olga was able to bring detailed knowledge of the data to every discussion of the development of the measure. She and David were the chief protagonists in the debate over including an item about eating ice cream. Olga won!

Introduction

A central thread of this book is the development and validation of the Dentine Hypersensitivity Experience Questionnaire (DHEQ) as a condition-specific measure for dentine hypersensitivity-related quality of life.

Several texts describe the construction of measures of health-related quality of life (HQoL), and all of them can be recommended as useful.[1–3] These texts set out a sequence of mapped stages, each with sufficient detail to help experienced researchers develop their own measures. However, like many aspects of research, even when there is an explicit protocol, the development of an HQoL measure requires many assumptions and judgments, some of which may be implicit. An implicit assumption may lead another researcher who makes a different assumption to an entirely different place. Assumptions are necessary and are made in all research. One way of minimizing their effect is to make them explicit so that they are visible and can therefore be challenged and revised.

Although seemingly quite minor, these assumptions and judgments may have profound consequences for the resulting measure. More generic texts may not have

Dentine Hypersensitivity. DOI: http://dx.doi.org/10.1016/B978-0-12-801631-2.00008-7

the space to outline the nature of these judgments, or the judgments themselves may be so dependent on the context (e.g., the condition or the population undergoing investigation) that they would not be relevant to a broad readership. The focus of this book is on one oral health problem and allows us to outline some of the judgments we made and may help readers with similar decisions in the future.

This section outlines the decisions made in the development of DHEQ using the stages in the development as a framework. These stages are slightly different from those described in Chapter 7, because some assumptions concerning our perspective as a group predated the development of the measure but are relevant here. Those stages are:

- Our perspective
- Explicitly determining the purpose of the measure
- Selection of a model
- The value of qualitative data
- Selection of domains
- Selection of descriptive system/response format and scaling
- Selection of items
- Reference period
- Panel testing

Clearly, this chapter builds on the work describing the construction of DHEQ by Boiko and colleagues, which is reproduced as Chapter 7 of this book.[4] In essence, this chapter presents the rationale or reasoning behind many of the decisions described in that work.

Our perspective

In some respects, the development of a measure of HQoL can be approached in two ways. One approach, sometimes called "top-down," places greater emphasis on the insights of health care workers experienced in caring for patients with a particular condition. The researchers may send lists of items and drafts to experts to compile and validate the questionnaire.[5] Conversely, the "bottom-up" approach emphasizes the lay experience of the condition to a greater extent.

Of course, these two approaches are not mutually exclusive, and most researchers rely on the expertise of both professionals and lay people. However, the bottom-up approach is most compatible with the perspective of our research team. Part of our raison d'être is to be person centered and to facilitate lay participation in health and health care. This perspective is widely accepted, is growing, and is part and parcel of the new public health movement.[6,7]

We hoped that our decision to start with the experiences of people with dentine hypersensitivity (DH) would yield a number of benefits. It would help measure the concepts, language, and terminology relevant to those people, which would in turn enhance its face and content validity, sensitivity, and responsiveness.[8] More immediate consequences of this perspective would be the way we explored the experiences of people with DH and the methods adopted to select items for the measure. All of these techniques would mean that the measure was likely to focus on the specific impacts of DH (i.e., to be condition specific).

Our person-centered perspective guides much of our work and is particularly evident in the work described here and in Chapters 7 and 10.

Explicitly determining the purpose of the measure

The instruction from our funders was to develop a measure of oral health-related quality of life (OHQoL) that could be used to evaluate products for DH. Their brief contained the following three key requirements: measure OHQoL; relate it to DH; and make it responsive to improvements in the condition.

The first stage was to define OHQoL. Several such definitions exist, and one criticism of research in this field is that the term is used without explicit definition.[9,10] As we have seen, quality of life, HQoL, and OHQoL are amorphous concepts that might include aspects of disease, symptoms, and/or physical functioning. Therefore, our adopted definition of OHQoL would determine exactly which aspects of the experience of DH we would measure.

Our criteria were that the definition would incorporate the most up-to-date concepts of OHQoL that were rooted in a detailed understanding of underlying theories of the concept and that the definition would yield a measure that was operationalizable and could be tested.

Based on these criteria, the definition we selected was from Locker and Allen:

> *The impact of oral disorders on aspects of everyday life that are important to patients and persons, with those impacts being of sufficient magnitude, whether in terms of severity, frequency, or duration, to affect an individual's perception of their life overall.*[9]

A key component of this definition is the inclusion of the idea that the impacts of oral disorders should be of sufficient magnitude "to affect an individual's perception of his or her life overall." Consequently, Locker and Allen argued that measures of HQoL (they were even unhappy with that phrase) should go beyond measuring specific impacts to incorporate questions about this more global appraisal, which would incorporate participants' individual beliefs, values, and concerns. This definition would mean that our measure would record both impacts and perceptions of the effect of DH on life overall.

The requirement of being useful for evaluation immediately made mathematical demands of the measure. To detect changes in a person's condition, a measure must be responsive. However, to be responsive in a meaningful way, measures need to distinguish real changes from the background noise of random measurement error. Consequently, evaluative measures also need to be stable when there is no change in the underlying condition.

In addition, there were implications of studying the impacts of a relatively minor condition such as DH. It was likely that the impacts of the condition would be modest and, consequently, there would be a correspondingly restricted scope for benefits brought about by treatment (compared with, for example, having a dental implant fitted or surgery for oral cancer). These considerations meant that the measure would need to detect those modest impacts (i.e., sensitivity) and capture small gradations of difference (i.e., precision).

Previous research had shown that a generic OHQoL measure (the Oral Health Impact Profile, OHIP) detected impacts of DH when comparing patients with the condition to unaffected patients.[11] However, the difference in mean scores between the two groups was less than 10% of the overall scale, suggesting that the measure was relatively insensitive to the condition and that there would be limited scope to respond to treatment.

There has been a long-standing debate about the relative merits of generic measures (that capture the impacts of several conditions) and those that are condition specific.[2,12,13] Generic measures may capture any unexpected benefits or adverse consequences of treatment and allow comparison of impacts between diseases. However, by focusing on the consequences of a particular condition, condition-specific measures exclude any unrelated experiences that would add to background noise and measurement error, therefore reducing reliability. This focus also keeps the questionnaire at a manageable length, therefore preventing undue burden to participants and allowing more questions to be asked about the relevant dimensions, which may also increase precision.

We concluded that the nature of the impacts of DH and consequences of treatment for DH would be relatively well-demarcated and that the advantages of brevity, specificity, and precision would far outweigh the likelihood of detecting unexpected consequences of treatment. We therefore opted to devise a measure specific to DH. Interestingly, since we made this decision, Bekes and Hirsch[14] have used the OHIP in an uncontrolled longitudinal evaluation of a product for DH in 713 patients attending 161 dental practices. There was a mean change of 13.5 OHIP points (approximately 7% of the range for OHIP), but the lack of a control group in this study prevents the disaggregation of treatment and placebo effects, which would be an important part of a randomized controlled trial. These findings can be compared with the impressive performance of DHEQ in three randomized controlled trials reported in Chapter 10 of this book.

Selection of a model

Most people researching OHQoL are dentists. Dentists are pragmatic people; they see an unrestorable tooth and take it out. The concepts involved are easily defined and there is relatively little use of theory in such decisions. Work in OHQoL is much more vague. The concepts are not as visible as a carious tooth and are not directly measurable. As Chapter 1 describes, OHQoL is amorphous and cannot be well-defined. It overlaps with some concepts and has indistinct relationships with others. Just as it was necessary to define OHQoL, it was equally necessary to define all the other concepts we were using. Similarly, we needed to consider how all these concepts were related to each other.

Starting with an existing theoretical model, we linked our study to current ideas about health and knowledge of the relationship between clinical states and the subjective experience of oral conditions. This would therefore provide a firm theoretical basis and a common understanding of our work. Furthermore, making explicit the hypothesized links between variables guides data analysis by indicating the analyses to be undertaken, and in restricting the analysis it reduces work and prevents type 1 error. Theoretical models also guide the interpretation of the results by suggesting in advance how variables may be related, in other words, whether the relationships are direct, whether they are mediated, or whether they might occur as a result of confounding.

In addition to these general necessities of using a conceptual model, there are specific benefits of adopting one when developing an OHQoL measure. The model can suggest areas for questioning, both when devising a topic guide for initial qualitative interviews and when developing subscales of items within the measure, which might correspond to levels in the model. In this way, the model may help ensure that the measure is comprehensive across the types and severity of impacts from the condition (i.e., content validity). In addition, in guiding the data analysis, the conceptual model becomes the "construct" against which the measure is validated (i.e., construct validation). It should be stressed, though, that in this form of validation both the measure and the theory are being tested and it is difficult to distinguish between the two. Thus, if construct validation does not support the validity of the measure, then careful thought is required to determine whether it is just the measure that is at fault or if it is both the measure and the theory that are at fault.

The selection of an appropriate model is therefore crucial. In addition to the fundamental misdirection from a weak model, such a model can lead researchers to devise questions that are irrelevant.[2]

We identified three models that might form the basis of our DH-specific OHQoL measure: Locker's model of oral health; the World Health Organization's International Classification of Functioning, Disability, and Health (ICF); and the Wilson and Cleary model linking clinical variables and quality of life.[15–17] We invested a considerable amount of time in choosing between them.

The Locker model specifically relates to oral health and was well-established. It had formed the basis of at least two generic OHQoL measures; the OHIP and the Oral Impacts on Daily Performance, which had served to validate it.[18–20] However, an appreciable body of work indicates that many things influence people's experiences with their own oral health; indeed, some of the strongest influences on OHQoL arise outside the mouth or the person, and these are absent from the Locker model.[21]

The ICF seems to have been developed, in part, to incorporate the perspectives of the disabilities movement that observed that it was the actions of the social and physical environments that restricted the ability of people with impairments to participate fully in life.[16] Consequently, it is a detailed and dynamic classificatory system that

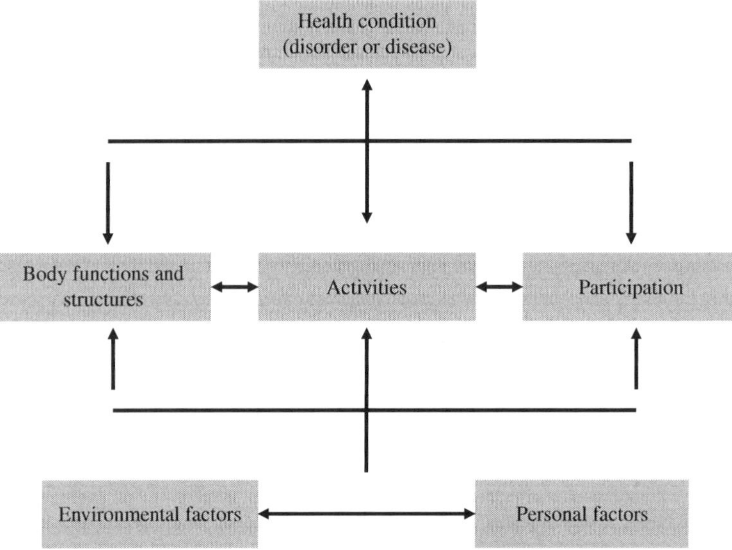

Figure 8.1 Interactions between the components of the ICF.[16]

sees the degree of participation as an interaction between a health condition (Figure 8.1), body functions, and structures, activities, and environmental and personal factors. Although more comprehensive, we debated at some length whether the incorporation of these dynamic and two-way relationships limited its usefulness as a framework for developing a measure. As noted, a key requirement of the framework was to provide a construct against which we could validate the measure. Relationships within the model can be hypothesized for the specific condition under investigation, which might have been addressed and which remains to be tested.

The Wilson and Cleary model linking clinical variables to quality of life incorporates environmental and individual factors, which would allow us to consider the effects of factors such as ethnicity, socioeconomic status, age, and personal sense of coherence.[17] In this regard, this model was considered compatible with Locker's but more comprehensive.[15]

Unlike the ICF, the Wilson and Cleary model hypothesizes relationships between levels, so that it and any measures derived from it are readily testable (Figure 8.2). Moreover, this model incorporated a level of general health perceptions, which we took to be analogous with the type of global assessment considered necessary by Locker and Allen.[9] In addition, the Wilson and Cleary model had been supported empirically in relation to oral health.[21,22]

Nonetheless, the debate about model selection remained unresolved. Consequently an interim decision was made to use the Wilson and Cleary model as a framework to assist collation of the qualitative data. We merely deferred a difficult decision. However, awareness of this debate did encourage the emergence of the stronger social dimension in the qualitative analysis.

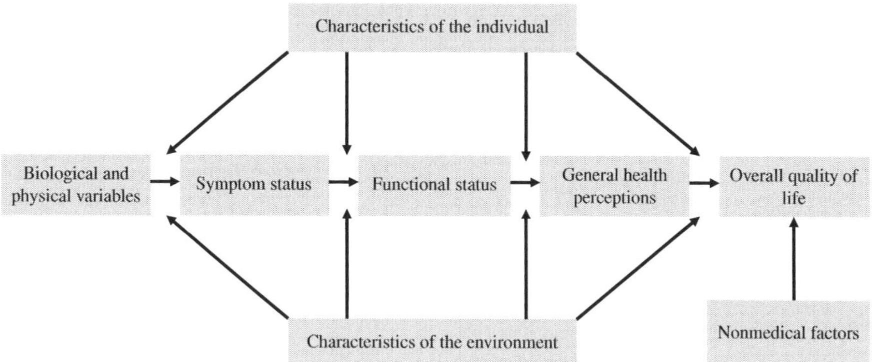

Figure 8.2 The Wilson and Cleary model. Linking clinical variables with HQoL: a conceptual model of patient outcomes.[17]

The value of qualitative data

As we have already noted, our person-centered bottom-up approach would determine the way we explored the experiences of people with DH. Specifically, we wanted affected people to tell us about their experiences comprehensively in their own words. Therefore, the natural choice was to collect qualitative data and to structure the data collection sufficiently to help participants think about their DH but without restricting them to checking lists of preselected words and phrases. We hoped that qualitative data would reveal the concepts, language, and terminology that were understandable and relevant to people with DH. Qualitative research allows the generation of data from participants giving their own interpretations and explanations during interviews. This is one area in which the multidisciplinary team was at an advantage. Rather than simply collecting a list of statements about the condition, our methodological and theoretical sensitivity allowed us to adopt a structure and logic for this task.

Purposive sampling was used to recruit people with different degrees of severity of DH, which was intended to yield insights into what is important and the degree of impact in mild and severe cases. Recruitment for the qualitative study therefore involved a snowball sample in which existing participants recruited other people with DH. However, previous market research by our funders had discovered that there were people who were reluctant to use the term "sensitivity" to describe their DH symptoms. Therefore, a recruitment agency gathered another group of participants on the basis that they experienced "twinges" and "discomfort" in their teeth.

Data were collected in interviews that were semistructured via a loose topic guide designed to help the interviewer cover ground identified as important while allowing exploration of new areas introduced by participants. Constant comparative analysis was employed as a technique in which data collection and analysis are conducted iteratively, so that ideas from previous interviews could be followed up in subsequent ones. The topic guide was devised from reviews of the DH literature and existing qualitative data from the funders. The guide avoided any specific wording or phrasing from the literature or existing OHQoL instruments. With the knowledge that not all people with DH considered the symptoms of DH as "sensitivity," this word was avoided.

Therefore, the qualitative study explored the daily experiences of people with DH (Chapter 6).[23] The findings showed the depth and complexity of pain associated with sensitivity, direct impacts on functional status and everyday activities such as eating, drinking, talking, toothbrushing, and social interaction, and less direct impacts on emotions and identity, all of which required subtle approaches to coping. This range of experiences supported the rationale for constructing a condition-specific measure to focus on these particular impacts.

There are many approaches to qualitative data analysis, but our chosen strategy of framework analysis was especially appropriate, because our conceptual model could be used for the analytic framework that we could then directly populate with the data.[24] This step made the method convenient and efficient. However, it is important not to "force" data into the model because this would prevent the emergence of new categories. In fact, deviant case analysis, which involves looking for data and categories that do not fit the model, is an important stage of framework analysis. The existence of such categories should prompt a revision of the model.

Selection of domains

As stated, the Wilson and Cleary model was used as the framework to organize the qualitative data. When matched to the data, this framework produced a robust set of domains, although the data also matched the ICF.

We needed a model that mapped the relationships between concepts discovered through the qualitative research. Both models are informed by the biopsychosocial approach but emphasize different things.

The Wilson and Cleary model links clinical variables with psychological and social outcomes and, to some extent, represents the biopsychosocial perspective quite well.[17] Thus, in this model DH results from biological and physical variables that interact with characteristics of the individual and of the environment to produce symptoms expressed in the form of pain. Pain is experienced in relation to its duration, intensity, tolerability, frequency, and extent. Pain results from various triggers. Symptoms may or may not lead to impacts on functional status expressed as impacts on eating, drinking, talking, toothbrushing, and social interaction. Again, this link is mediated by personal and environmental factors. Impacts on functional status can have further impacts on general health perceptions; in dentine sensitivity, these were expressed in terms of beliefs about feeling culpable and various lay and expert beliefs about the condition. Whether this does happen is heavily mediated by personal and environmental factors. The outcome in the Wilson and Cleary model is that these various impacts will have an impact on overall quality of life.[17]

However, the status of the ICF is somewhat ambiguous.[16] It is said to represent a shift from a "consequences of disease" model to a "consequences of health" model. This represents a desire for research of the consequences of both health and disease, with their consequences described as impacts on activities and participation. In ICF, DH would be seen as a health condition that interacts with personal and environmental factors to produce impacts expressed as pain and impacts on activities and participation. Impacts on pain appear to occur in the same way as that described by Wilson and Cleary, as a product of the interaction between the condition and personal and environmental factors. The pain also interacts with everyday activities, which in themselves can trigger the pain. These complex interactions impact on the person's ability to participate in social situations, eating meals with others, holding conversations, and attending dental appointments. The social psychological impact in this approach would be impacts on body functions and structures, activities, and participation.

Both models were designed to explain and explore the impact of chronic conditions on everyday life. Dentine sensitivity, for the most part in our data set, is understood within a health framework. It only becomes a chronic illness for a very small subset of people who have the condition. The Wilson and Cleary model is explanatory in that it seeks to explain the mechanisms of how a health condition impacts on quality of life. The ICF, however, is a consequences model, mapping the consequences of a health condition and how these consequences interact with personal and environmental variables. Its purpose is less explanatory.

Whichever model was used, additional domains arose from the data. This was another situation in which a multidisciplinary team of health workers and social and behavioral scientists working together was advantageous. In addition to the general literature on the biopsychosocial impacts of health conditions, the data were interpreted in light of the close understanding of the literature on chronic illness, coping, and illness beliefs.[25–29] The group's sensitivity to these ideas meant that themes such as coping and identity were represented in the final questionnaire.

After careful discussion, the qualitative analysis and theoretical sensitivity required some judgments to be made. However, in the end, a domain structure close to that of the Wilson and Cleary model was produced. With this approach, the need to cope could be regarded as a psychological impact of DH and could be considered in the functional limitation category of the model. Subsequent studies using DHEQ and other analyses based on this model substantiate this view.[4,30–33]

Readers will understand that the debate over the advantages of each of the two models was never fully resolved. Further work is required to consider the theoretical and empirical basis of these two approaches. Such thought might have profound implications for the way we think about the relationship between health and disease and their effects on everyday life.

Yet another approach would have been to generate many items from interviews and other sources and then use factor and Rasch analysis to restrict the item pool and identify the dimensions.[34] In our case, factor analysis was used to *confirm* the structure of DHEQ.[4] Only one change was required because factor analysis could not distinguish between the two subdomains of avoidance and approach coping in an interim version of the measure. Consequently, these two dimensions were aggregated in the final version. Nonetheless, the factor analysis supported the overall structure of DHEQ and reassuringly indicated that our bottom-up approach could produce a measure that had good mathematical properties.

Selection of descriptive system

It is a natural inclination to immediately think about the content of a questionnaire in terms of the subjects or topics of the items. However, the descriptive system of a questionnaire is central, particularly if the intention is to bring scores of individual items together into scales. Developing the descriptive systems involves selecting levels, a rating system, response formats, and scaling.

Attitudes and feelings can be assessed by directly inquiring about them. Direct estimation may be prone to bias (for instance, if the participants want to maximize the impact of their condition or have a desirability bias toward a product or team of researchers), but it is most straightforward to design and most likely to be clearly relevant to participants, which may encourage them to complete the questionnaire.

Scaling (creating an aggregate score from the responses to several items) is very important when attempting to measure vague concepts. The reliability of aggregate scores increases in relation to the number of items. This greater reliability decreases the variance in the scores and therefore makes it easier to detect differences between groups of participants (discrimination) and changes over time (responsiveness).

Precoded responses may be scored dichotomously (e.g., yes/no), ordinally (excellent, very good, good, fair, or poor), or continuously (on a scale of 1–10). Ordinal scales are often coded numerically and then are used as a continuous variable. The use of continuous judgments gives measures greater precision and does not lose statistical power. It seemed reasonable to use continuous judgments because we could not assume that products to manage DH would completely remove the signs or symptoms, but rather would alleviate them to a degree. We wanted to be able to capture this degree of change.

Our desire for greater precision suggested that we should use as many scale points as possible for each item. Streiner and Norman reviewed studies of the effect of differing numbers of scale points and found that reliability decreases below 7 points.[2] Many OHQoL scales have only 5-point scales, and we were unsure whether additional points would confuse participants. For instance, if DH is a relatively mild condition, then participants may not be able to judge between seven levels of a particular impact. In addition, a more precise scale might add to the burden of completing the questionnaire. Comparing versions of the questionnaire using both 5- and 7-point Likert scales in a small sample determined that participants were able to use 7-point scales without undue burden. The number of points on the scale was discussed at focus groups for panel testing. Although one or two participants found more points difficult, most could manage it. In quantitative testing, the scores for the two versions were similarly distributed, but the more precise scale would make any analyses more powerful.

We elected to have an odd number of scale points, therefore allowing participants the option of a "neutral response." This was because we thought some items might not apply to participants or because they simply could not reach a judgment for a particular item. Even numbers of points force participants to express an active judgment, which we thought unwise. A relatively high proportion of neutral responses ("neither agree nor disagree") in the cross-sectional validation study of DHEQ prompted an additional panel test with 11 participants who used neutral responses. These participants were asked to complete the questionnaire again, thinking out loud and commenting on its clarity as they progressed. This test suggested that participants were using the neutral response appropriately because they had mixed or neutral feelings about the item topic rather than not understanding it.

The question then arose regarding what characteristic of each impact to measure. Epidemiology often records the prevalence, frequency, or severity of a characteristic, and the choice should be determined by the condition undergoing investigation. At first glance, DH appeared suited for measuring the frequency of impacts. However, the interplay between symptoms and coping evident in our qualitative data suggested that these two domains might cancel each other out when measuring frequency of impact. For example, participants might not experience pain (no symptom) from a sensitive tooth if they avoided that tooth when eating (a coping impact), whereas if treatment rendered the pain tolerable they might use the tooth to eat (therefore removing the impact on coping) but might experience the pain (a symptom). Instead, participants were asked to agree or disagree with statements about possible impacts. Likert scales arranged between these two poles were therefore appropriate.

Selection of items

An immediate concern was how to refer to the pain experienced by people with DH. Several participants in the qualitative study had avowedly stated that they did not experience pain. In fact, previous research by the funder and our own qualitative data suggested that not everybody with DH described themselves as having "sensitive" teeth, and a considerable variety of descriptors were used instead. Therefore, we started the questionnaire with an item that asked participants to select a descriptor from a list of 21 or to use an open response and add their own word or phrase to describe what they felt. This item served to legitimize all the different words participants may use. Afterward we added the statement, "From now on in this questionnaire we are going to call what you feel as '*sensations in your teeth*' or '*sensations*'."

When devising the parts of the questionnaire that would form the scale for measuring impact, our patient-centered approach generated the items from the qualitative data. There are several guides on how to select items (see Streiner and Norman[2] for an example). General considerations include interpretability and reading level and being careful to avoid jargon, ambiguity, and double questions. Balancing these requirements to generate items about very specific aspects of experience is time-consuming and requires a certain amount of experience. Our most heated debate related to including an item about difficulty in eating ice cream, which was practically universal among the qualitative participants. Our concern was that if participants with even mild DH experienced this impact, then the item might not discriminate between people with different degrees of severity and would be unresponsive to treatment. Item impacts and mathematical tests of internal reliability subsequently supported its inclusion.

The first draft of the DHEQ had 48 items, which was deemed too many. It is typical to select too many items for a questionnaire and then eliminate redundant items during refinement. Once again, our person-centered perspective informed the method and we chose to reduce the item pool. Methods used to reduce the number of items fall into two broad approaches. Regression and factor analyses can be used to select items that are closely mathematically related, which will create a measure with high internal consistency. Instead, we opted for the item–impact method, which selects the items most relevant to affected people.[1] Using preliminary responses to the questionnaire from people with DH, we selected the items that reflected the most frequent and severe impacts, which would make the measure more sensitive and precise.

Finally, to avoid halo effects (whereby participants reach a global judgment about their condition and then quickly respond in the same way to all the items) or a response set (in which participants simply check all the boxes in the same way), the questionnaire was clearly divided into sections with subheadings and the format of the items was changed between sections.

Reference period

The reference period of a questionnaire is the period during which participants should recall their experiences. The reference period will be quoted many times in a questionnaire, with phrases such as "How much has this affected your everyday life in the last month" or "Think back for the last six months. How often have you...", for example. HQoL questionnaires typically use reference periods of a few weeks to 6 months. For the sake of clarity, it is wise to use the same reference period throughout the measure.

The reference period should be sufficiently long for enough participants to experience any impacts. Doing so will increase the reproducibility of the measure if impacts occur only sporadically. However, reference periods that are too long may misclassify if participants cannot remember their impacts. There is also a tendency to better recall events that happened more recently, creating recall bias.

An evaluative measure should be developed with all these considerations in mind and also with particular pragmatic considerations. Evaluations typically involve an assessment before and after the intervention. The intervention may take some time to exert an effect, and it is only after this time that the product can be evaluated. However, the reference period can only begin after this time, thereby adding to the duration of the study.

We felt a reference period of 1 month would be adequate to capture infrequent impacts or coping strategies, such as avoiding cold drinks, and reasonably practical in terms of study duration.

Panel testing

Panel tests are best treated like pilots. It is often best to conduct several small ones, learning from each and evaluating the results in another, rather than to conduct one big one. There were three panel tests of the DHEQ.

Before item reduction, we asked two groups of individuals with DH to complete the questionnaire and take part in focus groups to discuss any potential ambiguities and to ascertain if they knew what each question was implying, whether they understood it, and whether they would change or even remove any items. The focus groups allowed participants to agree or disagree with the comments of others and hinted to us whether comments were personal to one person or might be a more common perception. As we have already noted, 11 other participants were asked to complete the questionnaire while thinking out loud so that we could clarify why and how participants were using neutral responses. The panels confirmed the decision to adopt the 7-point scales and agreed with the content and format of the questionnaire

Conclusion

The work and the debates described in this chapter may surprise readers who imagined that questionnaires could be devised easily. We hope this report makes our thorough methodology apparent and justifies it to you. Our desire to be person centered is evident at many stages, as is the value of working in a multidisciplinary team with complementary perspectives. We believe this care was necessary for a measure with a firm theoretical and empirical foundation. In retrospect, these efforts have been rewarded. DHEQ has remarkable psychometric properties, as evidenced in the many studies that have evaluated it (Chapters 7, 9, 10, and 11). In addition, it is now being used to extend our understanding of aspects of health beyond DH.

References

1. Juniper E, Guyatt G, Jaeschke R. How to develop and validate a new health-related quality of life instrument. In: Spikler B, editor. *Quality of life and pharmacoeconomics in clinical trials*, 1996. Philadelphia, PA: Lippincott–Raven Publishers; 1996. p. 49–56.
2. Streiner DL, Norman GR. *Health measurement scales*. 3rd ed. Oxford: Oxford University Press; 2003.

3. Fayers P, Hays R, editors. *Assessing quality of life in clinical trials*. 2nd ed. Oxford: Oxford University Press; 2005.
4. Boiko OV, Baker SR, Gibson BJ, Locker D, Sufi F, Barlow APS, et al. Construction and validation of the quality of life measure for dentine hypersensitivity (DHEQ). *J Clin Periodontol* 2010;**37**:973–80. Available from: http://dx.doi.org/doi:10.1111/j.1600-051X.2010.01618.x.
5. Juniper EF, O'Byrne PM, Guyatt GH, Ferrie PJ, King DR. Development and validation of a questionnaire to measure asthma control. *Eur Respir J* 1999;**14**:902–7.
6. Guyatt G, Mitchell A, Irvine EJ, Singer J, Williams N, et al. A new measure of health status for clinical trials in inflammatory bowel disease. *Gastroenterology* 1989;**96**:804–10.
7. Baum F. *The new public health*. Oxford: Oxford University Press; 2002.
8. McColl E. Developing questionnaires. In: Fayers P, Hays R, editors. *Assessing quality of life in clinical trials*. 2nd ed. Oxford: Oxford University Press; 2005.
9. Locker D, Allen F. What do measures of oral health-related quality of life measure?. *Community Dent Oral Epidemiol* 2007;**35**:401–11.
10. Inglehart MR, Bagramian R. *Oral health-related quality of life*. London: Quintessence Publishing Co; 2011.
11. Bekes K, Johh MT, Schaller H-G, Hirsch C. Oral health-related quality of life in patients seeking care for dentin hypersensitivity. *J Oral Rehabil* 2009;**36**:45–51.
12. Dowie J. Decision validity should determine whether a generic or condition specific HRQOL measures is used in health care. *Health Econ* 2002;**11**:1–8.
13. Brazier J, Fitzpatrick R. Measures of health related quality of life in an imperfect world: a comment on Dowie. *Health Econ* 2002;**11**:17–9.
14. Bekes K, Hirsch C. What is known about the influence of dentine hypersensitivity on oral health-related quality of life?. *Clin Oral Investig* 2013;**17**(Suppl. 1):S45–51. Available from: http://dx.doi.org/doi:10.1007/s00784-012-0888-9.
15. Locker D. Measuring oral health: a conceptual framework. *Community Dent Health* 1988;**5**:3–18.
16. World Health Organisation. *International classification of functioning, disability and health*. Geneva: WHO; 2001.
17. Wilson IB, Cleary PD. Linking clinical variables with health-related quality of life: conceptual model of patient outcomes. *JAMA* 1995;**273**:59–65.
18. Slade GD, Spencer AJ. Development and evaluation of the Oral Health Impact Profile. *Community Dent Health* 1994;**11**:3–11.
19. Adulyanon S, Sheiham A. Oral impacts on daily performances. In: Slade GD, editor. *Measuring oral health and quality of life*. Chapel Hill, NC: University of North Carolina; 1997. p. 151–60.
20. Baker SR, Gibson B, Locker D. Is the oral health impact profile measuring up? Investigating the scale's construct validity using structural equation modelling. *Community Dent Oral Epidemiol* 2008;**36**:532–41. Available from: http://dx.doi.org/doi:10.1111/j.1600-0528.2008.00440.x.
21. Baker SR, Mat A, Robinson PG. What psychosocial factors influence adolescents' oral health? *J Dent Res* 2010;**89**:1230–5.
22. Baker SR, Pankhurst CL, Robinson PG. Testing relationships between clinical and non-clinical variables in xerostomia: a structural equation model of oral health-related quality of life. *Qual Life Res* 2007;**16**(2):97–308.
23. Gibson B, Boiko O, Baker S, Robinson PG, Barlow A, Player T, et al. The everyday impact of dentine sensitivity: personal and functional aspects. *Soc Sci Dent* 2010;**1**:11–20.

24. Ritchie L, Lewis J. *Qualitative research practice; a guide for social science students and researchers*. London: Sage Publications; 2005.
25. Bury M. Chronic illness as biographical disruption. *Sociol Health Illn* 1982;**4**:167–82.
26. Williams SJ. Chronic illness as biographical disruption or biographical disruption as chronic illness? Reflections on a core concept. *Sociol Health Illn* 2000;**22**:40–67.
27. Lazarus R, Folkman S. *Stress, appraisal and coping*. New York, NY: Springer; 1984.
28. Leventhal H, Meyer D, Nerenz D. The common sense representation of illness danger. In: Rachman S, editor. *Medical psychology*. New York, NY: Pergamon Press; 1980. p. 7–30.
29. Leventhal H, Benyamini Y, Brownlee S. Illness representations: theoretical foundations. In: Petrie K, Weinman J, editors. *Perceptions of health and illness*. Amsterdam: Harwood; 1997. p. 1–18.
30. Baker SR, Gibson BJ, Sufi F, Barlow A, Robinson PG. The Dentine Hypersensitivity Experience Questionnaire (DHEQ): a longitudinal validation study. *J Clin Periodontol* 2014;**41**:52–9. Available from: http://dx.doi.org/doi:10.1111/jcpe.12181.
31. Nammontri O, Robinson PG, Baker SR. Enhancing oral health via sense of coherence: a cluster-randomized trial. *J Dent Res* 2013;**92**:26–31. Available from: http://dx.doi.org/doi:10.1177/0022034512459757.
32. Krisdapong S, Prasertsom P, Rattanarangsima K, Sheiham A. Impacts on quality of life related to dental caries in a national representative sample of Thai 12- and 15-year-olds. *Caries Res* 2013;**47**:9–17. Available from: http://dx.doi.org/doi:10.1159/000342893.
33. Gururatana O, Baker SR, Robinson PG. Determinants of children's oral health related quality of life over time. *Community Dentistry Oral Epidmiol* 2014;**42**:206–15.
34. Cella DF, Dineen K, Arnason B, Reder A, Webster KA, Karabatsos G, et al. Validation of the functional assessment of multiple sclerosis quality of life instrument. *Neurology* 1996;**47**:129–39.

The dentine hypersensitivity experience questionnaire (DHEQ): a longitudinal validation study

Sarah R. Baker[1], Barry J. Gibson[1], Farzana Sufi[2], Ashley P.S. Barlow[2] and Peter G. Robinson[1]
[1]School of Clinical Dentistry, Claremont Crescent, University of Sheffield, Sheffield, UK,
[2]Glaxosmithkline Consumer Healthcare, Weybridge, Surrey, UK

Introduction

Person-reported outcome (PRO) measures are used to capture the psychosocial experience of pain, discomfort, and quality of life, thereby supplementing clinical indicators.[1] In oral health research, they identify aspects of life that are affected by oral conditions and that may be amenable to treatment.[2–4] PRO measures must therefore capture changes in quality of life. This is referred to as responsiveness: "*the accurate detection of change when it has occurred.*"[5]

Responsiveness can be examined for many reasons: (i) to detect change within individuals; (ii) to differentiate between subgroups of people who have improved, stayed the same, or worsened on an external anchor or referent (e.g., clinical endpoint or another PRO measure); and (iii) to detect treatment or intervention effects.[6] Furthermore, to aid interpretation of change, the minimally important difference (MID) is the smallest difference in a score that a person perceives as important.[7] Determining these aspects of responsiveness requires longitudinal studies that incorporate interventions of differing efficacy, together with an anchor or criterion.[6,8]

DH is a common oral health problem characterized by a sharp pain in response to an external stimulus that cannot be explained by any other dental disease.[9,10] It is related to the exposure of the dentine as a result of gingival recession or enamel erosion.[10] There has been little research of DH from the perspective of a patient. Patients' experiences of pain have been recorded in response to a stimulus within a clinical setting,[11,12] but there has been little consideration of the impact of DH on everyday life. Bekes et al.[13] found that a generic oral health-related quality of life (OHQoL) measure (OHIP-49) did not discriminate the impact of DH; although patients with hypersensitivity experienced more impacts and had poorer oral health than the general population, the difference in mean scores between the two samples was less than 10% of the overall scale.

Dentine Hypersensitivity. DOI: http://dx.doi.org/10.1016/B978-0-12-801631-2.00009-9

A qualitative study explored the daily experiences of people with DH and showed the complexity of pain associated with sensitivity and also the impacts on eating, drinking, talking, tooth brushing, and social interaction (see Chapter 6).[14] A condition-specific measure was developed to assess the impact of DH on the OHQoL (see Chapter 7).[15] This DHEQ detected functional limitations (e.g., slower eating), coping behaviors (e.g., warming food and drinks), emotional (e.g., anxiety), and social impacts (e.g., difficulties conversing) caused by DH and showed excellent reliability and validity in both a general population and a clinical sample.[15] It is now necessary to assess the responsiveness of DHEQ to change and to interventions.

The aim of this secondary analysis was to further evaluate DHEQ. There were two objectives: (i) to confirm its reliability and validity and (ii) to assess its responsiveness and determine the MID for DHEQ in clinical trials. We compared the ability of the measure to detect change in two types of trials: first, two trials comparing a test treatment against negative controls, in which we hypothesized *a priori* that we would detect a treatment effect using DHEQ; and second, a trial with an active control that anticipated no study group difference.

Methods

Participants

In total, 311 people aged 20–65 years participated in three clinical trials in the United Kingdom and the United States (Table 9.1). Each trial compared the efficacy of test and control dentifrices in providing relief from DH. Inclusion/exclusion criteria varied slightly across the three trials,[1] but inclusion criteria commonly included at least two nonadjacent sensitive teeth with a Schiff sensitivity score of ≥ 2 and a tactile threshold (Yeaple probe) of $\leq 20\,g$ force. Exclusion criteria included teeth with evidence of current or recent caries or reported treatment of decay in 12 months of screening, teeth with exposed dentine but with deep, defective, or facial restorations, teeth used as abutments for fixed or removable partial dentures, teeth with full crowns or veneers, orthodontic bands, or cracked enamel, and sensitive teeth with contributing etiologies other than erosion, abrasion, or recession of exposed dentine.

The studies were approved by an Institutional Review Board or Independent Ethics Committee in accordance with local requirements.

Clinical trial overview

All three trials were randomized, examiner-blinded, two-treatment arm, stratified, parallel design, single-site studies. Demographic, medical history, and medications were recorded, followed by oral examination. Participants returned at least 24 h

[1]Available on request.

Table 9.1 Participants' demographics across the three clinical trials at baseline

	Trial A	Trial B	Trial C
N	93	118	100
Control	46	58	49
Test	47	60	51
Gender			
Female	75	83	81
Male	18	35	19
Age			
Mean	42.90	36.19	45.21
SD, range	10.18, 20–60	10.71, 21–65	9.50, 20–65
Duration of DH			
<6 months	1	1	7
>6 to <12 months	8	14	6
>1 to <5 years	36	62	53
>5 to <20 years	33	41	29
>20 years	15	0	4
DH intensity[a]: Mean (SD)	6.25 (1.57)	6.46 (1.29)	5.80 (1.76)
Tactile: Mean (SD)	12.84 (3.46)	10.00 (0.00)	12.35 (3.05)
VAS: Mean (SD)	69.45 (13.07)	–	57.26 (22.07)
Schiff: Mean (SD)	2.76 (0.338)	2.71 (0.366)	2.69 (0.401)

[a]Potential range: 1 (not at all)–10 (extremely).

after screening for baseline assessment and completed DHEQ and standard tooth sensitivity assessments.

Participants were then randomized (1:1 allocation) according to baseline stratification and dispensed their study dentifrice, standard toothbrush, diary, brushing instructions, and timers. Participants brushed at home twice daily for 8 weeks (trials B and C) or 12 weeks (trial A). Participants returned for visits at weeks 4, 8 (trials B and C), and 12 (trial A).

The revised DHEQ has 48 items, of which 34 comprise an impact scale (Table 9.2, DHEQ is presented in full in Appendix 1). The scale contains five domains of *functional restrictions* (4 items), *coping* (12 items), *social impact* (5 items), *emotional impact* (8 items), and *identity* (5 items).[15] The items have coded responses on 7-point Likert scales: 1, strongly disagree; 2, disagree; 3, agree a little; 4, neither agree nor disagree; 5, disagree a little; 6, disagree; and 7, strongly disagree. A summary measure, "total score," was calculated as the sum of the scores per participant, with a possible range of 34–238. Domain scores were calculated in the same way.

Table 9.2 Mean (SD) scores, item impacts, and item-total correlations of DHEQ items at baseline for clinical trial A

		Mean	SD	Item impact	Item-total correlations (*r*)
1	Restrictions—pleasure out of eating	5.14	1.56	397.32[a]	0.737
2	Restrictions—cannot finish meal	3.92	1.62	168.95	0.759
3	Restrictions—longer to finish meal	5.06	1.61	375.45	0.752
4	Restrictions—problems eating ice cream	6.11	0.97	**584.73**[a]	0.572
5	Approach coping—modification of eating	5.55	1.36	**489.51**[a]	0.679
6	Approach coping—careful when breathing	5.61	1.16	**494.80**[a]	0.624
7	Approach coping—warming food/drinks	5.04	1.69	368.42	0.687
8	Approach coping—cooling food/drink	4.43	1.79	233.46	0.631
9	Approach coping—cutting fruit	4.72	1.78	289.34	0.616
10	Approach coping—putting a scarf over mouth	4.61	1.75	258.16	0.444
11	Avoidant coping—cold drinks/foods	5.08	1.56	**430.36**[a]	0.674
12	Avoidant coping—hot drinks/foods	4.06	1.73	183.11	0.532
13	Avoidant coping—contact with certain teeth	5.47	1.37	**453.46**[a]	0.603
14	Avoidant coping—change toothbrushing	4.46	1.53	249.31	0.459
15	Avoidant coping—biting in small pieces	5.15	1.51	387.80	0.610
16	Avoidant coping—other food	4.52	1.76	262.16	0.827
17	Social—longer than others to finish	4.29	1.78	217.07[a]	0.704
18	Social—choose food with others	3.99	1.72	158.80	0.837
19	Social—hide the way of eating	3.66	1.64	121.88	0.794
20	Social—unable to take part in conversations	2.73	1.50	32.49	0.623
21	Social—painful at the dentist	4.96	1.78	352.16[a]	0.485
22	Emotions—frustrated not finding a cure	4.77	1.55	297.65	0.669

(*Continued*)

Table 9.2 (Continued)

		Mean	SD	Item impact	Item-total correlations (*r*)
23	Emotions—anxious of eating contributes	5.10	1.50	384.03	0.723
24	Emotions—irritating sensations	5.83	1.05	**564.34**[a]	0.752
25	Emotions—annoyed I contributed	4.32	1.78	204.34	0.509
26	Emotions—guilty for contributing	3.82	1.69	144.01	0.550
27	Emotions—annoying sensations	5.76	1.02	**551.23**[a]	0.721
28	Emotions—embarrassing sensations	3.43	1.76	99.81	0.776
29	Emotions—anxious because of sensation	4.41	1.62	256.22	0.712
30	Identity—difficult to accept	3.28	1.74	95.45	0.602
31	Identity—different from others	3.26	1.71	101.39	0.716
32	Identity—makes me feel old	3.81	1.88	164.21[a]	0.478
33	Identity—makes me feel damaged	3.30	1.73	99.33	0.633
34	Identity—makes me feel unhealthy	3.65	1.79	164.62[a]	0.614

Figures in bold are those items with the greatest impact across the whole scale.
[a]Two items with the greatest impact within each domain (restrictions, coping behaviors, social impacts, emotional impacts, identity).

Participants also self-reported their global oral health as a score ranging from 1 (excellent) to 6 (very poor) on a 6-point scale. The effect of DH on life overall was measured using four items with ratings from 0 (not at all) to 4 (very much), creating a possible range of 0–16.

The *evaporative (air) sensitivity with Schiff sensitivity score* is determined by directing a 1-s application of air from a dental syringe to the tooth surface 1–2 mm coronal to the free gingival margin from a distance of 1 cm. The response was rated from 0 to 3 using the Schiff sensitivity scale.[16]

A *visual analogue scale (VAS)* rated participants' responses to the evaporative air test on a 100-mm scale scored from 0 (no pain) to 100 (intense pain).

The *tactile sensitivity* (*Yeaple probe*[17]) *score* was determined by directing the probe tip perpendicular to the buccal surface and slowly drawing it across the tooth surface with a force setting of 10 *g*. This is increased by 10 *g* with each successive challenge until either a "yes" response is recorded or the maximum is reached (20 *g* in trials B and C, 30 *g* in trial A, baseline settings only). The gram setting that elicited two consecutive "yes" responses was recorded as the tactile threshold.

Data analysis strategy

The analysis was conducted in two stages. In the first stage, cross-sectional validation was performed to further examine the reliability and validity of the DHEQ. First, *item impact* was calculated as mean score multiplied by the percentage of people who reported an impact (agree a little, agree, or strongly agree response on the item) to highlight the DH impacts that were most important for most people. Second, *internal consistency* of the DHEQ was assessed via item-total correlations, subscale-total correlations, and Cronbach's alpha coefficient. Third, *test–retest reliability* was assessed with intraclass coefficients (ICCs) using data from 41 control group participants who provided follow-up data at 12 weeks. Finally, *convergent validity* was examined by correlating DHEQ scores with global oral health ratings, self-reported quality of life, and clinical measures.

In the second stage of the analysis, the longitudinal validation was performed. First, we examined *responsiveness over time within individuals* by calculating the effect size (ES) for individual changes between baseline and follow-up as follows:

$$\text{Within-person individual effect size} = \frac{\text{posttest DHEQ score} - \text{pretest DHEQ score}}{\text{pretest (group) standard deviation}}$$

Within-person individual ES from 0.2 to 0.49 were classified as small, from 0.5 to 0.79 were classified as moderate, and more than 0.8 were classified as large.[18]

Second, we assessed *responsiveness using an external referent* in three ways in line with the literature.[6] First, participants were classified by their change in self-reported quality of life from baseline to final follow-up (i.e., better, same, or worse). Second, we assessed responsiveness to change indices, namely, Cohen's ES and the standardized response mean (SRM).[7] ES was calculated as the mean change between baseline and final follow-up divided by the baseline standard deviation (SD). SRM was calculated as the mean change between baseline and follow-up divided by the SD of the change score. Both the ES and SRM were calculated, stratified by category of change (worse, same, or better). Finally, to aid interpretation of DHEQ change scores for future clinical trials, we calculated the MID. The MID was defined as the mean change of the total scores in participants who reported any improvement in their self-reported quality of life (i.e., impact on life overall).

Third, we *calculated responsiveness to treatment.* If the DHEQ is responsive to treatments for DH, then it should be able to detect an effect of treatments with differing efficacies. Responsiveness was compared between trials with negative controls (trials A and B, in which our *a priori* hypothesis was that DHEQ change scores would differ between test and control groups) and an active control trial (trial C, in which it was anticipated to detect similar changes in both groups).

Three analyses were conducted as recommended by Buchbinder et al.[19]: (i) the observed treatment effect was calculated as the difference between mean change scores in the test and control groups; (ii) the relative percent improvement was calculated as

$$\frac{(\text{test change} - \text{control change})}{(\text{mean test baseline} - \text{mean control baseline})} \times 100;$$

and (iii) the standardized ES by trial arm were calculated as the observed treatment effect divided by the SD of the pooled change scores.

Results

Cross-sectional validation

The greatest impacts of DH on daily life were problems with eating ice cream, irritating sensations, annoying sensations, modification of eating, care when breathing, avoidance of cold drinks or foods, and avoiding contact with certain teeth (Table 9.2). With regard to *internal consistency*, nearly all item-total correlations were more than 0.4, indicating good internal consistency (Table 9.2). All correlations between domains and total scores were significant ($P < 0.05$) and consistent. The correlations between total score and the domains were as follows: social impact ($r = 0.905$); coping ($r = 0.911$); restrictions ($r = 0.876$); emotional impact ($r = 0.898$); and identity ($r = 0.756$). Internal consistency for the total score ($\alpha = 0.96$) and the domains of restrictions ($\alpha = 0.85$), coping ($\alpha = 0.91$), social impact ($\alpha = 0.84$), emotional impact ($\alpha = 0.89$), and identity ($\alpha = 0.90$) were all excellent.

Test–retest reliability was high for the total impact score at 0.77, but it varied for the following domains: identity (0.89), social impact (0.72), emotional impact (0.68), coping (0.61), and restrictions (0.43). All were significant ($P < 0.001$) except for the restrictions domain ($P < 0.01$). With regard to *convergent validity*, DHEQ total score was not correlated with global oral health ratings or the restrictions, coping, or emotions domains (Table 9.3). There were significant correlations between the identity and social impact domains and the global oral health rating. DHEQ total and all domain scores were strongly and significantly correlated with the scores for effect of DH on life overall. The total DHEQ and domain scores were significantly correlated with the mean tactile scores but not with the VAS and Schiff sensitivity assessment.

Longitudinal validation

Responsiveness over time within individuals

DHEQ total and domain scores in all three trials are summarized in Table 9.4. There were significant decreases in total scores and most domain scores across all trials. The identity domain consistently showed the lowest change, followed by the

Table 9.3 Pearson's correlation coefficients between DHEQ total and domain scores and self-reported global oral health, quality of life and clinical measures at baseline in clinical trial A

	Self-reported		Clinical		
	Global oral health	QoL	Tactile	VAS	Schiff[a]
DHEQ total	0.190	0.793***	−0.34***	0.16	0.20
Restrictions	0.149	0.677***	−0.32**	0.11	0.10
Coping	0.042	0.700***	−0.30**	0.20	0.28**
Social	0.292**	0.720***	−0.27**	0.10	0.12
Emotional	0.166	0.718***	−0.34***	0.13	0.11
Identity	0.287**	0.640***	−0.24*	0.09	0.17

[a] Spearman's correlation coefficient.
$^{*}P<0.05$, $^{**}P\leq 0.01$, $^{***}P\leq 0.001$.

Table 9.4 Participants DHEQ total, domain (mean, SD) and change scores across clinical trials

	Baseline	8-week	12-week	Pre-post change[a]	ES
Trial A					
Total	151.54 (36.05)	–	141.41 (38.34)	10.13***	0.28
Restrictions	20.01 (4.84)	–	19.13 (5.22)	0.88	0.18
Coping	58.26 (13.53)	–	54.12 (15.98)	4.14***	0.31
Social	19.42 (6.46)	–	17.71 (6.60)	1.70***	0.26
Emotional	36.82 (9.16)	–	34.73 (9.00)	2.10*	0.23
Identity	17.02 (7.45)	–	15.71 (7.24)	1.31**	0.18
Trial B					
Total	150.85 (36.79)	130.48 (43.49)	–	20.38***	0.56
Restrictions	20.10 (5.07)	17.18 (5.59)	–	2.92***	0.58
Coping	57.01 (13.15)	48.77 (15.97)	–	8.24***	0.63
Social	19.66 (6.39)	17.22 (6.99)	–	2.44***	0.38
Emotional	37.55 (9.92)	31.30 (10.62)	–	6.25***	0.64
Identity	16.80 (8.60)	15.91 (7.99)	–	0.90	0.10
Trial C					
Total	136.69 (30.96)	108.84 (39.17)	–	27.84***	0.86
Restrictions	19.15 (4.49)	14.66 (5.52)	–	4.49***	0.99
Coping	52.64 (11.98)	41.79 (15.67)	–	10.86***	0.90
Social	16.65 (6.29)	13.21 (5.98)	–	3.43***	0.54
Emotional	34.90 (9.25)	27.87 (11.49)	–	7.03***	0.75
Identity	14.57 (6.88)	12.69 (6.67)	–	1.88***	0.27

[a]Paired *t*-test. $^{*}P<0.05$, $^{**}P\leq 0.01$, $^{***}P\leq 0.001$. Higher score, worse OHQoL. Trials A and B, negative control; trial C, active control.

social domain. Within-person individual effects for total DHEQ scores were large in the active control trial (trial C, 0.86) and small to moderate in the negative control trials (0.28 and 0.56 for trial A and trial B, respectively). The coping domain showed the largest ES in all trials (0.31, 0.63, and 0.90 in trial A, trial B, and trial C, respectively), followed by the emotional domain (0.23, 0.64, and 0.75). The identity domain showed the smallest effect (0.18, 0.10, and 0.27).

Responsiveness using an external referent

A total of 58, 78, and 84 participants reported improved DHEQ total scores, with 22, 30, and 12 reporting deterioration in trials A, B, and C, respectively. Only nine participants reported no change from baseline to follow-up across the three trials.

To examine responsiveness using an external criterion, participants were classified by their change in self-reported quality of life from baseline to final follow-up (i.e., better, same, or worse). One-way analysis of variance was used to compare change in DHEQ scores across these three groups. There were main effects of change groups in all three trials ($F = 12.34$, 8.24, and 14.02; all $P < 0.001$). If the DHEQ is responsive to change, then it should show changes for participants who reported improvement in quality of life, and these should be substantially greater than those demonstrating stability in the external referent. The results were compatible with these expectations; those participants who reported an improvement in quality of life at follow-up (the "better" group) had more improvement in DHEQ scores (i.e., less DHEQ impacts over time) than those in the "same" and "worse" groups (Table 9.5). In the active control trial (trial C), all groups showed positive change scores (improvement in OHQoL) regardless of whether categorized as "better," "same," or "worse." These findings are compatible with expectations because all participants received active treatment.

Using responsiveness to change indices, we found that in the negative control trials, both the ES and SRM could be classified as large ES in the better group (>0.80),[18] with very small ES in both same and worse groups. In the active control trial, all ES were large (Table 9.5).

Finally, the MID was calculated as the mean change of the total scores for participants who reported any improvement in their self-reported quality of life (i.e., impact on life overall). For the negative control trials, the MID range was between 22 (trial A, $n = 41$) and 29 points (trial B, $n = 71$), whereas for the active control trial the MID was larger at 39 points (trial C, $n = 55$).

Responsiveness to treatment

DHEQ detected a treatment effect in the first negative control trial (trial A). Participants who received the active treatment reported greater reduction in DHEQ scores at follow-up than controls (Table 9.6). The intervention effect was moderate (Cohen's classification = 0.47). Contrary to our hypothesis, changes in both groups were similar in the second negative control trial (trial B). The relative improvement in this trial was 2.1%, with a very small ES (0.12). As hypothesized, changes in

Table 9.5 DHEQ total change scores, ES, and SRM according to QoL outcome category at follow-up

	n	Change (SD)	SRM	ES
Trial A				
Better	41	22.39 (22.58)	0.99	0.73
Same	17	2.53 (23.22)	0.11	0.05
Worse	26	−4.23 (21.92)	0.19	0.12
Trial B				
Better	71	29.44 (34.73)	0.85	0.84
Same	13	3.31 (21.39)	0.16	0.09
Worse	25	3.52 (27.09)	0.13	0.08
Trial C				
Better	55	39.38 (28.67)	1.37	1.31
Same	25	16.64 (14.37)	1.16	0.49
Worse	15	28.54 (7.37)	3.87	0.81

DHEQ scores recorded so higher score = improvement in OHQoL. Trials A and B, negative control; trial C, active control.

Table 9.6 DHEQ total change scores, treatment effect, and standardized ES of treatment among trial participants

	Test (SD) change	Control (SD) change	*F*	Observed treatment effect	Relative % improvement	SES
Trial A	15.93 (26.09)	4.05 (23.45)	4.80*	11.88	7.68	0.47
Trial B	22.43 (34.16)	18.21 (33.86)	0.42	4.22	2.05	0.12
Trial C	24.35 (24.98)	31.49 (32.35)	1.47	−7.14	−5.14	−0.25

Trials A and B, negative control; trial C, active control.
$^*P < .05$

DHEQ scores were similar in the test and control groups in the active control trial. The relative improvement was 5.1% in the control group compared with test group, with a small ES (− 0.25).

Discussion

The application of condition-specific person-reported measures, such as the DHEQ, as outcome measures in randomized clinical trials is relatively new in dentistry. As such, it is particularly important that any newly developed, or indeed existing, PRO measure meets the gold standard guidelines for establishing responsiveness or sensitivity to change, a key component of any evaluative measure.[7] Using the methods recommended in recent guidelines and by the clinical significance consensus group,[6,8] we demonstrated that the DHEQ is highly responsive to changes in functional and personal experiences of DH for individuals over an 8- to 12-week timeframe. There was significant improvement in DHEQ scores in all three trials. These findings indicate that the DHEQ is valuable for use in longitudinal evaluation of people with DH. Interestingly, the coping domain showed the greatest change, compatible with these items being transient or amenable to change (e.g., warming certain foods or drinks, avoiding cold drinks). In contrast, the identity and social domains were least responsive, perhaps because they are less likely to change over the short term (e.g., makes me feel old/damaged). This differential pattern of change was reflected in the ES for the individual domain scores. Overall, for the total DHEQ score, the ES were small to moderate in the negative control trials (0.26–0.56) and large in the active control trial (0.86). This finding is expected; if both control and test groups received antisensitivity dentifrice, then they would be expected to show the greatest improvement in OHQoL (Table 9.4).

Using changes in self-reported quality of life as the external referent, the DHEQ was sensitive to improvement in quality of life status. ES and SRM (range: 0.73–1.31) for the "improved" group represent large effects according to Cohen's classification, which suggests robust responsiveness. Similarly, the small SRM and ES in the stable group in the negative control trials (range: 0.05–0.09) provide further support for the stability of the DHEQ.

When physiological or clinical measures are tested, there is often a gold standard for comparison. In most cases relating to PRO measures such as OHQoL, however, there is no gold standard. Given that we do not have a gold standard in relation to DH, we could not examine clinically significant change in DHEQ; that is, "a difference in score that is large enough to have an implication for the patient's treatment or care."[8] This has important implications for determining the MID, a statistic or cut-point that is vital for interpreting OHQoL scores in cross-sectional comparisons of groups and in longitudinal evaluation of treatment strategies. Here, we chose to use participants' self-reported overall quality of life. In accordance with the clinical significance consensus group, we would argue that the persons themselves should be the ultimate judges of their health status. Thus, a person's self-report should be a better anchor for estimating MIDs of quality of life measures.[8] In this study, the MID was estimated at 22, 29, and 39 in the three trials. These results mean that if the measures were to be used in further clinical trials, then a DHEQ total change score between 22 and 29 would be required in negative control trials, with a larger change score of 39 in active control trials. As with all PRO measures, the MID is not an immutable characteristic; it will vary across populations and treatments.[6] Nevertheless, the MID presented here is important for the interpretation of statistically significant results in future DH-related clinical trials as well as for determining future sample sizes for clinical trials using DHEQ.

The findings suggest that DHEQ is responsive to treatments, supporting our *a priori* hypotheses in two out of the three trials. In the first, participants using an active dentine antisensitivity dentifrice reported significantly fewer impacts of DH on their daily lives at 12-week follow-up compared with those who received no antisensitivity ingredient (Table 9.6). The responsiveness to treatment was supported by the active control trial; improvements in test and control groups were similar, again supporting our hypothesis that participants should improve regardless of group. In contrast, DHEQ did not detect an effect in the second negative control trial (trial B). The explanation for this is unclear because the inclusion criteria and study design were comparable in both negative control trials. However, controls in trial B showed much greater change scores than those in trial A (18.21 vs. 4.05, respectively) and the SDs were much greater in test and control groups (34.16 and 33.86, respectively) compared to those in trial A (26.09 and 23.45). These changes in the control group influence responsiveness statistics. Furthermore, it is important to note that the trials were powered to detect changes in Schiff scores as the primary outcome, rather than DHEQ.

The clinical significance consensus group commented that evaluations of PROs, particularly quality of life, in clinical trials and the assessment of clinically significant thresholds for change are complex.[8] Strategies to determine change have not kept pace with the explosion in quality of life measures in medicine and dentistry. Although we used several analytic strategies, DHEQ, like all such measures, should be constantly reviewed, incorporating new data and psychometric techniques as they emerge. We used both anchor-based (i.e., tied to an external referent) and distribution-based methods (i.e., linked to a statistical parameter such as SD) as outlined at the consensus meeting (Table 9.3)[8]. Despite their widespread use, responsiveness statistics are derived using SDs from a given sample and are therefore population specific and context specific. We mitigated this to some degree by calculating responsiveness across three trials incorporating both active and negative control arms. Nevertheless, the MID identified here may not be appropriate outside of clinical trials, such as with people with mild DH in the general population. In addition, we calculated the MID based on actual changes in quality of life rather than global ratings of change in health. The latter, most commonly used method for determining MID was not available in the dataset. Given that we have no previous information regarding what constitutes a minimal change in the anchor (the effect of DH on life overall), we selected a threshold of *any* change. Although this strategy is acceptable, it is a conservative approach and does risk overestimating the MID.[6]

The DHEQ was developed by meticulously following recommended methods and adopting a robust theoretical framework.[20,21] The present study supports findings from previous cross-sectional validation and a study in China (see Chapters 7 and 11).[15,22] This body of work demonstrates the DHEQ to be a reliable and valid measure of the experience of DH, and to be a useful addition to the growing condition-specific OHQoL instruments for facilitating our understanding of the biopsychosocial impacts of oral conditions. DHEQ is therefore the measure of choice because of its direct reference to the problems associated with sensitive teeth. It can therefore help to explore the nuances of impacts in relation to DH specifically, which, in turn, has possible implications for our future understanding of a condition that is sometimes considered an enigma, in addition to serving as a tool for use in clinical trials.[23]

References

1. Jokovic A, Locker D, Stephens M, Kenny D, Tompson B, Guyatt G. Validity and reliability of a questionnaire for measuring child oral-health-related quality of life. *J Dent Res* 2002;**81**:459–63.
2. Baker SR, Pankhurst CL, Robinson PG. Utility of two oral health-related quality of life measures in patients with xerostomia. *Community Dent Oral Epidemiol* 2006;**34**: 351–62.
3. Pearson NK, Gibson BJ, Davis DM, Geilier S, Robinson PG. The effect of a domiciliary denture service on oral health related quality of life: a randomised controlled trial. *Br Dent J* 2007;**203**:E3.
4. Awad MA, Feine JS. Measuring patient satisfaction with mandibular prostheses. *Community Dent Oral Epidemiol* 1998;**26**:400–5.
5. DeBruin AF, Diederiks JPM, De Witte LP, Stevens FCJ, Philipsen H. Assessing the responsiveness of a functional status measure: the sickness impact profile versus the SIP68. *J Clin Epidemiol* 1997;**50**:529–40.
6. Revicki D, Hays RD, Cella D, Sloan J. Recommended methods for determining responsiveness and minimally important differences for patient-reported outcomes. *J Clin Epidemiol* 2008;**61**:102–9.
7. Guyatt GH, Walter S, Norman G. Measuring change over time: assessing the usefulness of evaluative instruments. *J Chronic Dis* 1987;**40**:171–8.
8. Wyrwich KW, Bullinger M, Aaronson N, Hays RD, Patrick DL, Symonds T, The Clinical Significance Consensus Meeting Group. Estimating clinically significant differences in quality of life outcomes. *Qual Life Res* 2005;**14**:285–95.
9. Cummins D. Dentin hypersensitivity: from diagnosis to a breakthrough therapy for everyday sensitivity relief. *J Clin Dent* 2009;**20**:1–9.
10. Ayad F, Ayad D, Zhang YP, DeVizio W, Cummins D, Mateo LR. Comparing the efficacy in reducing dentin hypersensitivity of a new toothpaste containing 8/0% arginine, calcium carbonate, and 1450 ppm flouride to a commercial sensitive toothpaste containing 2% potassium ion: an eight-week clinical study on Canadian adults. *J Clin Dent* 2009;**20**:10–6.
11. Rees JS, Addy M. A cross-sectional study of dentine hypersensitivity. *J Clin Periodontol* 2002;**29**:997–1003.
12. Al-Wahadni A, Linden GJ. Dentine hypersensitivity in Jordanian dental attenders—a case control study. *J Clin Periodontol* 2002;**29**:688–93.
13. Bekes K, Johh MT, Schaller H-G, Hirsch C. Oral health-related quality of life in patients seeking care for dentin hypersensitivity. *J Oral Rehabil* 2009;**36**:45–51.
14. Gibson B, Boiko OV, Baker SR, Robinson PG, Barlow A, Player T , et al. The everyday impact of dentine sensitivity: personal and functional aspects. *Soc Sci Dent* 2010;**1**:11–21.
15. Boiko OV, Baker SR, Gibson BJ, Locker D, Sufi F, Barlow APS, et al. Construction and validation of the quality of life measure for dentine hypersensitivity (DHEQ). *J Clin Periodontol* 2010;**37**:973–80.
16. Schiff T, Dotson M, Cohen S, De Vizio W, McCool J, Volpe A. Efficacy of a dentifrice containing potassium nitrate, soluble pyrophosphate, PVM/MA copolymer, and sodium fluoride on dentinal hypersensitivity: a twelve week clinical study. *J Clin Dent* 1994;**5**:87–92.
17. Polson AM, Caton JG, Yeaple RN, Zander HA. Histological determination of probe tip penetration into gingival sulcus of humans using an electronic pressure-sensitive probe. *J Clin Periodontol* 1980;**7**:479–88.

18. Cohen J. *Statistical power analysis for the behavioural sciences*. 2nd ed. NJ: Lawrence Erlbaum Associates; 1998.
19. Buchbinder R, Bombardier C, Yeung M, Tugwell P. Which outcome measure should be used in rheumatoid arthritis clinical trials? *Arthritis Rheum Arthritis Care Res* 1995;**38**:1221–2.
20. Juniper EF, Guyatt GH, Jaeschke R. *How to develop and validate a new health-related quality of life instrument. Quality of life and pharmacoeconomics in clinical trials*. Philadelphia, PA: Lippincott-Raven; 1996.
21. Wilson I, Cleary P. Linking clinical variables with health-related quality of life: a conceptual model of patient outcomes. *JAMA* 1995;**273**:59–65.
22. He SL, Wang JH, Wang MH. Development of the Chinese version of the Dentine Hypersensitivity Experience Questionnaire. *Eur J Oral Sci* 2012;**120**:218–23.
23. Johnson RH, Zulgar-Nairn BJ, Koval JJ. The effectiveness of an electroniosing toothbrush in the control of dentinal hypersensitivity. *J Periodontol* 1982;**53**:353–9.

Derivation of a short form of the dentine hypersensitivity questionnaire*

Carolina Machuca[1], Sarah R. Baker[1], Farzana Sufi[2], Steve Mason[2], Ashley P.S. Barlow[2] and Peter G. Robinson[1]
[1]School of Clinical Dentistry, Claremont Crescent, University of Sheffield, Sheffield, UK,
[2]GlaxoSmithKline, Consumer Health Care, Weybridge, UK

Introduction

Oral health care has changed from considering only the clinical aspects of oral conditions to placing greater emphasis on the subjective experience of the mouth. Subjective experience is especially important in relation to dentine hypersensitivity (DH) because it is a "short, sharp pain arising from exposed dentin in response to stimuli typically thermal, evaporative, tactile, osmotic, or chemical and which cannot be ascribed to any other form of dental defect or disease."[1] This diagnosis of exclusion indicates that the condition does not have unique clinical signs. It therefore gives primacy to the affected person or persons to detect and evaluate treatment for the condition.

One formal way to assess the subjective experience of oral conditions is to measure oral health-related quality of life (OHQoL); the "*Impacts of oral disorders on everyday life that are important to people and of sufficient magnitude to affect perception of their life overall.*"[2] Generic OHQoL measures detect the impacts of a broad range of oral conditions and are therefore useful for assessing and comparing impacts in general populations.

However, generic measures tend to not detect the specific and nuanced problems of particular oral diseases. A generic OHQoL measure, the Oral Health Impact Profile, was found to be relatively insensitive to some aspects of DH.[3,4] Accordingly, an OHQoL measure specific for the impacts of DH was developed based on interviews with people with the condition.[5,6]

The Dentine Hypersensitivity Experience Questionnaire (DHEQ) contains a main scale of 34 items that records impacts in five subscales: functional restrictions, coping, emotions, identity, and social impact. Participants respond to each item on 7-point Likert scales labeled as *strongly disagree*, *disagree*, *disagree a little*, *neither agree nor disagree*, *agree a little*, *agree*, and *strongly agree*, which are coded 1–7, respectively.

*Previously published as: Machuca C, Baker SR, Sufi F, Mason S, Barlow A, Robinson PG. Derivation of a short form of the Dentine Hypersensitivity Experience Questionnaire. Journal of Clinical Periodontology 2014;41:46–51.

Dentine Hypersensitivity. DOI: http://dx.doi.org/10.1016/B978-0-12-801631-2.00010-5

Several summary scores can be calculated. The prevalence can be calculated by the proportion of people who broadly agree (agree a little, agree, or strongly agree) with one or more items. The extent of impact is the number of items broadly agreed on by each person and the total score is calculated as the sum of the item codes.

DHEQ also enquires about different aspects of pain on visual analogue scales and contains a global oral health rating that asks participants to rate their oral health as excellent, good, fair, poor, or very poor. Finally, a smaller scale of four items considers the effect of DH on participants lives overall. DHEQ has been found to have excellent face and content validity, reliability, validity, and responsiveness to change.[6,7]

However, the original version has 48 items. Although measures with more items provide more precise and reliable data for use in research, they can take a long time to administer and complete, making their clinical application more burdensome for clinicians, researchers, and participants. Shorter forms of questionnaires may improve response rates, reduce missing data, reduce the cost of data collection, and facilitate their use in broader segments of the population, such as the elderly or larger samples.[8] Therefore, the aim of this project was to derive and evaluate a short form of the DHEQ.

Methods

Data from three previous studies of people with DH intended to develop and validate the long form of the DHEQ were used: the original cross-sectional validation ($n = 160$) and two randomized control trials ($n = 193$).[6,7] Sample size calculations were difficult in the absence of any preliminary data about the size of the short forms or their psychometric properties, but samples smaller than this have been used to devise and evaluate short forms of OHQoL questionnaires.[9] The three data sets ($N = 353$) were pooled and the merged data were split into two halves randomly. Short forms were derived using the first half and evaluated using the second half of the data.

Development

Two methods were used to derive the short forms, the item-impact and the regression methods. The item-impact method selects items that are most important to people with the condition.[9–11] After a questionnaire has been administered, item impacts are calculated as the product of the frequency of individuals experiencing a particular impact and the mean score for that item. The top two and three ranked items in each of the five subscales were selected to create item-impact-derived 10-item and 15-item short forms (SFI-10 and SFI-15, respectively).

The regression technique was applied to the same data. Total DHEQ score for the long form was used as the dependent variable, and all of the items were tested as independent variables in a single multiple regression model. The items were ranked by their contribution to the coefficient of variation (R^2), and the two and three items making the highest contribution in each subscale were selected for the regression-derived 10-item and 15-item short forms (SFR-10 and SFR-15, respectively).

Evaluation

The properties of the short forms were tested using the second half of the data. Scores for all four short forms were calculated by summing the response codes to their items.

Face and content validity were assessed using descriptive statistics, including measures of central tendency, the range of variability, and any floor and ceiling effects.

Internal consistency was assessed via Cronbach's α, α with each item deleted, and corrected item totals. Test–retest reliability was assessed using the Intraclass Correlation Coefficient (ICC) of the one-way random effect parallel model.

To evaluate the precision of the short forms, relative validity (RV) was assessed as the ratios of F statistics for the four short forms and the original form of DHEQ.

Sensitivity was evaluated as the proportion of people who broadly agreed with each item in each short form. Criterion validity was evaluated by correlating the long form with each of the short forms. Construct validation involved correlating the short form scores with global ratings of oral health and with the effect of DH on overall well-being.

Responsiveness was tested using the short forms in the two trials that aimed to assess the effectiveness of desensitizing toothpastes. The trials used accepted guidelines for the design and conduct of DH studies, being randomized, double blind (trial A), or examiner blind (trial B), two-treatment arm, stratified (by maximum baseline Schiff sensitivity score), parallel design, and single-site study in participants with at least two sensitive teeth.[12] At screening, participants provided written informed consent. Demographics, medical history, and medication use were recorded, followed by an oral examination, which included a Schiff sensitivity assessment. Those participants meeting the eligibility criteria returned a minimum of 24h after screening for the baseline assessment. At baseline, participants completed the DHEQ, and tooth sensitivity assessment using an evaporative air stimulus. Participants in trial A also completed a visual analogue scale to record the severity of their pain. Eligible participants were then randomized to treatment according to baseline stratification, assigned their study dentifrice, and provided with toothbrush, diary, brushing instructions, and timers. Participants used their allocated dentifrice at home twice per day for the next 12 weeks (trial A) or 8 weeks (trial B). DHEQ data were collected at baseline and at 12 weeks in trial A and at baseline and week 8 in trial B. Analysis of covariance (ANCOVA) using the long form of DHEQ had indicated a treatment effect in trial A but not in trial B. The ANCOVA was repeated for each of the short forms in both trials.

Results

Development

The item-impact and regression techniques were each used to produce 10-item and 15-item measures using the first half of the merged data. Table 10.1 summarizes

Table 10.1 Item impacts and items selected for short forms of DHEQ

	Mean score for item	% of people with impact	Item-impact	Included in 10- (*) and 15- (†) item-impact short forms	Included in 10- (*) and 15- (†) item regression short forms
Restrictions: problems eating ice cream	6.0	94.4	1007.7	*	
Restrictions: pleasure out of eating	4.7	70.1	585.7	*	†
Restrictions: longer to finish meal	4.5	61.0	486.3	†	*
Restrictions: cannot finish meal	3.2	28.8	163.7		*
Coping: modification of eating	5.1	80.8	735.2	*	
Coping: avoiding certain teeth	5.2	77.4	708.2	*	
Coping: careful when breathing	4.8	70.1	598.3	†	
Coping: avoiding cold drinks/ foods	4.9	67.8	581.7		
Coping: warming food/drinks	4.7	66.7	550.7		
Coping: biting in small pieces	4.5	62.1	497.8		
Coping: change toothbrushing	4.2	49.7	371.9		
Coping: avoiding other food	4.1	43.5	313.2		*
Coping: cutting fruit	3.8	42.9	290.3		†
Coping: cooling food/drink	3.8	40.1	267.6		*
Coping: putting a scarf over mouth	3.6	37.9	241.5		
Coping: avoiding hot drinks/foods	3.6	33.9	216.3		
Social: painful at the dentist	4.2	51.4	382.5	*	†
Social: longer than others to finish	3.7	39.5	258.6	*	
Social: choose food with others	3.6	39.0	248.7	†	*
Social: hide the way of eating	3.1	26.0	142.9		*
Social: unable to take part in conversations	2.2	7.3	28.9		
Emotions: annoying sensations	5.4	87.0	837.0	*	
Emotions: irritating sensations	5.4	88.7	849.8	*	
Emotions: anxious of eating contributes	4.7	67.2	557.4	†	*
Emotions: frustrated not finding a cure	4.4	49.7	383.5		
Emotions: annoyed with self for contributing	4.1	44.1	318.2		
Emotions: anxious because of sensations	3.8	43.5	295.8		†
Emotions: guilty for contributing	3.5	32.2	201.0		
Emotions: embarrassing sensations	2.9	18.6	95.1		*
Identity: makes me feel unhealthy	3.5	40.7	253.0	*	*
Identity: makes me feel old	3.4	36.2	217.0	*	
Identity: makes me feel damaged	3.0	28.2	149.4	†	
Identity: difficult to accept	2.9	21.5	110.1		†
Identity: different from others	2.7	20.9	99.9		*

the item impacts and indicates which items were included in each short form. The SFI-10 and SFR-10 short forms had only one question in common: "The sensations make me feel as though I am unhealthy," which was in the identity subscale section. The proportion of participants who broadly agreed with each item in the SFI-10 ranged from 36.2% to 94.4%, whereas items in the SFR-10 had a prevalence range from 18.6% to 67.2%.

SFI-15 and SFR-15 had six items in common, including all three items in the social subscale. The prevalence range for items in the SFI-15 was 28.4% to 94.4%, whereas the SFR-15 items had a prevalence range of 18.6% to 43.5%.

Evaluation

The scores obtained from all participants using the second half of the data indicate that all the short forms detected wide variation, suggesting reasonable content validity (Table 10.2). The item-impact short forms produced mean scores 12 points higher than those of the regression-derived forms. There were no floor and ceiling effects.

The item-impact short forms were more sensitive. The prevalence of impact using SFI-10 and using SFR-10 was 68% and 37%, respectively. Using SFI-15 and SFR-15, the prevalence was 93% and 61%, respectively.

All four short forms had a Cronbach's α more than 0.84 and three had Cronbach's αs more than 0.9, indicating excellent internal consistency (Table 10.3).

Table 10.2 SFI-10, SFR-10, SFI-15, and SFR-15 scores among 176 people with self-reported DH

Short form	Range of possible values	Mean score (SD)	Range of scores	% with minimum score	% with maximum score
SFI-10	10–70	47.36 (10.5)	10–68	0.56	1.14
SFR-10	10–70	35.73 (12.8)	10–68	1.14	0.56
SFI-15	15–105	67.79 (16.6)	15–102	0.56	1.14
SFR-15	15–105	55.50 (18.7)	15–103	0.56	0.56

Table 10.3 Reliability statistics of SFI-10, SFR-10, SFI-15, and SFR-15

Short form	Cronbach's α ($n = 176$)	Range of α if items deleted	Range of corrected item total correlations
SFI-10	0.840	0.813–0.838	0.385–0.718
SFR-10	0.904	0.888–0.909	0.417–0.718
SFI-15	0.906	0.895–0.906	0.405–0.741
SFR-15	0.924	0.916–0.925	0.454–0.782

When individual items were deleted, α remained stable. The corrected item total correlations were all positive and more than 0.38, supporting the internal reliability of the short forms.

The ICC was calculated using data from 93 participants in the 12-week clinical trial of desensitizing toothpastes. ICCs more than 0.9 indicate that measures are stable over time. The ICCs for SFR-10, SFI-15, and SFR-15 were 0.925, 0.921, and 0.939, respectively; only SFI-10 had an ICC of less than 0.9 (i.e., 0.895).

The RV coefficients, expressed as the ratios of F statistics, were used to compare the precision of the short forms with the original DHEQ (Table 10.4). The item-impact 10-item and 15-item short forms demonstrated 18% and 13% greater precision than the original questionnaire, whereas the regression short forms showed less precision.

All four short forms correlated with the long form (all $r > 0.93$, $P < 0.001$) and were highly positively correlated with ratings of overall well-being, but not with global ratings of oral health (Table 10.5 and Figure 10.1).

Responsiveness was tested using the short forms in the two trials of desensitizing toothpastes. The long form of DHEQ had indicated a treatment effect in trial A but not in trial B. As shown in Table 10.6, ANCOVA did not detect a treatment effect using any of the short forms, although all four approached significance in trial A.

Table 10.4 RV coefficients for the original and four short forms of DHEQ

Form	F_{test}	RV
DHEQ	74.278	
SFI-10	87.804	1.18
SFR-10	39.462	0.53
SFI-15	83.614	1.13
SFR-15	39.653	0.53

Table 10.5 Correlations between short form scores and oral health and overall well-being global ratings

Short form	Oral health		Overall well-being	
	r_s	P	r_s	P
SFI-10	0.119	0.117	0.735[a]	<0.001
SFR-10	0.103	0.174	0.732[a]	<0.001
SFI-15	0.082	0.281	0.760[a]	<0.001
SFR-15	0.130	0.086	0.729[a]	<0.001

[a]Spearman's Correlation is significant at the 0.01 level (2-tailed).

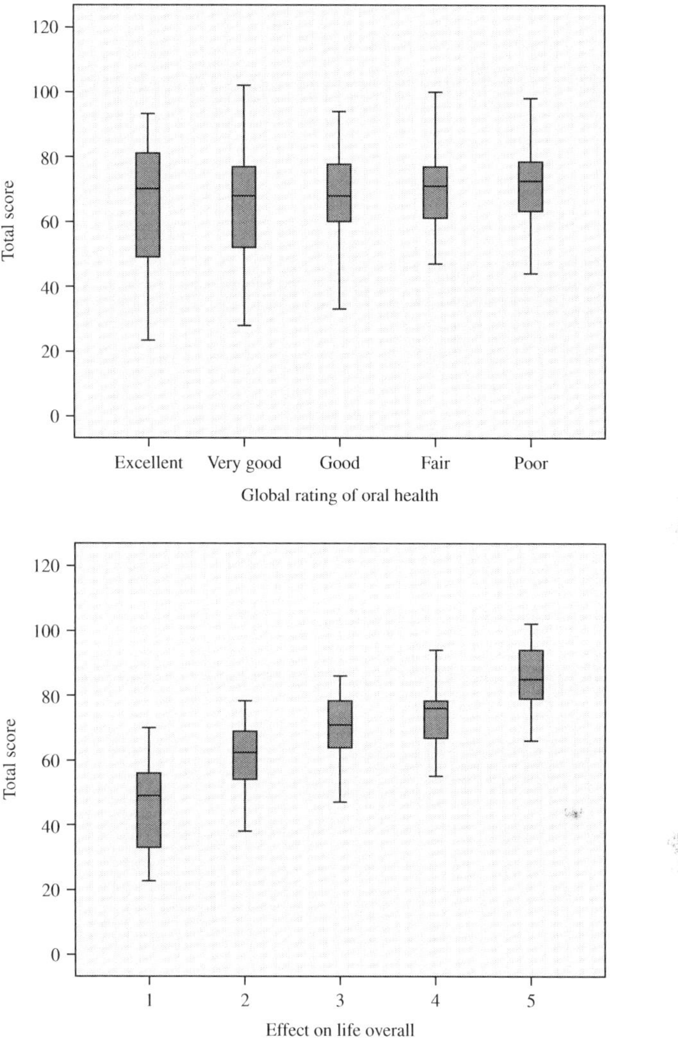

Figure 10.1 SFI-15 scores in relation to global ratings of oral health and effect on life overall.

Discussion

The aim of this study was to derive and evaluate short forms of the DHEQ. All four short forms performed well. Face and content validity appeared to be acceptable. They all had Cronbach's α more than 0.8, indicating excellent internal consistency; however, depending on the use of the test and the cost of misinterpretation, higher values might be required. Nunnally[13] argued that α more than 0.9 are required

Table 10.6 ANCOVA of between-group outcome effects in two trials of desensitizing toothpastes

Source	Type III sum of squares	df	Mean square	F	Significance
Trial A: 12-week follow-up					
SFI-10	156.655	1	156.655	2.777	0.100
SFI-15	386.552	1	38.52	3.371	0.070
SFR-10	213.254	1	213.254	3.547	0.063
SFR-15	417.028	1	417.028	3.038	0.085
Trial B: 8-week follow-up					
SFI-10	127.270	1	127.270	1.377	0.244
SFI-15	291.571	1	291.571	1.547	0.217
SFR-10	87.450	1	87.450	1.244	0.268
SFR-15	106.174	1	106.174	0.702	0.404

for use in individuals and, in the present data, three short forms exceeded this value. The stability of all four short forms, as indicated by the high ICCs, indicates that they could be used to distinguish levels of impact in samples of individuals receiving treatment for DH. Their correlations with overall well-being ratings demonstrated that the instruments discriminated between the effects of DH on the lives of participants overall.

Several studies have demonstrated that people with DH do not seem to consider it a serious dental condition and few seek professional help for it.[4,14–16] DH shares characteristics with some chronic illnesses, where living with the condition involves adopting mechanisms for adjusting and integrating DH into everyday life.[5] Those findings may explain the lower correlations of the short forms with the global rating of oral health (Table 10.5 and Figure 10.1). DH may not be regarded by people as a real oral health problem.

In evaluative measures, the interest is focused on recording within-participant changes over time as assessed by the analysis of variance (through the F_{test} value) and the RV. Conversely, this responsiveness may be offset by poor stability (low ICCs), which may mask real changes in OHQoL. Although all the short forms demonstrated high stability, none showed a significant effect after treatment in trial A (Table 10.6). That is, consistent with the long form, no treatment effect was expected or found in trial B, but a treatment effect was possible but not found in trial A.[7] However, in trial A there was a tendency for all the short forms to respond, and it should be kept in mind that the power of this analysis was reduced because the evaluation of the short forms was conducted using only half of the data. Item-impact-derived short forms have better responsiveness for use in evaluative research, but the data in this study indicate that the original long form of DHEQ may be superior for this purpose.[17]

The choice between the item-impact and regression approaches to develop short forms has been discussed several times.[9,11,18] The choice appears to be dependent on the philosophy of the investigators, because the method chosen is less important than the content and properties of the instrument. The 10-item and 15-item short forms had only one and six items in common, respectively. These differences between the statistical and person-centered approaches support the consideration of several different approaches to develop short forms.[9,11]

A measure developed to detect impacts on people's daily lives should be based on what is most important to them. This philosophy supports the use of the item-impact method, in which lay people are key figures judging the most important items. This information can then be combined with psychometric analysis to aid the selection. The combination of both perspectives should ensure that the final derived short forms meet both the philosophical and mathematical criteria for a good measure.[9]

Several detailed issues warrant consideration in the choice of short form. First, the item-impact method was used in the derivation of the original DHEQ long form.[6] Therefore, the regression short forms are quite sensitive. Nonetheless, the regression short forms contained more low-frequency items and were less sensitive. The item-impact short forms detect a greater prevalence of impacts because the technique selects items that are more frequently experienced by the people who are affected. Short forms that contain more high-frequency items detect the full impact of conditions. This greater sensitivity of the item-impact short forms to detect impacts is a distinct advantage. Second, the slightly higher αs obtained with the regression forms may be a mathematical artefact (the part–whole correlation effect) because of similarities between the methods used to derive these short forms (regression) and to evaluate them (correlation). For this reason, the apparently superior internal reliability of the regression-derived short forms should be judged with care. Overall, because of their greater sensitivity, precision, and responsiveness, the item-impact-derived short forms are superior.

In general, measures with more items have greater reliability, which supports the use of the 15-item version.[19] Although the properties of both item-impact short forms were similar, the 15-item version had greater internal reliability and precision; therefore, it is preferable and is provided in Appendix 2. The additional time required to complete a 15-item questionnaire is small for the extra advantage gained. Furthermore, the excellent psychometric properties of the 15-item short form suggest it can be used with individual patients. This would allow the DHEQ to move from being purely a research tool to one that can be used to detect and monitor DH and to assess its response to treatment in individual patients.

This study has the limitation that is based on data obtained in the United Kingdom and the United States. Expanding the characteristics of the sample and using the DHEQ in different communities and different languages is recommended to explore the applicability of both long and short versions of the questionnaire.

In conclusion, item-impact-derived short forms of DHEQ were more sensitive and precise. The excellent psychometric properties of the 15-item short form suggest it can be used with individual patients. The DHEQ 15-item short form derived using the item-impact method is better than the other short forms for assessing the subjective aspects of people experiencing DH.

References

1. Canadian Advisory Board on Dentin Hypersensitivity. Consensus-based recommendations for the diagnosis and management of dentin hypersensitivity. *J Can Dent Assoc* 2003;**69**:221.
2. Locker D, Allen F. What do measures of 'oral health-related quality of life' measure?. *Community Dent Oral Epidemiol* 2007;**35**:401–11.
3. Slade G, Spencer A. Development and evaluation of the oral health impact profile. *Community Dent health* 1994;**11**:3–11.
4. Bekes K, John M, Schaller H, Hirsch C. Oral health-related quality of life in patients seeking care for dentin hypersensitivity. *J Oral Rehabil* 2009;**36**:45–51.
5. Gibson B, Boiko O, Baker S, Robinson P, Barlow A, Player T , et al. The everyday impact of dentine sensitivity: personal and functional aspects. *Soc Sci Dent* 2010;**1**:11–20.
6. Boiko O, Baker S, Gibson B, Locker D, Sufi F, Barlow A, et al. Construction and validation of the quality of life measure for dentine hypersensitivity (DHEQ). *J Clin Periodontol* 2010;**37**:973–80.
7. Baker S, Gibson B, Sufi F, Barlow A, Robinson P. The dentine hypersensitivity experience questionnaire: a longitudinal validation study. *J Clin Periodontol* 2014;**41**:52–9.
8. Dillman D. Mail and internet surveys. Hoboken, New Jersey: Wiley; 2007. p. 18.
9. Jokovic A, Locker D, Guyatt G. Short forms of the Child Perceptions Questionnaire for 11–14-year-old children (CPQ11–14): development and initial evaluation. *Health Qual Life Outcomes* 2006;**4**:4.
10. Guyatt G, Bombardier C, Tugwell P. Measuring disease-specific quality of life in clinical trials. *CMAJ* 1986;**134**:889.
11. Locker D, Allen P. Developing short-form measures of oral health-related quality of life. *J Public Health Dent* 2002;**62**:13–20.
12. Holland G, Narhi M, Addy M, Gangarosa L, Orchardson R. Guidelines for the design and conduct of clinical trials on dentine hypersensitivity. *J Clin Periodontol* 1997;**24**:808–13.
13. Nunnaly J. *Psychometric theory*. New York, NY: McGraw-Hill; 1978.
14. Orchardson R, Gillam D. Managing dentin hypersensitivity. *J Am Dent Assoc* 2006;**137**:990–8.
15. Bartold P. Dentinal hypersensitivity: a review. *Aust Dent J* 2006;**51**:212–8.
16. Gillam D, Seo H, Bulman J, Newman H. Perceptions of dentine hypersensitivity in a general practice population. *J Oral Rehabil* 1999;**26**:710–4.
17. Allen P. Assessment of oral health related quality of life. *Health Qual Life Outcomes* 2003;**1**:40.
18. Juniper E, Guyatt G, Streiner D, King D. Clinical impact versus factor analysis for quality of life questionnaire construction. *J Clin Epidemiol* 1997;**50**:233–8.
19. Streiner D, Norman G. *Health measurement scales: a practical guide to their development and use*. 2nd ed. Oxford: Oxford University Press; 2008.

Development of the chinese version of the dentine hypersensitivity experience questionnaire*

S.L. He and J.H. Wang
Chongqing Key Laboratory for Oral Diseases and Biomedical Sciences; Department of Pediatric Dentistry, The Affiliated Hospital of Stomatology, Chongqing Medical University, Chongqing, China

Despite its relatively recent emergence over the past few decades, oral health-related quality of life (OHQoL) has important implications for the clinical practice of dentistry and dental research. OHQoL is a multidimensional construct that includes a subjective evaluation of the individual's oral health, functional well-being, emotional well-being, expectations and satisfaction with care, and sense of self. It has wide-reaching applications in survey and clinical research. OHQoL is increasingly acknowledged as an important health outcome in clinical trials and oral healthcare programs.

Over the past decades, a variety of generic instruments have been developed to comprehensively assess the impact of oral disorders on patient quality of life. However, as other chapters in this book reveal, generic instruments are too broad to detect small but important changes related to specific oral disorders such as dentine hypersensitivity (DH). Disease-specific instruments may be more sensitive to impacts and therefore show the changes in those impacts better than generic scales.

The Dentine Hypersensitivity Experience Questionnaire (DHEQ) assesses aspects of OHQoL related to DH.[4] If dentistry is to embrace the lay perspective of health, then there is an increasing need for cross-cultural adaptation of OHQoL instruments for use in other languages and cultures. Linguistic differences may vary between small grammatical constructions and the complete absence of words or ideas to be translated between languages. Moreover, our cultures represent the different ways in which we view our world. Thus, cultures that view the world differently may manifest different impacts of the mouth on their everyday life. For example, someone who lives in a culture that does not eat in public is unlikely to feel embarrassed because of difficulties eating. Because of these variations in social and economic structures, culture, and language, OHQoL instruments must be

*Previously published as: Development of the Chinese version of the Dentine Hypersensitivity Experience Questionnaire (DHEQ). *European Journal of Oral Science*, 2012; **22**:218–23.

Dentine Hypersensitivity. DOI: http://dx.doi.org/10.1016/B978-0-12-801631-2.00011-7

cross-culturally adapted to adequately capture impacts while retaining their psychometric properties.

This translation, adaptation, and validation of an instrument or scale for cross-cultural research is time-consuming and requires careful planning and adoption of rigorous methodological approaches to derive a reliable and valid measure of relevance to the target population. Therefore, we conducted a study to translate the DHEQ into Chinese, to evaluate its cross-cultural adaptation, and to test its reliability and validity among Chinese people. The major findings of our study are presented here.

DH is a common oral symptom defined as a "short, sharp pain arising from exposed dentin in response to stimuli typically thermal, evaporative, tactile, osmotic, or chemical and which cannot be ascribed to any other form of dental defect or pathology."[1] It is a highly prevalent oral condition, affecting 25.5–34.1% of Chinese people. This oral disorder not only affects patients' physical health but also affects their social function and psychosocial well-being. Therefore, there is a need to comprehensively assess the impact of DH on daily life. OHQoL is a multidimensional construct quantifying the extent to which oral disorders affect functioning, psychosocial well-being, sense of self, expectations, and satisfaction with care. It has important implications for the clinical practice of dentistry and dental research.

Commonly used instruments to assess OHQoL include the Oral Health Impact Profile (OHIP) and the Oral Impacts on Daily Performance (OIDP).[2,3] Whereas these instruments may comprehensively assess the multidimensional impact of oral disorders, they may be less useful for assessing the burden of specific oral disorders (e.g., dental esthetics, oral malodor, impairment of swallowing function) on OHQoL. Such generic measures can be too broad to accurately assess the links between specific oral conditions and OHQoL. Bekes and colleagues found that the particular impacts of DH were not fully captured through OHIP-49. Although the impact of DH has drawn attention in the past, no specific instrument has been available to assess its effect on people.

A DHEQ that assesses aspects of OHQoL related to DH was recently proposed by Boiko et al.[4] This condition-specific measure showed good psychometric properties in the United Kingdom and the United States. DHEQ could be used to evaluate negative impacts specifically related to DH and could aid in the development of effective interventions and health policies for DH. However, this questionnaire cannot be directly used in non-English-speaking countries. Rigorous psychometric evaluation must be accomplished before it can be used in other areas. Therefore, this study aimed to translate the DHEQ into Chinese, to evaluate its cross-cultural adaptation, and to test its reliability and validity among Chinese people.

Materials and methods

DHEQ is an English language OHQoL instrument developed by the University of Sheffield, UK. It consists of 50 items (two items were deleted in the revised English version) grouped into five domains to describe the pain (items 1–6), three visual analogue scales to measure pain (items 7–9), impact scales (items 10–45), a scale to record effects on life overall (items 46–49), and a global oral health rating (item 50). The impact scale has five subscales, "restrictions," "adaptation," "social impact," "emotional impact," and "identity," with responses given on 7-point Likert scales ranging from "strongly agree" to "strongly disagree" (equivalent to scores of 1–7). For items 46–49, responses were given on a 5-point Likert scale with responses ranging from "not at all" to "very much" (equivalent to scores of 1–5). For the last item, a 6-point Likert scale was used, ranging from "excellent" to "very poor" (equivalent to scores of 1–6). DHEQ is provided in full in Appendix 1.

The DHEQ was translated into Chinese using the forward–backward process proposed by Del Greco et al.[5] The process included several major steps:

1. Two independent translators first translated the DHEQ from English to Mandarin. Standard Chinese, also known as Mandarin, is an official language of the People's Republic of China. Both translators were fluent in English and Chinese and had background knowledge of dentistry.
2. Then, the two independent versions were back-translated from Mandarin to English by a professional English teacher and two bilingual dental specialists, none of whom knew the original questionnaire.
3. The translated and back-translated versions were compared and discussed by an expert panel consisting of two dental specialists with extensive knowledge of OHQoL assessment who were fluent in both English and Mandarin. A preliminary Chinese DHEQ version was then produced.
4. The preliminary Chinese DHEQ was pilot tested on a convenience sample of 20 patients.
5. After the test, emerging problems were discussed. The Chinese version was considered final when there were no substantial differences.

The study sample comprised a total of 110 patients with DH who had been previously recruited for a randomized clinical trial of the effect of desensitizing toothpaste at the West China College of Stomatology, Sichuan University. The Schiff cold air sensitivity scale was used to assess patients' responses to air blast hypersensitivity assessment. Teeth were evaluated for hypersensitivity by directing a 1-s blast of air onto the hypersensitive cervical area, from a distance of 1 cm, using cold air from a dental unit at 60 psi (± 5 psi) and a temperature of 70°F (± 3°F). The degree of hypersensitivity was recorded in accordance with the Schiff cold air sensitivity scale. The scale was scored as follows:

0, subject does not respond to air stimulus;
1, subject responds to air stimulus but does not request discontinuation of stimulus;
2, subject responds to air stimulus and requests discontinuation or moves from stimulus; and
3, subject responds to air stimulus, considers stimulus to be painful, and requests discontinuation of the stimulus.

Participants had at least one sensitive tooth with a response ≥ 1 on the Schiff scale. Patients with any orthodontic appliances or who were unable to understand the DHEQ questions were excluded from the study. Two patients were excluded because of this latter criterion.

The study was approved by the Ethics Committee of West China College of Stomatology, Sichuan University. Written informed consent was obtained from all participants. A detailed explanation was provided before participants completed the Chinese version of DHEQ. DHEQ was completed in the waiting room. If they had any questions, then they could consult research assistants at any time. To evaluate the test–retest reliability, 30 participants were randomly selected to make a return visit after 2 weeks.

Statistical analysis

Two types of reliability were adopted to assess the DHEQ. Internal consistency was evaluated by calculating Cronbach's alpha for the multi-item subscales and test–retest reliability was determined via intraclass correlation coefficients (ICCs) using data from the 30 participants who completed DHEQ again after a 2-week interval. Cronbach's alpha of 0.70 or greater is considered acceptable for comparisons between groups. Descriptors for ICC denoting poor to fair, moderate, good, and excellent agreement correspond to scores of <0.40, 0.41–0.60, 0.61–0.80, and more than 0.80, respectively.[6]

Validity was assessed in three ways. Construct validity was determined using exploratory factor analysis. However, a Bartlett's test of sphericity coefficient and Kaiser–Meyer–Olkin (KMO) test must first be conducted to determine whether there are sufficient significant correlations among items to perform this analysis. Factor loadings more than 0.40 were considered significant.

Discriminative validity was tested by comparing the subscale scores against the degree of hypersensitivity recorded with the Schiff cold air sensitivity and by investigating the correlation between DHEQ subscale scores and the global rating of oral health. Statistical analyses were conducted using SPSS 20.0 (SPSS, Chicago, IL).

Results

A total of 110 participants were recruited from a university-affiliated clinic for this study. All the DHEQ questionnaires were completed fully. The mean age of the participants was 47.2 ± 12.5 years (range: 20–67) and 72.7% were female. A total of 17 participants (15.5%) had first-degree hypersensitivity diagnosed; 58 (52.7%) displayed second-degree hypersensitivity; and 35 (31.8%) exhibited third-degree hypersensitivity. Table 11.1 presents the means, standard deviations of the pain scale (items 1–6), the overall effect on life scale (items 46–49), and the global oral health rating (item 50). Table 11.3 shows that the items related to "problems with eating ice cream (90%)," "avoiding contact with certain teeth (87%)," and "pleasure out of eating (86%)" were reported most frequently on DHEQ.

Table 11.2 shows the internal consistency of the multi-item scales. Cronbach's alpha for the total impact score of DHEQ was 0.96, and values for the subscales

Table 11.1 Mean scores for the pain scale, the effect on life overall scale, and the global oral health rating

	No. of items	Range	Mean score (SD)
Pain scale (VAS)			
Intensity	1	1–10	6.25 (1.91)
Bothersomeness	1	1–10	5.48 (2.05)
Tolerability	1	1–10	4.82 (1.86)
Effect on life overall	4	1–20	2.77 (0.97)
Global oral health rating	1	1–6	4.45 (0.97)

Table 11.2 Internal consistency and test–retest reliability of the individual subscales

Subscale	No. of items	Internal consistency ($n = 110$)	Test–retest reliability ($n = 30$)
		Internal consistency (Cronbach's alpha)	ICC (95% CI)
Total score	36	0.955	0.85 (0.66–0.99)
Subscales			
Restrictions	5	0.69	0.73 (0.43–0.96)
Adaptation	12	0.89	0.90 (0.77–0.99)
Social impact	5	0.80	0.89 (0.75–0.99)
Emotional impact	8	0.87	0.88 (0.69–0.99)
Identity	6	0.84	0.83 (0.62–0.99)

ranged from 0.69 for "restrictions" to 0.89 for "adaptation." All subscales exceeded the minimum reliability standard of 0.70, except the functional restrictions subscale, whose value of 0.69 nearly reached the threshold. The corrected item-total correlations ranged from 0.129 (item 13) to 0.760 (item 11). Only the corrected item-total correlation of item 13 ("I am uncertain when I am going to have these sensations in my teeth") failed to reach the recommended minimum correlation of 0.20. In addition, when deleting this item, alpha increased slightly.

Test–retest reliability was calculated for the 30 participants who repeated the test after 2 weeks. ICC values were 0.73 and 0.90 for functional restrictions and adaptation, respectively. ICC values for DHEQ indicated good to excellent agreement.

The result of the KMO test was 0.75 and the result of Bartlett's test of sphericity was 3466.1 (degrees of freedom [df] = 630, $P < 0.001$), demonstrating sufficient significant correlations to perform factor analysis (Table 11.3). All items had factor loadings more than 0.40. Factor analysis extracted eight factors (a figure consistent with the original English version), which together accounted for 74% of the variance. Items in most subscales loaded on only one factor for

Table 11.3 Range, mean scores, corrected item-total correlations and factor analysis results for the Chinese version of the DHEQ

Item	Mean	SD	% of people who had impact	Corrected item-total correlation	Cronbach's alpha if item deleted	Factor loading
Restrictions						
10. Pleasure out of eating	2.60	1.17	86	0.573	0.954	0.603
11. Cannot finish meal	3.56	1.61	56	0.760	0.952	0.777
12. Longer to finish meal	3.22	1.39	66	0.727	0.953	0.752
13. Uncertainty when	3.61	1.47	61	0.129	0.957	0.523
14. Problems with eating ice cream	2.37	1.18	90	0.240	0.955	0.580
Adaptation						
15. Modification of eating	2.78	1.28	84	0.533	0.954	0.569
16. Careful when breathing	3.16	1.37	75	0.560	0.954	0.598
17. Warming foods/drinks	3.02	1.36	72	0.541	0.954	0.578
18. Cooling foods/drinks	3.14	1.46	74	0.649	0.953	0.682
19. Cutting fruit	3.39	1.59	64	0.724	0.953	0.756
20. Putting a scarf over mouth	3.66	1.68	54	0.573	0.954	0.600
21. Avoiding cold foods/drinks	2.75	1.32	75	0.529	0.954	0.565
22. Avoiding hot foods/drinks	3.24	1.48	66	0.710	0.953	0.739
23. Avoiding contact with certain teeth	2.71	1.07	87	0.482	0.954	0.520
24. Change tooth brushing	2.66	1.20	85	0.449	0.954	0.478
25. Biting in small pieces	3.10	1.53	67	0.685	0.953	0.727
26. Avoiding other food	2.82	1.41	77	0.632	0.953	0.677

(*Continued*)

Table 11.3 **(Continued)**

Item	Mean	SD	% of people who had impact	Corrected item-total correlation	Cronbach's alpha if item deleted	Factor loading
Social						
27. Longer than others to finish	3.63	1.62	58	0.731	0.952	0.765
28. Choose food with others	2.98	1.30	77	0.605	0.953	0.645
29. Hide the way of eating	3.70	1.46	50	0.633	0.953	0.667
30. Unable to take part in conversations	4.50	1.63	31	0.679	0.953	0.694
31. Painful at the dentist	4.17	1.74	43	0.450	0.955	0.477
Emotions						
32. Frustrated not finding a cure	3.12	1.49	70	0.534	0.954	0.566
33. Anxious of eating contributes	3.15	1.44	70	0.667	0.953	0.699
34. Irritating sensations	3.29	1.40	64	0.702	0.953	0.726
35. Annoyed with myself for contributing	3.32	1.47	63	0.727	0.953	0.742
36. Guilty for contributing	4.10	1.77	39	0.625	0.953	0.646
37. Annoying sensations	2.78	1.32	80	0.337	0.955	0.709
38. Embarrassing sensations	3.69	1.63	50	0.688	0.953	0.705
39. Anxious because of sensations	3.50	1.46	61	0.662	0.953	0.672
Identity						
40. Difficult to accept	4.04	1.70	51	0.647	0.953	0.673
41. Part of my life	3.11	1.50	65	0.431	0.955	0.434
42. Different from others	4.26	1.70	39	0.618	0.953	0.633
43. Makes me feel old	3.23	1.68	67	0.714	0.953	0.735
44. Makes me feel damaged	3.81	1.69	49	0.678	0.953	0.696
45. Makes me feel unhealthy	3.50	1.64	65	0.689	0.953	0.720

Table 11.4 Discriminative validity of subscale scores according to the Schiff cold air sensitivity scale

Subscale	Schiff cold air sensitivity scale score			*H* value*
	1 ($n = 17$, mean (SD))	**2 ($n = 58$, mean (SD))**	**3 ($n = 35$, mean (SD))**	
Total score	157.24 (40.21)	122.00 (21.80)	97.63 (28.08)	33.50
Subscales				
Restrictions	20.82 (5.14)	15.31 (3.71)	12.83 (3.29)	27.38**
Adaptation	49.47 (16.10)	36.57 (7.67)	29.89 (9.04)	26.77**
Social impact	23.88 (6.16)	20.07 (4.89)	14.80 (4.32)	31.56**
Emotional impact	34.41 (8.87)	27.57 (6.74)	22.31 (8.66)	22.03**
Identity	28.65 (6.72)	22.48 (6.11)	17.80 (7.20)	25.90**

*Kruskal–Wallis test.
**$P < 0.001$.

Table 11.5 Convergent validity of the DHEQ: correlations between subscale scores with global oral health rating

Subscale	r_s[a]	95% confidence interval	*P*
Total score	−0.57	−0.70 to −0.41	0.00
Subscales			
Restrictions	−0.35	−0.51 to −0.16	0.00
Adaptation	−0.50	−0.64 to −0.31	0.00
Social impact	−0.53	−0.67 to −0.36	0.00
Emotional impact	−0.53	−0.66 to −0.40	0.00
Identity	−0.59	−0.71 to −0.42	0.00

[a]Spearman's rank correlation coefficient.

the DHEQ (i.e., adaptation, social, or identity), but the items in the emotions subscale were split into two and those in the restrictions subscale were divided into three factors. There were significant differences between the total and subscale scores of the DHEQ, categorized by the Schiff cold air sensitivity scale (Table 11.4). Total and subscale scores of the DHEQ had significant negative correlations with global oral health status (r_s ranged from −0.59 to −0.35), thus indicating good validity (Table 11.5).

Discussion

To our knowledge, this study provides the first introduction of the DHEQ in a non-English-speaking country and also constitutes the first evaluation of the psychometric properties of the DHEQ in a non-English-speaking country. In recent years, a large number of studies assessing OHQoL have been conducted.[7–10] Foreign instruments must be adapted before they can be used for speakers of different languages in other cultures. Therefore, we performed a cross-cultural adaptation of the original DHEQ and evaluated the reliability and validity of the Chinese version of the DHEQ according to international studies. The results presented here demonstrate the utility of the Chinese version of the DHEQ to measure the biopsychosocial impact of DH.

For internal consistency, Cronbach's coefficient alpha to test reliability exceeded 0.70 for all measures except for the functional restrictions subscale. The corrected item-total correlations were all well above the recommended level of 0.2, except for one item: "I am uncertain when I am going to have these sensations in my teeth." Deleting this item increased alpha for the entire DHEQ slightly, indicating that the item did not provide useful data and should be deleted. The same item was also deleted from the revised English version of the DHEQ. This item was negatively worded and therefore may have introduced method bias. A positively worded item may have obtained more satisfactory results. Alternatively, for this item, the predictability of sensations simply may not form part of the impact of DH. Nevertheless, these results demonstrate that the Chinese version of the DHEQ has good internal consistency.

Test–retest reliability for the overall scale and subscales demonstrated good to excellent agreement, although these values were slightly lower than those reported for the English version of the DHEQ. These findings indicate that the DHEQ is a reliable and stable instrument for assessing the impacts of DH.

In factor analysis, all items had factor loadings of more than 0.40, indicating that all items had strong relationships with their factors. The item "Having these sensations in my teeth is now just a part of my life," which had the lowest factor loading (0.434), was deleted from the revised English version on the basis of factor analysis. Two subscales (restrictions and emotions) had items that loaded onto two or three factors. A possible reason for this may be that the power is insufficient to detect expected associations between variables. Therefore, our results should be verified with a larger sample.

Differences in the overall and subscale scores of DHEQ were found among participants with different levels of hypersensitivity. That is, the severity of DH was inversely associated with OHQoL. This is an important finding because it may reveal a new function for both scales: the ability to discriminate between participants with different levels of hypersensitivity. Furthermore, significant correlations were observed between DHEQ subscale scores and global oral health rating. The correlations were higher than those obtained in the English validation of the DHEQ. Participants who reported higher levels of impact had poorer overall oral health than those who reported lower levels. Overall, the scales showed good convergent validity.

However, one limitation of the present study should be considered. All participants were recruited from patients attending a clinic; thus, the results cannot be extrapolated to people with mild DH. Further research could evaluate the psychometric properties of the DHEQ in a sample of the general population.[11]

In conclusion, this study confirmed the reliability and validity of the Chinese version of the DHEQ. It was appropriate to use the questionnaire to assess the OHQoL of participants in China who experience DH.

China has the largest population in the world. However, the assessment of OHQoL in China started in the beginning of the twenty-first century. OHQoL has a multitude of substantive applications in the field of dentistry, health care, and dental research as we move from bench to applied science and person-centered approaches to measure treatment needs and efficacy of care. Patient-oriented outcomes like OHQoL will enhance our understanding of the relationship between oral health and general health, and will demonstrate to clinical researchers and practitioners that improving the quality of a patient's well-being goes beyond simply treating dental maladies. OHQoL research can be used to inform public policy and help eradicate oral health disparities.

According to Qiang Gao, the former minister of Ministry of Health of PR China, at present, the development of China's healthcare system is faced with important opportunities as well as problems. It is an important historical mission for us to strengthen healthcare system reform and increase governmental input for more preventive services, products, and programs to establish examples of successful prevention programs.

Oral health care is a global concern that breaks down national boundaries. New scientific findings and technologies can arise anywhere in the world. The globalization of the health system will surely affect the area of oral health. Success in preventing and controlling oral disease in China is increasingly dependent on the ability to share knowledge and expertise with others around the world in a free and open manner. Dentistry in China must be fully involved in international organizations and activities for research, education, clinical practice, product development and distribution, and health promotion. In addition, China will benefit from international cooperation and collaboration (including OHQoL instrument adaptation).

Acknowledgment

The authors express their appreciation to Dr. Olga Boiko of the University of Sheffield for providing the original DHEQ and thank all the participants of the study.

References

1. Holland GR, Narhi MN, Addy M, Gangarosa L, Orchardson R. Guidelines for the design and conduct of clinical trials on dentine hypersensitivity. *J Clin Periodontol* 1997;**24**:808–13.
2. Slade GD. Derivation and validation of a short-form oral health impact profile. *Community Dent Oral Epidemiol* 1997;**25**:284–90.
3. Slade GD, Spencer AJ. Development and evaluation of the oral health impact profile. *Community Dent Health* 1994;**11**:3–11.

4. Boiko OV, Baker SR, Gibson BJ, Locker D, Sufi F, Barlow AP, et al. Construction and validation of the quality of life measure for dentine hypersensitivity (DHEQ). *J Clin Periodontol* 2010;**37**:973–80.
5. Del Greco L, Walop W, Eastridge L. Questionnaire development. 3: Translation. *Can Med Assoc J* 1987;**136**:817–8.
6. Bartko JJ. The intraclass correlation coefficient as a measure of reliability. *Psychol Rep* 1966;**19**:3–11.
7. He SL, Wang JH, Wang MH, Deng YM. Validation of the Chinese version of the Halitosis Associated Life-quality Test (HALT) questionnaire. *Oral Dis* 2012;**18**:707–12.
8. He SL, Wang JH, Li M. Validation of the Chinese version of the Summated Xerostomia Inventory (SXI). *Qual Life Res* 2013;**22**:2843–7.
9. Li M, He SL. Reliability and validity of the Chinese version of the chronic oral mucosal diseases questionnaire. *J Oral Pathol Med* 2013;**42**:194–9.
10. Zhao Y, He SL. Development of the Chinese version of the Oro-facial Esthetic Scale. *J Oral Rehabil* 2013;**40**:670–7.
11. Foster Page LA, Thomson WM, Jokovic A, Locker D. Validation of the Child Perceptions Questionnaire (CPQ 11–14). *J Dent Res* 2005;**84**:649–52.
12. Bekes K, John MT, Schaller HG, Hirsch C. Oral health-related quality of life in patients seeking care for dentin hypersensitivity. *J Oral Rehabil* 2009;**36**(1):45–51.

Part Three

Psychology and the Measurement of Pain and Impact

Response shift and oral health quality of life in dentine hypersensitivity

12

Marta Krasuska, Sarah R. Baker and Peter G. Robinson
School of Clinical Dentistry, Claremont Crescent, University of Sheffield, Sheffield, UK

Introduction

This chapter introduces the concept of response shift, which refers to the fact that, over time, the meaning of self-evaluation of quality of life changes. It also provides an overview of two related studies exploring response shift in oral health quality of life (OHQoL) in people with dentine hypersensitivity.

OHQoL can be understood as "impacts of oral conditions on everyday life that are important to people and of sufficient magnitude to affect perception of their life overall."[1] OHQoL is increasingly used in dentistry to capture subjective experiences of oral health, as an outcome measure in clinical trials, and to evaluate oral health care.

The use of OHQoL adds value by introducing a person-centered perspective into dentistry. In doing so, it allows for representation of the impact of oral conditions on quality of life. Nevertheless, it is not without challenges. This chapter focuses on the challenge that occurs when a person's subjective understanding of OHQoL changes, thus undermining assessments of OHQoL over time. For example, when evaluating effects of treatment, response shift threatens the validity of the evaluations.

What is response shift?

The concept of "response shift" provides a way of understanding how the meaning of quality of life can change over time and how this affects the subjective assessments of quality of life. Response shift has been defined as a "change in the meaning of one's self-evaluation of quality of life as a result of change in the person's internal standards of measurement (recalibration); change in the person's values, i.e., change in the relative importance of component domains of quality of life (reprioritization); or redefinition of quality of life (reconceptualization).[2(p1508)]

The term "response shift" was first introduced into quality of life research in a series of articles published in *Social Science and Medicine* in 1999.[2–7] The concept

Dentine Hypersensitivity. DOI: http://dx.doi.org/10.1016/B978-0-12-801631-2.00012-9

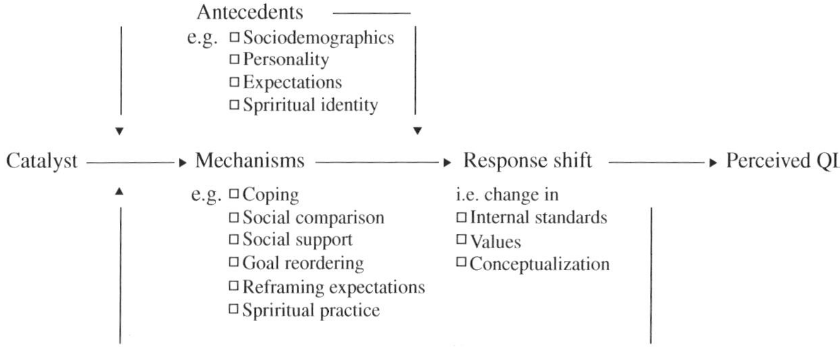

Figure 12.1 Sprangers and Schwartz theoretical model of response shift and quality of life. *Source*: Reprinted from Sprangers and Schwartz[2(p1509)]. Copyright 1999, with permission from Elsevier.

was meant to explain apparently paradoxical findings regarding the quality of life of people with chronic illnesses and disabilities. For example, people with poor health and disabilities were often found to report quality of life comparable with that reported by healthy individuals.[8,9] Likewise, the quality of life of people with life-threatening conditions was found to be stable over time, despite their deteriorating health.[10] Finally, assessments of quality of life by relatives, close friends, and lay caregivers acting as proxies underestimate the quality of life of people with chronic conditions.[11–13]

In addition to introducing the definition of response shift, Sprangers and Schwartz[2] also proposed a theoretical model of response shift and quality of life (Figure 12.1). This model explains how response shift occurs using five components:

1. Catalyst, a change in health that initiates the process of response shift
2. Mechanisms, behavioral, cognitive, and affective processes
3. Response shift, including recalibration, reprioritization, and reconceptualization
4. Antecedents, stable dispositional characteristics of a person
5. Perceived quality of life, including physical, psychological, and social domains.

One or more types of response shift can occur in response to the same catalyst.

In 2004, Rapkin and Schwartz updated the Sprangers and Schwartz model by adding an additional component. *Appraisal* was defined as a cognitive process involved in the self-assessment of one's quality of life on an item or a dimension in a quality of life questionnaire.[14] Within the updated model, appraisal can mediate between mechanisms and response shift. Alternatively, changes in appraisal alone can result in response shift.

Since the introduction of the concept of response shift into quality of life research, numerous studies have been published exploring it. These studies have identified response shifts in relation to serious chronic and life-threatening conditions such as cancer, arthritis, HIV/AIDS, and stroke, and also to some extent in the general population.[15–21]

A number of methods have been developed to assess response shift. The most commonly used methods include:

- Retrospective assessment (i.e., the THEN TEST)[18,22,23]
- Statistical methods, mainly using structural equation modeling[24,25]
- Individualized measures, whereby an individual is asked to select and rate the value of different domains of quality of life[26,27]
- Qualitative interviews.[28–30]

Despite the large number of studies to date, there is no agreement regarding what processes lead to response shift and how to best measure and account for it.[31–33]

Response shift has also been investigated in dentistry in relation to OHQoL. A change in participants' internal standards of measurement (recalibration) of OHQoL was found in partially edentulous patients receiving prosthodontic treatment.[34,35] In both cases, adjusting for response shift increased the apparent magnitude of the treatment effect. Recalibration, reprioritization, and reconceptualization of OHQoL were also found in edentulous individuals after receiving dentures in a clinical trial of conventional versus implant-retained dentures.[36] Adjusting for response shift resulted in the apparent treatment effect reaching statistical significance. Gregory et al.[37] investigated the meaning of oral health to people with visibly missing teeth or damaged front teeth. Their qualitative study found that the relevance of oral health changed over time and was determined by a number of factors, including participants' aspirations toward their oral health, the accessibility of oral care, and the perceived norms in society. Finally, Gibson et al.[38] found that individuals' views of dentine hypersensitivity and its impact on their everyday life changed over time, suggesting response shift.

Response shift in dentine hypersensitivity

As other chapters indicate, dentine hypersensitivity is characterized by episodes of pain in the teeth in response to external stimuli. However, the impacts of dentine hypersensitivity on the person go beyond the experience of pain to include impacts on eating, drinking, talking, toothbrushing, and social interactions.[38] These impacts cannot be captured solely using clinical indicators. Therefore, it is important to use the subjective measures to assess the impacts of dentine hypersensitivity on everyday life.

Response shift is very relevant to dentine hypersensitivity. First, the central role of the affected person in identifying the subjective experience of pain may make the condition susceptible to response shift. Second, evaluations of treatments, either in formal research or in less formal appraisals of products purchased for the conditions, might be undermined by response shift. For example, a participant in a trial might score OHQoL on a scale at baseline; however, if something happens (such as an illness of a family member), then the dentine hypersensitivity may assume less importance and its impacts on the quality of life would diminish, even if the product being tested was ineffective. Third, the availability of a precise and robust measure of OHQoL for dentine hypersensitivity (the DHEQ) allows detailed examination of response shift. Finally, response shift has predominantly been studied in severe conditions, so an investigation of the phenomenon in DH will elucidate its role in other minor conditions.

To obtain a better understanding of response shift and its influence on the assessment of OHQoL in people with dentine hypersensitivity, two studies were conducted. The first investigated response shift within the framework of a clinical trial for treatments of dentine hypersensitivity. The second qualitatively explored response shift and its underlying mechanisms through in-depth interviews with people with self-diagnosed dentine hypersensitivity.

Recalibration in a randomized controlled trial for treatments of dentine hypersensitivity

The first study explored one type of response shift (i.e., recalibration) that has been found to undermine assessments of treatment effects.[17,36,39] Recalibration represents changes in internal standards of measurement, and therefore this study also investigated the influence of recalibration on the assessment of treatment effects. More specifically, the study investigated the average magnitude and direction of recalibration, recalibration at the individual level, and the areas of OHQoL sensitive to recalibration. It also compared two methods for accounting for recalibration on assessments of change in OHQoL.

This study was nested within a clinical trial of mouthwashes for DH. The trial was an 8-week, randomized, controlled, four-treatment arm (three active treatment and one placebo arm), parallel design trial sponsored by GlaxoSmithKline that took place between March and July in Hamburg, Germany. Participants were reimbursed for participating.

The DHEQ[40] was used as an outcome measure to complement the more clinical Schiff and tactile sensitivity scores.[41]

DHEQ has excellent validity and reliability and has increasingly been used as an outcome in clinical trials to complement clinical indicators.[40,42] Thirty-four items assess the impacts of dentine hypersensitivity on quality of life within five impact subscales: limitations (4 items), coping (12 items), social impacts (5 items), emotional impacts (8 items), and identity (5 items).[40] The items are answered on a 7-point Likert scales with scores ranging from 1 (Strongly disagree) to 7 (Strongly agree). DHEQ is given in full in Appendix 1.

Two approaches, the THEN TEST and IDEALS, were incorporated into the DHEQ to account for recalibration. Both the THEN TEST and IDEALS rely on the design of the questionnaire to measure recalibration. In the THEN TEST, the traditional study design involving self-assessment of one's quality of life at baseline ("pre") and at follow-up ("post") is supplemented with a retrospective reassessment ("then") of the baseline score at follow-up.

In IDEALS, individuals also complete the questionnaire at baseline and at follow-up ("actual"). However, each administration of the questionnaire is supplemented with another version that inquires about how they would want things to be

Table 12.1 Example of analogous THEN TEST and IDEALS items used in DHEQ

THEN TEST	IDEALS
Baseline: Thinking about yourself over the last month to what extent would you agree or disagree with the following statement: *"Having sensations in my teeth takes a lot of pleasure out of eating and drinking."* ("pre")	Baseline: Thinking about yourself over the last month, to what extent would you agree or disagree with the following statement: *"Having sensations in my teeth takes a lot of pleasure out of eating and drinking."* ("$\text{actual}_{\text{baseline}}$") What would your ideal be? *"Having sensation in my teeth takes a lot of pleasure out of eating and drinking."* ("$\text{ideal}_{\text{baseline}}$")
Follow-up: Thinking about yourself over the last month to what extent would you agree or disagree with the following statement: *"Having sensations in my teeth takes a lot of pleasure out of eating and drinking."* ("post") How do you now think you were at the time of Screening assessment: *"Having sensations in my teeth takes a lot of pleasure out of eating and drinking."* ("then")	Follow-up: Thinking about yourself over the last month, to what extent would you agree or disagree with the following statement: *"Having sensations in my teeth takes a lot of pleasure out of eating and drinking"* ("$\text{actual}_{\text{follow-up}}$") What would your ideal be? *"Having sensations in my teeth takes a lot of pleasure out of eating and drinking."* ("$\text{ideal}_{\text{follow-up}}$")

ideally ("ideal"). An example of both designs applied to an item from the DHEQ is presented in Table 12.1.

If measured using the THEN TEST, a discrepancy between the baseline assessment ("pre") and the retrospective assessment ("then") indicates the magnitude and direction of recalibration. An upward shift in the internal standards of measurement is assumed when people retrospectively assesses themselves as worse than when they assessed themselves at baseline. A downward shift is assumed when people retrospectively assesses themselves as better than they had been at baseline.

In the IDEALS design, the magnitude and direction of recalibration is indicated by shifts in the "ideal" score between baseline and follow-up. An upward shift in the internal standards of measurement is assumed when a follow-up "ideal" score is higher than the baseline "ideal" score. A downward shift is assumed when a follow-up "ideal" score is lower.

The THEN TEST and IDEALS approaches also allow for calculation of the change in quality of life adjusted for recalibration. For the THEN TEST, the "then" and "post" scores are used to calculate the adjusted change in OHQoL scores. For the IDEALS, the difference between the "actual" and "ideal" scores is compared between baseline and follow-up scores.

In the trial, participants were given the THEN TEST or IDEALS versions of DHEQ at random, which they completed at screening and at the end of the trial (12–14 weeks later). Clinical status was assessed using the Schiff and tactile sensitivity scores.

Recalibration occured in the participants. On average, among participants receiving active treatment, the THEN TEST detected a downward shift in the internal standards during the trial. By contrast, the IDEALS indicated an upward shift.

The direction of the shift in the internal standards was in the expected direction for the IDEALS but not for the THEN TEST. For the IDEALS, an upward shift meant that, on average, the participants' expectations toward their oral health increased. Such a change helps to maintain a stable, satisfactory quality of life in response to changing life circumstances, such as an improvement or deterioration in health.[43] Additionally, it is possible that participating in a clinical trial for treatments for dentine hypersensitivity might have increased the participants' expectations toward their oral health.[37]

However, the THEN TEST detected a downward shift, which meant that at the end of the trial, participants retrospectively assessed themselves as better than they had assessed themselves to be at the beginning of the trial. This direction of recalibration was the opposite of that expected and contradictory to the literature.[23,34–36]

One possible reason why participants retrospectively reassessed themselves as being better is that they overestimated the impacts of dentine hypersensitivity at the outset. This could have been attributable to the effort justification bias, a tendency to increase the attractiveness of a goal or an outcome if a considerable effort is required to obtain it.[44] Alternatively, participants might have overestimated their level of impacts at screening to ensure that they qualified to participate and receive the reimbursement. Another possible reason is that while participating in a trial, participants started to pay more attention to their teeth and, when retrospectively assessing their OHQoL, realized they were experiencing fewer impacts before the trial had begun.

Finally, the similarity in the direction and magnitude of recalibration found in the three active treatment groups and the placebo group suggests that recalibration might have been part of a placebo effect.

At an individual level, many participants did not undergo recalibration and, of those who did, some shifted their internal standards downwards and others shifted upwards (whether assessed by the THEN TEST or the IDEALS). Similar heterogeneous shifts have been found in other studies of recalibration at the individual level.[18,22,23,45,46] Several factors might explain this heterogeneity. Different people might react to the same catalyst differently, so that only a subset of individuals undergoes recalibration, and within this subset some may shift their internal standards downwards and others may shift their internal standards upwards. The other reason might be that different people might need different amounts of time to accommodate changes in health status; some people might have needed more time than the duration of the trial provided. Personality traits such as extraversion, resilience, and rigidity might also play a role in how people experience recalibration.[47]

Emotional impacts were most sensitive and identity was least sensitive to recalibration. Such variation in sensitivity to recalibration may arise if subscales that involve more subjective assessments are more sensitive.[48] In addition, floor and ceiling effects may also leave little room for recalibration in some subscales. Both factors may have played a role.

Adjusting for recalibration reduced the values, magnitude, and significance of participants' changes in OHQoL during the trial, whether using the THEN TEST or IDEALS. Additionally, the changes in OHQoL adjusted using the THEN TEST switched, indicating deterioration rather than improvement during the trial. Interestingly, neither the unadjusted change nor the adjusted change correlated with the Schiff and tactile sensitivity scores.

These findings are inconsistent with those from other studies. Participants receiving health interventions had adjusted changes that pointed to greater improvements in quality of life than the unadjusted changes.[18,49] The reason for these unexpected findings is different for the THEN TEST and the IDEALS. For the THEN TEST, participants retrospectively rated themselves as better, whereas in other studies of people receiving treatment, participants tended to retrospectively rate themselves as worse. This resulted in the counterintuitive results.

The IDEALS method incorporates the recalibration into the score. If an improvement in oral health is followed by increased expectations towards it, then the value of adjusted change is lower than the value of unadjusted change.

Clinical criteria are often weakly correlated with self-reported OHQoL.[50–52] The weak relationship found in this study can be explained by the fact that the clinical assessment focused exclusively on the symptom of pain, whereas OHQoL encompasses emotional and social impacts and coping.[40] Additionally, the clinical assessment in a clinical setting might not correspond with the pain experienced on a day-to-day basis. Finally, the evaluation of pain in clinical trials for dentine hypersensitivity is not without problems.[41]

In comparisons between the two methods, the THEN TEST is a well-established method for recalibration assessment, used either alone[53–55] or in combination with other methods.[26,56–58] It has been validated against other methods such as structural equation modeling[17,59] and individualized methods.[60] Guidelines on how to conduct the THEN TEST studies have also been formulated.[61] However, this method has been criticized for its susceptibility to bias, including recall, effort justification, and social desirability bias.[44] These biases undermine the validity of the THEN TEST. Additionally, the implicit theory of change, postulating that people infer about their past state from their current state rather than directly recall their past state, was proposed as an alternative to recalibration to explain discrepancies between initial and retrospective assessments.[32]

The IDEALS method, on the contrary, has not been widely used to assess recalibration in the area of quality of life, although it has been applied to education and management research.[62,63] The potential threat to the validity of the IDEALS approach is that individuals might indicate the highest possible score as their "ideal" at baseline, leaving no room for an upward shift in the internal standards. However, such ceiling effects did not occur in this study because participants were willing to accept some degree of impact as their "ideal" at baseline. Another potential threat arises if participants do not understand the question about their "ideal." There was some evidence that not all participants correctly understood the IDEALS assessment.

In this study, the THEN TEST appeared to be more sensitive to recalibration than the IDEALS, because it indicated greater recalibration and detected more cases of recalibration at an individual level. However, this apparent sensitivity and detection of shifts in the opposite direction might result from its susceptibility to bias.

This list of potential biases in the THEN TEST leaves the IDEALS as a promising approach for assessing recalibration, but it requires further validation.

Response shift in people with dentine hypersensitivity: a longitudinal qualitative study

Our qualitative study explored response shift and its underlying mechanisms in 20 people with dentine hypersensitivity.

Each participant took part in two in-depth interviews over 6–12 months. During each interview, participants also completed the DHEQ. Participants were purposefully sampled to recruit individuals who had only recently recognized they had sensitive teeth and those who had the condition for some time. The areas explored during the first interview included the influence of dentine hypersensitivity on quality of life, how this could change, possible catalysts, mechanisms, and antecedents, and appraisal. The follow-up interview also explored changes since the first interview and participants' reflections on taking part in the study.

Data were analyzed in accordance with the framework approach with the addition of a constant comparative method.[64,65] A theoretical framework was based on the original models of response shift[2,14,66] but was revised in response to new categories evident in the data.

This study also found that response shift occurred in people with dentine hypersensitivity. Two main themes, "response shift in the context of adaptation" and "response shift in the context of self-assessment" emerged.

Participants displayed two types of response shift: reprioritization and reconceptualization. Reprioritization involved a change in the relative importance of dentine hypersensitivity for the person's quality of life:

> *I'm so busy, because I've got two young children, that I haven't got time to be anxious about something that is going to hurt.*
>
> *(Agnes, baseline)*

Reconceptualization concerned the inner concept of dentine hypersensitivity, with shifts along a continuum between dentine hypersensitivity conceptualized as "*a big negativity*" or "*a big problem*" at one end and as something "*normal*" or "*part of a normal life*" at the other.

The most important mechanisms involved in the occurrence response shift, and adaptation identified in the study were acceptance, problem-oriented coping, habits, and downward social comparison. Two of these mechanisms were new to the response shift model: acceptance and habits.

Acceptance is "allowing, tolerating, embracing, experiencing, or making contact with a source of stimulation that previously provoked escape, avoidance, or aggression."[67(p215)]. It is an "active and aware embracement of thoughts, feelings, and memories without unnecessary attempts to change their frequency or form, especially when doing so would cause harm."[68] Acceptance helps people adjust to chronic health conditions, especially chronic pain.

Acceptance encompassed accepting the pain in teeth and acknowledging that a cure for DH was unlikely:

> *I've accepted it a lot more because (...) I think if there was something you could do about it you'd pay that money and have it done definitely. But the only thing you can do is to put as many things in place to try and counteract it as you possibly can, because, you know, there is no cure for it. You know, that it is something you've got to deal with personally.*
>
> *(Doreen, baseline)*

Problem-oriented coping strategies are directed toward resolving or minimizing problems. In this study, strategies were used either to minimize the pain from DH (e.g., eating on one side of the mouth or avoiding cold drinks) or to prevent the condition from progressing (e.g., avoiding acidic food or using dentine hypersensitivity-specific oral care products).

Habit formation was the second new mechanism to emerge from the analysis. Habits are behaviors that are triggered automatically by cues present in the person's everyday life.[69] The formation of habits represents a shift from "conscious voluntary control over the behavior to a lower order behavioral control that is scarcely available to consciousness."[70]

Habit formation related to problem-oriented coping include oral hygiene routines (e.g., toothbrushing, flossing), eating, and drinking:

> *If I think of eating normally I think of like eating at both sides of the mouth. And I think now I eat more towards the right hand side because the most sensitive teeth are on the left but I just, as I say I just sort of do it automatically now so it has kind of become normal.*
>
> *(Olivia, follow-up)*

Finally, downward social comparison, when people compare themselves with those who are worse off than themselves, was an important mechanism of adaptation to DH. Downward social comparison was performed largely in relation to pain:

> *The pain is probably the same as you said, but it's just the way I see it is different I guess. (...) I think it's a consequence of our discussion, you see, like thinking that some people are worse off than me made me think oh yeah that's fine. I guess. (...)*

> *Cause last time we were talking made me realise some people are much worse off than that.' 'like some people they can't even eat anything too hot or anything too cold you know like they just can't do it cause it's too painful. I guess mines not that bad.*
>
> *(Olivia, follow-up)*

Context was very important for adaptation to DH and for the occurrence of response shift. Context included the immediate social environment (family and friends), dentists, dentine hypersensitivity oral care products (toothpastes and mouthwashes), and advertisements for these products. Context influenced response shift through the mechanisms mentioned already, such as social comparison, problem-oriented coping, and acceptance as well as social support.

Three types of catalysts initiated adaptation to DH: the onset of the condition, external events, and participating in the study.

Onset of the condition was an expected catalyst because of its place within the response shift model (Figure 12.1) and as a result of being documented in a number of other studies.[30,57,71]

The two other catalysts were less expected because they are not explicitly mentioned in the model. External events precipitated reprioritization of the importance of DH for the person's quality of life by placing it in a wider context. Thus, the effect of DH on quality of life depends not only on the condition, but also on other things important to the person at the particular time.

Finally, as shown by Olivia (previous quote), participating in the study also made participants reflect on their condition and its role in their lives. Participation prompted some to discuss DH with others or to seek further information about it. Completing the DHEQ questionnaire during the interview was also a source of social comparison and information and therefore prompted reflection on dentine hypersensitivity among participants.

The results of this study suggest that psychosocial interventions (taking part in this study can be seen as one) can foster adaptation to dentine hypersensitivity and supported the recommendations in Chapter 3.

Conclusion

The findings of these two studies confirmed that people with dentine hypersensitivity experience response shift. Response shift is part of adaptation to DH and affects assessment of the effects of treatment.

There is variability in the way people experience response shift in terms of timing, magnitude, direction, mechanisms involved in its occurrence, and the catalysts that can initiate response shift.

The results showed that response shift in relation to DH can occur at any point in time over the duration of the condition. It can be a result of the onset of the condition, important life events (not related to dentine hypersensitivity), or an intervention (such as participating in a clinical trial or a psychological study). When occurring within the frame of a clinical trial, it can also be part of a placebo effect.

In order to gain a full understanding of response shift and how people adapt to dentine hypersensitivity it is important to place the condition in the wider context of people's lives.

Two methods of measuring and accounting for response shift within a clinical trial were compared in the first study. The THEN TEST (the retrospective assessment) is a well-established method but might be susceptible to bias. In comparison, IDEALS is a new but promising method for assessing and accounting for response shift.

References

1. Locker D, Allen F. What do measures of "oral health-related quality of life" measure?. *Community Dent Oral Epidemiol* 2007;**35**:401–11.
2. Sprangers MAG, Schwartz CE. Integrating response shift into health-related quality of life research: a theoretical model. *Soc Sci Med* 1999;**48**:1507–15.
3. Schwartz CE, Sprangers MAG. Methodological approaches for assessing response shift in longitudinal health-related quality-of-life research. *Soc Sci Med* 1999;**48**:1531–48.
4. Gibbons FX, Buunk BP. Individual differences in social comparison: development of a scale of social comparison orientation. *J Pers Soc Psychol* 1999;**76**:129–42.
5. Schwartz CE, Sendor RM. Helping others helps oneself: response shift effects in peer support. *Soc Sci Med* 1999;**48**:1563–75.
6. Wilson IB. Clinical understanding and clinical implications of response shift. *Soc Sci Med* 1999;**48**:1577–88.
7. Daltroy LH, Larson MG, Eaton HM, Phillips CB, Liang MH. Discrepancies between self-reported and observed physical function in the elderly: the influence of response shift and other factors. *Soc Sci Med* 1999;**48**:1549–61.
8. Albrecht GL, Devlieger PJ. The disability paradox: high quality of life against all odds. *Soc Sci Med* 1999;**48**:977–88.
9. Brickman P, Coates D, Janoffbulman R. Lottery winners and accident victims—is happiness relative. *J Pers Soc Psychol* 1978;**36**:917–27.
10. Bach JR, Tilton MC. Life satisfaction and well-being measures in ventilator assisted individuals with traumatic tetraplegia. *Arch Phys Med Rehabil* 1994;**75**:626–32.
11. Sprangers MAG, Aaronson NK. The role of health-care providers and significant others in evaluating the quality-of-life of patients with chronic disease—a review. *J Clin Epidemiol* 1992;**45**:743–60.
12. Spiller JA, Alexander DA. Domiciliary care: a comparison of the views of terminally ill patients and their family caregivers. *Palliat Med* 1993;**7**:109–15.
13. Sneeuw KCA, Aaronson NK, deHaan RJ, Limburg M. Assessing quality of life after stroke—the value and limitations of proxy ratings. *Stroke* 1997;**28**:1541–9.
14. Rapkin BD, Schwartz CE. Toward a theoretical model of quality-of-life appraisal: implications of findings from studies of response shift. *Health Qual Life Outcomes* 2004;**2**:14.

15. Schwartz CE, Bode R, Repucci N, Becker J, Sprangers MAG, Fayers PM. The clinical significance of adaptation to changing health: a meta-analysis of response shift. *Qual Life Res* 2006;**15**:1533–50.
16. Schwartz CE, Rapkin BA. Understanding appraisal processes underlying the thentest: a mixed methods investigation. *Qual Life Res* 2012;**21**:381–8.
17. Mayo NE, Scott SC, Ahmed S. Case management poststroke did not induce response shift: the value of residuals. *J Clin Epidemiol* 2009;**62**:1148–56.
18. Razmjou H, Schwartz CE, Yee A, Finkelstein JA. Traditional assessment of health outcome following total knee arthroplasty was confounded by response shift phenomenon. *J Clin Epidemiol* 2009;**62**:91–6.
19. Westerman MJ, The A-M, Sprangers MAG, Groen HJM, van der Wal G, Hak T. Small-cell lung cancer patients are just "a little bit" tired: response shift and self-presentation in the measurement of fatigue. *Qual Life Res* 2007;**16**:853–61.
20. Ogden J, Lo J. How meaningful are data from Likert scales? An evaluation of how ratings are made and the role of the response shift in the socially disadvantaged. *J Health Psychol* 2012;**17**:350–61.
21. Galenkamp H, Huisman M, Braam AW, Deeg DJH. Estimates of prospective change in self-rated health in older people were biased owing to potential recalibration response shift. *J Clin Epidemiol* 2012;**65**:978–88.
22. Osborne RH, Hawkins M, Sprangers MAG. Change of perspective: a measurable and desired outcome of chronic disease self-management intervention programs that violates the premise of preintervention/postintervention assessment. *Arthritis Rheum Arthritis Care Res* 2006;**55**:458–65.
23. Hinz A, Barboza CF, Zenger M, Singer S, Schwalenberg T, Stolzenburg JU. Response shift in the assessment of anxiety, depression and perceived health in urologic cancer patients: an individual perspective. *Eur J Cancer Care* 2011;**20**:601–9.
24. King-Kallimanis BL, Oort FJ, Visser MRM, Sprangers MAG. Structural equation modeling of health-related quality-of-life data illustrates the measurement and conceptual perspectives on response shift. *J Clin Epidemiol* 2009;**62**:1157–64.
25. Ahmed S, Bourbeau J, Maltais F, Mansour A. The Oort structural equation modeling approach detected a response shift after a COPD self-management program not detected by the Schmitt technique. *J Clin Epidemiol* 2009;**62**:1165–72.
26. Dempster M, Carney R, McClements R. Response shift in the assessment of quality of life among people attending cardiac rehabilitation. *Br J Health Psychol* 2010;**15**:307–19.
27. Sharpe L, Butow P, Smith C, McConnell D, Clarke S. Changes in quality of life in patients with advanced cancer—evidence of response shift and response restriction. *J Psychosom Res* 2005;**58**:497–504.
28. Beeken RJ, Eiser C, Dalley C. Health-related quality of life in haematopoietic stem cell transplant survivors: a qualitative study on the role of psychosocial variables and response shifts. *Qual Life Res* 2011;**20**:153–60.
29. Korfage IJ, Hak T, de Koning HJ, Essink-Bot M-L. Patients' perceptions of the side-effects of prostate cancer treatment—a qualitative interview study. *Soc Sci Med* 2006;**63**:911–9.
30. Sinclair VG, Blackburn DS. Adaptive coping with rheumatoid arthritis: the transforming nature of response shift. *Chronic Illn* 2008;**4**:219–30.
31. Barclay-Goddard R, Epstein JD, Mayo NE. Response shift: a brief overview and proposed research priorities. *Qual Life Res* 2009;**18**:335–46.

32. Norman GR. Hi! How are you? Response shift, implicit theories and differing epistemologies. *Qual Life Res* 2003;**12**:239–49.
33. Ubel PA, Peeters Y, Smith D. Abandoning the language of "response shift": a plea for conceptual clarity in distinguishing scale recalibration from true changes in quality of life. *Qual Life Res* 2010;**19**:465–71.
34. Reissmann DR, Remmler A, John MT, Schierz O, Hirsch C. Impact of response shift on the assessment of treatment effects using the oral health impact profile. *Eur J Oral Sci* 2012;**120**:520–5.
35. Kimura A, Arakawa H, Noda K, Yamazaki S, Hara ES, Mino T, et al. Response shift in oral health-related quality of life measurement in patients with partial edentulism. *J Oral Rehabil* 2012;**39**:44–54.
36. Ring L, Hofer S, Heuston F, Harris D, O'Boyle CA. Response shift masks the treatment impact on patient reported outcomes (PROs): the example of individual quality of life in edentulous patients. *Health Qual Life Outcomes* 2005;**3**:55.
37. Gregory J, Gibson B, Robinson PG. Variation and change in the meaning of oral health related quality of life: a "grounded" systems approach. *Soc Sci Med* 2005;**60**:1859–68.
38. Gibson B, Boiko O, Baker SV, Robinson PG, Barlow A, Player T, et al. The everyday impact of dentine hypersensitivity: personal and functional aspects. *Soc Sci Dent* 2010;**1**:11–22.
39. Joore MA, Potjewijd J, Timmerman AA, Anteunis LJC. Response shift in the measurement of quality of life in hearing impaired adults after hearing aid fitting. *Qual Life Res* 2002;**11**:299–307.
40. Boiko OV, Baker SR, Gibson BJ, Locker D, Sufi F, Barlow APS, et al. Construction and validation of the quality of life measure for dentine hypersensitivity (DHEQ). *J Clin Periodontol* 2010;**37**:973–80.
41. Curro FA, Friedman M, Leight RS. Design and conduct of clinica trials on dentine hypersensitivity. In: Addy M, Embery G, Edgar WM, Orchardson R, editors. *Tooth wear and sensitivity: clinical advances in restorative dentistry*. London: Martin Duntiz Ltd; 2000.
42. Baker SR, Gibson BJ, Sufi F, Barlow A, Robinson PG. The dentine hypersensitivity experience questionnaire (DHEQ): a longitudinal validation study. *J Clin Periodontol* 2014;**41**:52–9.
43. Carver CS, Scheier MF. Scaling back goals and recalibration of the affect system are processes in normal adaptive self-regulation: understanding "response shift" phenomena. *Soc Sci Med* 2000;**50**:1715–22.
44. Hill LG, Betz DL. Revisiting the retrospective pretest. *Am J Eval* 2005;**26**:501–17.
45. Wagner JA. Response shift and glycemic control in children with diabetes. *Health Qual Life Outcomes* 2005;**3**:38.
46. Mayo NE, Scott SC, Dendukuri N, Ahmed S, Wood-Dauphinee S. Identifying response shift statistically at the individual level. *Qual Life Res* 2008;**17**:627–39.
47. Schwartz CE, Sprangers MAG. Reflections on genes and sustainable change: toward a trait and state conceptualization of response shift. *J Clin Epidemiol* 2009;**62**:1118–23.
48. Schwartz CE, Rapkin BD. Reconsidering the psychometrics of quality of life assessment in light of response shift and appraisal. *Health Qual Life Outcomes* 2004;**2**:16.
49. Bitzer EM, Petrucci M, Lorenz C, Hussein R, Doerning H, Trojan A, et al. A comparison of conventional and retrospective measures of change in symptoms after elective surgery. *Health Qual Life Outcomes* 2011;**9**:1–9.

50. Locker D, Slade G. Association between clinical and subjective indicators of oral health status in an older adult population. *Gerodontology* 1994;**11**:108–14.
51. Baker SR, Mat A, Robinson PG. What psychosocial factors influence adolescents' oral health? *J Dent Res* 2010;**89**:1230–5.
52. Baker SR, Pankhurst CL, Robinson PG. Testing relationships between clinical and non-clinical variables in xerostomia: a structural equation model of oral health-related quality of life. *Qual Life Res* 2007;**16**:297–308.
53. Andrykowski MA, Donovan KA, Jacobsen PB. Magnitude and correlates of response shift in fatigue ratings in women undergoing adjuvant therapy for breast cancer. *J Pain Symptom Manage* 2009;**37**:341–51.
54. Chin K, Fukuhara S, Takahashi K, Sumi K, Nakamura T, Matsumoto H, et al. Response shift in perception of sleepiness in obstructive sleep apnea–hypopnea syndrome before and after treatment with nasal CPAP. *Sleep* 2004;**27**:490–3.
55. Balain B, Ennis O, Kanes G, Singhal R, Roberts SNJ, Rees D, et al. Response shift in self-reported functional scores after knee microfracture for full thickness cartilage lesions. *Osteoarthritis Cartilage* 2009;**17**:1009–13.
56. Blair H, Wilson L, Gouick J, Gentleman D. Individualized vs. global assessments of quality of life after head injury and their susceptibility to response shift. *Brain Inj* 2010;**24**:833–43.
57. Korfage IJ, de Koning HJ, Essink-Bot M-L. Response shift due to diagnosis and primary treatment of localized prostate cancer: a then-test and a vignette study. *Qual Life Res* 2007;**16**:1627–34.
58. Hoefer S, Pfaffenberger N, Renn D, Platter M, Ring L. Coronary intervention improves disease specific health-related quality of life but not individualised quality of life: a potential response shift effect? *Appl Res Qual Life* 2011;**6**:81–90.
59. Gillison F, Skevington S, Standage M. Exploring response shift in the quality of life of healthy adolescents over 1 year. *Qual Life Res* 2008;**17**:997–1008.
60. Ahmed S, Mayo NE, Wood-Dauphinee S, Hanley JA, Cohen SR. The structural equation modeling technique did not show a response shift, contrary to the results of the then test and the individualized approaches. *J Clin Epidemiol* 2005;**58**:1125–33.
61. Schwartz CE, Sprangers MAG. Guidelines for improving the stringency of response shift research using the thentest. *Qual Life Res* 2010;**19**:455–64.
62. Schmitt N, Pulakos ED, Lieblein A. Comparison of 3 techniques to assess group-level beta-change and gamma-change. *Appl Psychol Meas* 1984;**8**:249–60.
63. Visser MRM, Oort FJ, Sprangers MAG. Methods to detect response shift in quality of life data: a convergent validity study. *Qual Life Res* 2005;**14**:629–39.
64. Ritchie J, Lewis J, editors. *Qualitative research practice: a guide for social science students and researchers*. London, Thousand Oaks, New Delhi: SAGE Publications Ltd.; 2003.
65. Corbin J, Strauss A. *Basics of qualitative research: techniques and procedures for developing grounded theory*. 3rd ed. London, Thousand Oaks, New Delhi: SAGE Publications Inc.; 2008.
66. Barclay-Goddard R, King J, Dubouloz C-J, Schwartz CE, Response Shift Think Tank Working Group. Building on transformative learning and response shift theory to investigate health-related quality of life changes over time in individuals with chronic health conditions and disability. *Arch Phys Med Rehabil* 2012;**93**:214–20.
67. Cordova JV. Acceptance in behavior therapy: understanding the process of change. *Behav Anal* 2001;**24**:213–26.

68. Hayes SC, Luoma JB, Bond FW, Masuda A, Lillis J. Acceptance and commitment therapy: model, processes and outcomes. *Behav Res Ther* 2006;**44**:1–25.
69. Wood W, Neal DT. A new look at habits and the habit-goal interface. *Psychol Rev* 2007;**114**:843–63.
70. Graybiel AM. Habits, rituals, and the evaluative brain. *Annu Rev Neurosci* 2008;**31**:359–87 Palo Alto: Annual Reviews.
71. Ahmed S, Mayo NE, Wood-Dauphinee S, Hanley JA, Cohen SR. Response shift influenced estimates of change in health-related quality of life poststroke. *J Clin Epidemiol* 2004;**57**:561–70.

Development of condition-specific scales for reporting the pain of dentine hypersensitivity

13

Lisa J. Heaton[1], *Ashley P.S. Barlow*[2] *and Susan E. Coldwell*[1]
[1]Department of Oral Health Sciences, University of Washington, Seattle, WA, [2]GlaxoSmithKline Consumer Healthcare, Weybridge, UK

Introduction

This chapter describes the development of a set of labeled magnitude (LM) scales to measure the pain of dentine hypersensitivity (DH). The aim of this scale development was to create scales that were sufficiently sensitive to quantitatively discriminate between different levels of pain associated with DH. Limitations in the use of more common measures, such as the visual analog scale (VAS), are described in the measurement of DH pain. The qualitative and quantitative means of developing these scales are described, including focus groups, magnitude estimation, and follow-up interviews. This chapter also describes the use of these scales in clinical research and potential future directions for the scales. A full description of this work is provided in the work by Heaton et al.[1]

Measurement of subjective pain

Assessing pain is a complex task, largely because of its subjective multidimensional nature.[2,3] The most commonly used scales involve participants rating pain intensity using verbal descriptors (category scales) or whole numbers (e.g., 0–10), or by marking a point on a line (VAS). Many researchers adopt category and number scales because they are easy to use, intuitive, and require minimal instructions. Such scales offer a limited number of descriptive choices, sometimes forcing individuals to rate different pain intensities under the same category because the choices available do not correlate with the individual's self-perceived assessment of the pain experience. Category scales bring the added problem that there may be uneven spacings in magnitude between words, such that the perceived intensity of a shift from, for example, "weak" to "mild" pain is less than from "strong" to "intense" pain.[4]

Another commonly used measure is the VAS, which typically has two verbal descriptors labeling each end of a 100-mm linear scale that spans the entire range of

Dentine Hypersensitivity. DOI: http://dx.doi.org/10.1016/B978-0-12-801631-2.00013-0

pain experience from, for example, "no pain" to "worst pain imaginable." Participants are free to mark anywhere along the line. However, individuals do not always use VAS scales consistently because they lack labels along the line to guide responses. Furthermore, because the scales span the entire range of pain experience from none to extremely high, the low end of the pain range is compressed to approximately the bottom one-third of the scale in an effort to capture the full range of pain intensity on the line. Therefore, individuals in clinical studies of low to moderate intensity pain conditions such as DH are likely to use only the lower one-third of the entire scale, leading to the possibility of floor effects and low discrimination in these clinical studies.

In an attempt to combine the benefits of both category scales and VAS, researchers began to combine verbal descriptors and numerical measurements into category ratio scales of pain. Heft and Parker[4] suggested combining category ratio scales with line scales by using labels placed at distances "reflecting the spacings between words as the subjects perceive them." Their graphic rating scale of pain intensity has six descriptors ranging from "faint" to "intense" aligned along a horizontal line at intervals determined by magnitude estimation of perceived differences between the words. Green subsequently developed a similar vertical scale with six descriptors ranging from "barely detectable" to "strongest imaginable" aligned at distances determined by magnitude estimation.[5]

Our group developed four different category ratio scales, also known as LM scales, specifically targeted at measuring the pain of DH. We began this process by conducting interviews and focus groups with people experiencing DH to identify the terms they themselves use to describe the pain of DH. Magnitude estimation studies were next conducted using people with DH and pain-free volunteers to determine relative magnitude of each term and, thus, its appropriate position along the new labeled scales. Both experimental and paper-and-pencil tests were used to confirm the sensitivity of the scales to low and moderate pain. The steps comprising this work are summarized here and are described more completely in the work by Heaton et al.[1]

Development of scales to assess DH pain

All the protocols describing the research activities were reviewed by the Institutional Review Board at the University of Washington. All participants signed informed consent after a detailed discussion of the study objectives, requirements, and risks/benefits with a member of the research team.

Water stimulation with VAS and focus groups

The initial stage of LM scale development involved establishing a collection of words that individuals with DH use to describe the pain of DH. Twenty-six individuals recruited from flyers advertising for people with DH were consented and then screened at University of Washington dental clinics to confirm their diagnosis. After completion of a water stimulation task, they participated in focus groups. In the water stimulation task, a dental assistant applied cold (4°C) and room temperature (25°C) water to up to three teeth diagnosed with DH and stimulated one at a time with 0.5 ml of water. The assistant applied the water of one temperature to a particular tooth, waited for 10 min, and then applied the water of the second temperature to the same tooth. After the water stimulation, subjects participated in focus groups facilitated by a trained moderator. Participants were asked to provide terms to describe the pain of DH at both its worst and lowest levels of pain. The focus group discussions were recorded and transcribed verbatim. Transcripts were reviewed to identify consistent themes with regard to descriptive terms associated with DH.

Four grouped themes were common among the words used to describe the pain of DH. *Intensity* included such terms as "sharp," "faint," and "stabbing." *Duration* terms included "temporary," "lingering," and "sudden." *Tolerability*, the theme describing how well participants were able to endure the pain of DH, included terms such as "annoying," "uncomfortable," and "unbearable." *Description* terms included those that encompassed both intensity and duration. For example, "twinge" suggested experiencing a low level of pain for a short time, whereas "ache" was suggestive of a higher level of pain for a longer period of time.

Magnitude estimation task

The second stage in the scale development process was to determine where the words from the focus groups should be located on the scales. The objective was to assign a magnitude to each term relative to the others. For this part of the scale development, 20 individuals with DH from the earlier phase of the study and 24 healthy individuals without pain completed a magnitude estimation task with the pain terms obtained from the previous focus groups.

In the magnitude estimation task, individuals were instructed to assign numeric values to words based on their magnitude or perceived intensity. To assist with this, one word in each scale was selected as the modulus and assigned a value of 12.[6] Providing a modulus in this way gives participants a benchmark or anchor point against which they could compare the remaining words and give relative ratings.[7] For example, "dull" was selected as the modulus for the intensity scale and was assigned a value of 12. Individuals were instructed to assign further values proportionally such that terms they perceived as three times as strong as the modulus term were assigned the number 36, whereas those half as strong were assigned the value 6, and so forth.

Individuals with and without pain did not differ significantly in their magnitude estimates of the pain terms. In the one case in which the two groups differed on a term ("unbearable"), the mean rating from the DH group was used because the final scales are designed for use by people with DH diagnosed. To produce a standard line version of each scale, the resulting magnitude estimates for the descriptive terms were subsequently multiplied by a constant to produce ratings that spanned a 100-mm scale. The resulting four LM scales with corresponding values for each term in millimetres (mm) are shown in Figure 13.1.

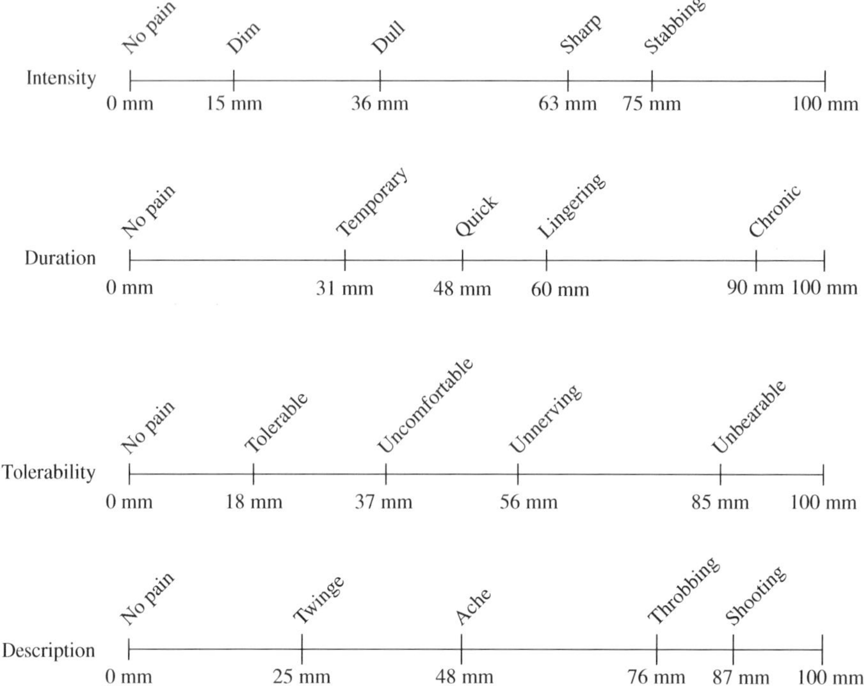

Figure 13.1 LM scales resulting from magnitude estimation task. *Note*: mm markings are for reference only and are not displayed when scales are given to participants. Scales are represented to scale at 100 mm.

Assessing the LM scales using water stimulation

The next stage in the development process was to test how well individuals with DH could distinguish between two water temperatures by using the newly developed LM scales. The use of water at cold and at room temperatures was chosen to represent two stimuli likely to provoke a higher pain response and lower pain response, respectively. Twenty participants from the first phase of the study first completed a four-item written oral pain-related calibration task containing the four LM scales to help standardize interpretation of oral pain (including the touch of a pill on the tongue, canker sore, biting your tongue, and a toothache) and to ensure they understood how to use the scales. After the calibration task, each participant completed a water stimulation task with both cold (4°C) and room temperature (25°C) water as previously described. Participants then completed the four LM scales, rating their sensory response to the water stimulus on their sensitive teeth. We found that participants gave higher ratings for the cold water than for the room temperature water for each of the four scales. For example, the means were 60.2 mm (standard deviation [SD] = 14.1 mm) and 35.3 mm (SD = 20.2 mm) for the intensity scale on the most sensitive teeth of the participants.

Scale orientation and rating of non-oral pain scenarios

It was important to determine whether people completed the LM scales differently if the scales were presented horizontally or vertically. We also used this as an opportunity to observe how the scales performed when used to rate non-oral pain. The scales were also compared directly to a VAS.

Forty adults completed paper-and-pencil measures, rating the pain associated with eight hypothetical pain scenarios on both the LM scales and VAS. Half (20 participants) of the participants completed the LM scales presented vertically ("no pain" at the bottom and the highest pain descriptor near the top), whereas the other half (20 participants) completed the scales horizontally ("no pain" at the left and the highest pain descriptor to the right). Participants also completed VAS measures presented in the horizontal orientation, as it is typically presented in studies. The eight pain scenarios rated included both low-level (paper cut, splinter, sunburn, bruise) and high-level (broken bone, chronic back pain, migraine headache, childbirth) non-oral pain sensations.

Ratings on the LM scales were equivalent whether they were presented horizontally or vertically (e.g., mean intensity scores for a splinter were 42.2 mm [SD = 18.7 mm] and 39.9 mm [SD = 17.7 mm], respectively). Although ratings on the VAS for the low-level pain scenarios were placed within the bottom one-third of the scale on average across all four low-level scenarios, participants gave higher mean pain ratings for three of the four LM scales for the low-level pain scenarios compared with the VAS. Specifically, intensity, duration, and description produced significantly higher ratings than did VAS (e.g. mean intensity scores for a splinter were 41.1 mm [SD = 18.0 mm] for LM and 24.3 mm [SD = 16.1 mm] for VAS), whereas ratings on the tolerability scale did not differ significantly from VAS. When examining the high-level pain scenarios, participants gave significantly higher ratings on the LM duration, tolerability, and description scales compared to the VAS, whereas intensity did not differ significantly from VAS.

Comparison of VAS and LM scales using a water stimulation task

Finally, to directly compare the LM scales and VAS regarding measurement of DH pain, we asked participants with DH to complete both types of scales after cold and room temperature water stimulation of their sensitive teeth. Twenty-eight participants completed the water stimulation task as previously described and completed pain assessments using the four LM scales and one VAS. Order of scale presentation and whether LM scales or VAS were completed first were counterbalanced across participants. Likewise, the order of presentation of water temperatures was randomized and counterbalanced for each participant.

Participants consistently rated the pain associated with placement of the cold water significantly higher (i.e., more painful) than the room temperature water on each of the four LM scales as well as on the VAS. The description LM scale showed the greatest difference between the temperatures, and the mean of the description difference score was significantly larger than that of the mean VAS difference score (32.3 mm [SD = 24.6 mm] and 24.3 mm [SD = 22.7 mm], respectively).

Preference interviews

After completing the water stimulation, 23 of the participants in the water stimulation task completed individual interviews about their experiences using both the LM scales and VAS. Of the 23 interview participants, 3 did not express a preference for one scale over the other, 7 expressed a clear preference for the VAS, whereas 13 preferred the LM scales. Participants who preferred the VAS noted that it was more familiar to them and that the words presented on the LM scales were sometimes confusing. These individuals were more likely to describe the LM scales as "categorical" and noted that the categories did not always fit their pain experiences. Participants with a preference for the LM scales reported that the descriptors allowed them to be more precise in their measurements, and that they found the VAS to be vague and open-ended.

Application of the LM scales in clinical research

Since development and validation of the LM scales, they have been used in both observational and interventional clinical research. The LM scales were used to assess the frequency and severity with which DH presents in private dental practice. Cuhna-Cruz et al.[8] surveyed 37 dental practices in a dental practice-based research network for the prevalence of DH in their patients. They found that 12% of patients in these practices experienced DH. After identifying teeth with DH in their patients, dentists in this study used an air/water syringe to apply a 1-s air puff to the sensitive teeth. Patients then completed the LM scales and VAS. Ratings on three of the four LM scales (intensity, duration, and description) were higher than the VAS ratings and thus involved a greater amount of the 100-mm scales than the VAS. Results from this study supported the initial hypothesis raised during development that individuals use a larger proportion of the LM scales than VAS when rating the pain of DH.

Barlow et al.[9] used both the LM scales and VAS in a randomized, parallel group, controlled, clinical trial to examine changes in DH pain response after use of either an experimental toothpaste with 5% potassium nitrate or a negative control toothpaste. Study subjects completed the LM scales and VAS after an air blast stimulus to their sensitive teeth at baseline, 1 week, and 2 weeks after using either toothpaste. Between-treatment differences were detected using the LM scales; the VAS ratings did not produce any differences between the toothpastes. The authors concluded that the LM scales were better able to detect treatment differences between the desensitizing and control toothpastes than the standard VAS measure.

Conclusions

An appreciable amount of work has gone into the development and assessment of a set of four LM scales specific for the pain associated with DH. Although pain intensity scales such as the standard VAS have proven useful in the measurement of high-intensity pain conditions such as postoperative pain,[10,11] the nature of low-level episodic pain may not be fully captured by these standard measures. The goal of the current set of studies was to develop scales that would encourage participants to use a greater portion of the 100-mm scales, thereby avoiding the floor effects seen when using the VAS with such low-level pain conditions, and to broaden the scope of the pain assessment beyond "intensity" to cover other condition-specific dimensions that characterize the DH pain response.

Both the LM scales and VAS showed good distinction between cold water and room temperature water. In comparing the VAS and LM scales on eight hypothetical pain scenarios, participants utilized a greater portion of the 100-mm measures of the LM scales than the VAS, particularly when rating low-level pain, such as a bruise, paper cut, and splinter. Using the LM scales for low-level pain may thus reduce floor effects seen with the VAS. No significant differences emerged on ratings provided on vertically and horizontally presented LM scales. This suggests that the LM scales may be presented in either orientation. When presented in conjunction with the VAS, however, users may want to present all scales on the horizontal orientation for ease of use by participants.

One limitation in the developmental work was that many of the same participants with DH participated in multiple phases of the study. However, the scales have subsequently been used in clinical studies with new participants,[8,9] with similar findings as with the developmental work but among unique and diverse patient populations.

The LM scales, alone or in combination with the standard VAS, can provide information regarding the intensity, duration, tolerability, and description of the pain associated with this condition, allowing less of an opportunity for the floor effects seen when low-grade pain is measured with the VAS. These scales permit individuals with DH to rate broader and condition-specific aspects of their pain, allowing researchers and clinicians to better understand not only the DH pain condition but also other low-level pain conditions and, ideally, to evaluate the effectiveness of treatment-based management strategies.

References

1. Heaton LJ, Barlow AP, Coldwell SE. Development of labeled magnitude scales for the assessment of pain of dentin hypersensitivity. *J Orofac Pain* 2013;**27**(1):72–81.
2. Beecher HK. The subjective response and reaction to sensation; the reaction phase as the effective site for drug action. *Am J Med* 1956;**20**(1):107–13.
3. Merskey H, Bogduk N, editors. *Classification of chronic pain: description of chronic pain syndromes and definitions of pain terms*. Seattle: IASP Press; 1994.

4. Heft MW, Parker SR. An experimental basis for revising the graphic rating scale for pain. *Pain* 1984;**19**(2):153–61.
5. Green BG, Shaffer GS, Gilmore MM. Derivation and evaluation of a semantic scale of oral sensation magnitude with apparent ratio properties. *Chem Senses* 1993;**18**:683–702.
6. Heaton LJ, Garcia LJ, Gledhill LW, Beesley KA, Coldwell SE. Development and validation of the Spanish Interval Scale of Anxiety Response (ISAR). *Anesth Prog* 2007;**54**(3):100–8.
7. Heft MW, Gracely RH, Dubner R, McGrath PA. A validation model for verbal description scaling of human clinical pain. *Pain* 1980;**9**(3):363–73.
8. Cunha-Cruz J, Wataha JC, Heaton LJ, Rothen M, Sobieraj M, Scott J, et al. Northwest Practice-based Research Collaborative in Evidence-based DENTistry The prevalence of dentin hypersensitivity in general dental practices in the northwest United States. *J Am Dent Assoc* 2013;**144**(3):288–96.
9. Barlow AP, Jeffrey P, Heaton LJ, Amini P, Gallob J, Coldwell SE. Clinical evaluation of dentine hypersensitivity using labelled magnitude scales. *J Dent Res* 2010;**89**(B):2713 <www.dentalresearch.org>.
10. Breivik EK, Skoglund LA. Comparison of present pain intensity assessments on horizontally and vertically oriented visual analogue scales. *Methods Find Exp Clin Pharmacol* 1998;**20**(8):719–24.
11. Gagliese L, Weizblit N, Ellis W, Chan VW. The measurement of postoperative pain: a comparison of intensity scales in younger and older surgical patients. *Pain* 2005;**117**(3):412–20.

The role of illness beliefs and coping in the adjustment to dentine hypersensitivity

14

Jenny M. Porritt[1], *Farzana Sufi*[2], *Ashley P.S. Barlow*[2] *and Sarah R. Baker*[1]
[1]School of Clinical Dentistry, Claremont Crescent, University of Sheffield, Sheffield, UK,
[2]GlaxoSmithKline Consumer Healthcare, Weybridge, UK

Introduction

Dentine hypersensitivity is a common oral health problem affecting adults worldwide with many general dental practice patients reporting sensitivity in their teeth.[1–3] The condition is characterized by a short, sharp pain to an external stimulus (e.g., cold) that cannot be explained by any other dental pathology and is trigged by the exposure of the dentine as a result of gingival recession or enamel erosion.[4] Emerging research has revealed that dentine hypersensitivity can cause pain and functional limitations (e.g., difficulties eating, drinking, and toothbrushing), which can affect the quality of life of those individuals who suffer from the condition.[5] However, there is evidence to suggest that there is a huge variation in how people respond to the condition.[5]

The way in which psychological factors contribute to an individual's experience of dentine hypersensitivity could highlight possible areas for intervention.[5] The self-regulation model (SRM) of illness developed by Leventhal and colleagues proposes that individuals are active problem solvers and suggests that central cognitive constructs called illness beliefs guide coping in response to a health threat.[6–8] Illness beliefs are a set of thoughts a person has developed about a specific health problem.[9] The self-regulation framework is based on a parallel processing framework in which illness beliefs comprise both cognitive representations and emotional aspects of the illness that are processed simultaneously in the response to an internal and/or external stimuli.[10] The original SRM, developed in the 1980s, hypothesized that illness beliefs comprised five dimensions: identity; timeline; cause; consequence; and control/cure. Dimensions such as personal responsibility, disruptiveness of the condition, illness coherence, and emotional impacts of the health condition have since been proposed.[11,12]

Previous research has provided support for the SRM through the discovery that illness beliefs are strongly associated with health-related outcomes for a variety of different conditions such as heart disease, asthma, and cancer.[13–15] Perceptions that an illness is curable/controllable are significantly related to more positive well-being, social functioning, and vitality, whereas perceptions that a health condition

Dentine Hypersensitivity. DOI: http://dx.doi.org/10.1016/B978-0-12-801631-2.00014-2

has severe consequences and a chronic timeline are often associated with worse psychological well-being and social functioning.[16] However, it is important to recognize that the way illness beliefs influence oral health-related outcomes may be different than the role they play in people's responses to life-threatening conditions or more debilitating illnesses. McCarthy et al.,[17] however, found that illness beliefs might play a key role in how individuals adjust after oral surgery (third molar extraction). Shorter timeline expectations of healing predicted a speedier return to work, and preoperative expectations that recovery could be controlled positively predicted the patient's quality of healing. Interestingly, medical factors were not significant predictors of any of the oral health outcomes assessed. Further evidence of the importance of illness beliefs in predicting oral health outcomes of patients with a range of conditions (e.g., temporomandibular disorder (TMD), burning mouth syndrome) was generated from the results of a longitudinal study by Galli et al.[18]

The SRM postulates that coping may mediate the relationship between illness beliefs and health outcomes.[16] A meta-analytic review revealed that those individuals who perceived they had control over their health condition used more problem-focused coping strategies than those who perceived low controllability over their illness.[16] In contrast, beliefs that an illness is highly symptomatic, has a chronic timeline, and has severe consequences are all associated with the use of avoidance coping strategies.[16] However, there is also some evidence that coping strategies and illness beliefs may be distinct mechanisms that exert their influence on health outcomes independently.[19] Health anxiety has also been associated with the failure to engage in protective coping strategies, and it is argued that the avoidance behaviors or "passive" strategies (e.g., reassurance-seeking, self-monitoring of symptoms) that individuals with health anxiety tend to use often prolong pain and exacerbate anxiety over the long term.[20,21]

There is limited research on the relationship between coping strategies and oral health outcomes; however, a study conducted by Turner et al.[22] supports the detrimental role that passive coping may have on daily functioning outcomes in TMD patients. It should, however be recognized that although frequent use of a specific coping strategy may be useful for one group of patients, the frequent use of that same strategy may be counterproductive for individuals experiencing a different health condition.[23]

To date, no research has examined the relationship between health anxiety, illness beliefs, pain-related coping strategies, and health outcomes in individuals with dentine hypersensitivity. Therefore, the current study aimed to determine the clinical and psychological factors that influence how people cope and adjust to dentine hypersensitivity.

Method

Participants

This study formed part of a larger piece of research that used daily diaries to investigate the day-to-day impacts of dentine hypersensitivity. Previous studies that have used daily health diaries have typically had relatively small sample sizes (<100) because of the practical considerations associated with this method of data collection.[24,25] Therefore, this study adopted a similar approach and aimed to recruit approximately 100 participants. Participants were sampled purposively from staff and students at a large university in the United Kingdom. A screening questionnaire was administered to 280 individuals who expressed an interest in the study and 184 individuals completed this questionnaire. Only those adults who met the strict inclusion criteria were able to participate ($N = 101$). This included people who had attended routine dental appointments and had experienced dentine hypersensitivity on a frequent basis (minimum of several sensations per week). Individuals who had received periodontal surgery within the past 6 months (or had teeth scaled/root-planed within the past 3 months) and those experiencing dry mouth or serious/painful health conditions were excluded.

Design

On obtaining informed written consent, participants were directed to the online survey, which comprised five questionnaires (assessing illness beliefs, health anxiety, coping strategies, OHRQoL and HRQoL) and collected information on clinical (e.g., duration and frequency of condition and number of sites affected) and demographic data (e.g., gender, age). One month after baseline, participants completed the online questionnaires again, primarily to collect information on follow-up health outcomes (OHRQoL and HRQoL). Participants were paid a small financial incentive for completing the baseline and follow-up measures. Ethical approval for the study was obtained from the University of Sheffield Research Ethics Committee.

Measures

Clinical variables

The length of time the individuals had experienced DH and the frequency of their pain sensations were assessed using two items, "How long have you been experiencing any sensations in your teeth?" and "How often do you have any sensations in your teeth?", from DHEQ.[26] Responses were made on a 6-point scale (0 = none to 5 = more than 20 years) and an 8-point scale (0 = never to 7 = several times per day), respectively.

Pain-related coping strategies

The Vanderbilt Pain Management Inventory (VPMI) was used to assess how individuals respond to pain sensations caused by their dentine hypersensitivity.[27] The questionnaire consists of 18 items and participants rated the frequency with which they used active and passive coping strategies when their pain reached a moderate or greater level of intensity on a 5-point scale (1 = never do when in pain to 5 = very frequently do when in pain). The active coping scale measures use of strategies that involve efforts of the patients to function despite the pain or that actively distract themselves from their pain. The passive coping scale assesses tendencies to restrict activities because of the pain and/or dependence on others for help with pain control.

Health anxiety

Health anxiety was measured using the Short Health Anxiety Inventory (SHAI).[28] The questionnaire is composed of 14 questions that consist of a group of four statements: "I do not worry about my health;" "I occasionally worry about my health;" "I spend much of my time worrying about my health;" and "I spend most of my time worrying about my health." Responses are ranked on a 4-point scale (0–3), with total scores ranging from 0 to 42 and higher scores reflecting higher levels of health anxiety.

Illness beliefs

Illness beliefs were assessed using 38 items from the Illness Perception Questionnaire-Revised (IPQ-R)[29] and the following dimensions were measured: consequences; acute/chronic and cyclical timelines; emotional representations; illness coherence; personal control; and treatment control (sensitivity toothpaste). An example item from the measure was "My symptoms come and go in cycles" (cyclical timeline). Responses were on a 5-point Likert scale ranging from 1 (strongly disagree) to 5 (strongly agree). High scores reflect increased perceived consequences, increased control, chronicity of condition, increased illness coherence, increased emotional impacts, and the cyclical nature of the condition.

OHRQoL

OHRQoL for dentine hypersensitivity was assessed using the Revised DHEQ.[26] This measure comprises 34 items and five subscales (identity, restrictions, coping adaptations, social impacts, and emotional impacts). Responses use a 7-point Likert scale (1 = strongly disagree to 7 = strongly agree) and total scores were calculated for subscales and the overall OHRQoL. Higher scores reflect an increased number of negative impacts.

HRQoL

HRQoL was measured by the EuroQoL-5D measure (EQ-5D).[30] The measure composes five dimensions (mobility, self-care, usual activities, pain/discomfort, and anxiety/depression), and each dimension has three levels (e.g., 1 = "I have no problems with self-care;" 2 = "I have some problems washing or dressing myself;" and 3 = "I am unable to wash or dress myself"). EQ-5D health states were converted into a single index by applying a formula that attaches values (weights) to each level of the dimensions. The value sets used in this study were those derived for EQ-5D in the United Kingdom using the EQ-5D visual analogue scale.

Analysis

Structural equation modeling (SEM) using AMOS 18.0 was used to test the proposed model (Figure 14.1). The path analysis technique used measures to the

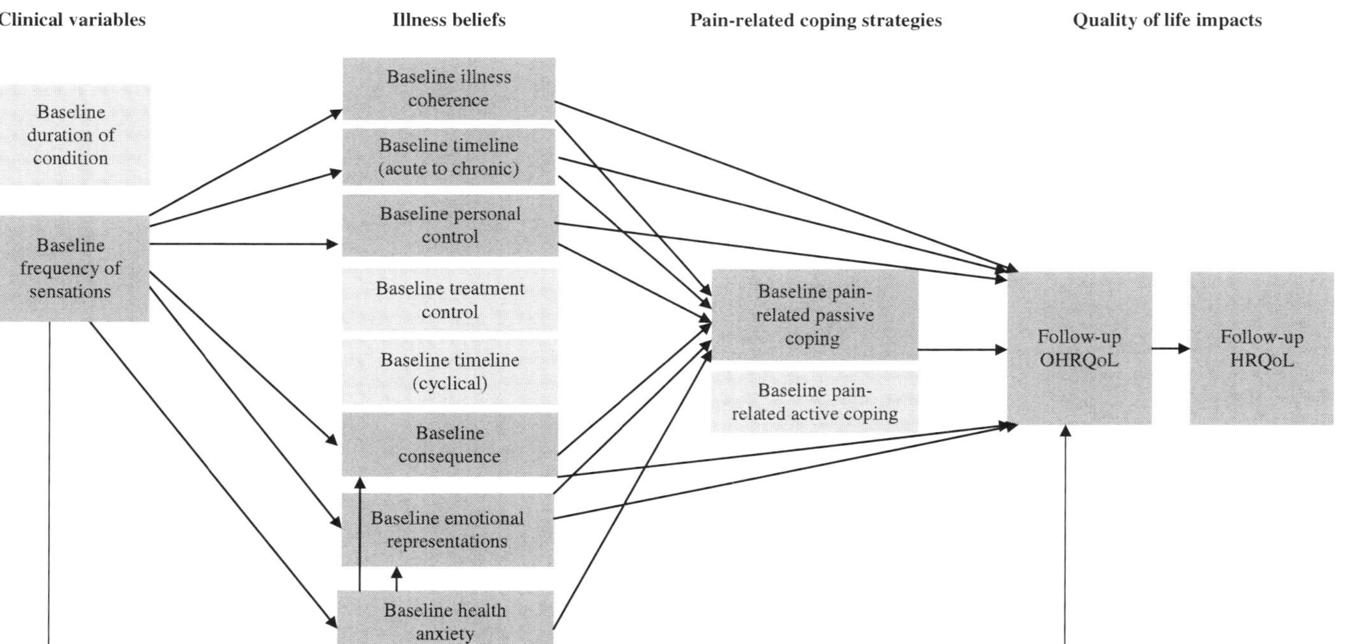

Figure 14.1 Direct pathways hypothesized between clinical variables, illness beliefs, pain-related coping and follow-up quality of life impacts experienced by adults with dentine hypersensitivity tested within model 1.

Note: Variables in pale grey not entered into final model because these were nonsignificant predictors of the primary outcome variable (follow-up OHRQoL).

extent that the model fit a data set and allowed testing of interrelationships between a range of variables simultaneously. A bootstrapping technique was conducted using the data because this procedure has been advocated as the best approach when sample sizes are small to medium (<200).[31] In addition, the bias-corrected 95% confidence interval (CI) bootstrap percentiles were used because these have been shown to be more accurate when dealing with smaller sample sizes and mediation effects.[31,32] A preselection criterion was used for the path analysis and only baseline predictors of follow-up OHRQoL (DHEQ) that had $P < 0.20$ were entered into the model (based on Spearman and Pearson correlations). Maximum likelihood was used and adequacy of overall model fit was assessed using five fit indices including the following: chi-square test statistic, which should not be significantly different from the observed data; chi-square divided by degrees of freedom (CMIN/df), which should be lower than 2.0; root mean-squared error of approximation (RMSEA), which should be less than 0.08; incremental fit index (IFI), which should be more than 0.95; and standardized root mean square residual (SRMR), which should be less than 0.08.[33–35] The error variances between illness beliefs were allowed to correlate freely. Missing data were replaced by the item's median score to generate total scores. However, if more than 50% of the values for any given questionnaire were missing, then total scores were not calculated. Within the SEM analysis, the regression imputation technique handled this missing data.

Results

Participants

One hundred and one respondents fulfilled all of the study's inclusion criteria. The majority of participants were female ($N = 69$), and their ages ranged between 18 and 63 years (mean age: 26.3 years; standard deviation (SD) = 8.6). Thirty-six participants were single or widowed, 20 were married, 29 were in a relationship, and 16 were cohabiting. Of the participants who provided information regarding their ethnic background, 92 were white British/other, two were Indian, three were from multiple heritages, and one had an Asian background. All 101 participants completed the baseline and follow-up questionnaire (100% response rate). However, at follow-up, two participants were excluded from the data analysis because of missing data.

An SRM of dentine hypersensitivity

A summary of clinical variables appears in Table 14.1. Means and SD for health anxiety, illness beliefs, and coping are outlined in Table 14.2, and bivariate associations between clinical variables, illness beliefs, coping, and health-related outcomes are presented in Table 14.3. On the basis of the preliminary analysis, active pain-related coping strategies, duration of condition, beliefs associated with illness

Table 14.1 Summary of 101 participants' clinical information relating to their dentine hypersensitivity at baseline and follow-up

Variable	Number of people at baseline	Number of people at follow-up
Duration of condition		
< 6 months	5	–
6 months–1 year	9	–
1–5 years	37	–
5–20 years	41	–
> 20 years	9	–
Frequency of condition		
Once a month	1	0
Several times a month	5	4
Once a week	6	1
Several times a week	33	29
Once a day	14	18
Several times a day	42	49
Cause of sensations		
Cold fluids	94	95
Ice cream	91	86
Cold foods	74	88
Cold air	66	61
Sweet things	47	47
Having teeth cleaned at the dentist	44	43
Hot fluids	39	42
Metals touching my teeth	34	32
Hot foods	28	27
Toothbrushing	33	38
Acidy fruits	21	23
Sticky foods	6	14
Tooth-whitening products	6	6
Salty foods	0	0
Duration of sensations		
A few seconds	62	59
About a minute	23	24
Several minutes	13	15
About half an hour	1	2
Longer than half an hour	2	1

(*Continued*)

Table 14.1 **(Continued)**

Variable	Number of people at baseline	Number of people at follow-up
Areas of mouth affected		
Left side	57	57
Right side	52	62
Top front	62	69
Top back	37	45
Bottom front	51	57
Bottom back	44	43

timeline (cyclical), and treatment control were not entered into the SEM because these variables did not meet the preselection criteria used in this study ($P < 0.20$; data not shown). A total of 22 direct pathways were hypothesized within the model based on the SRM of Leventhal et al.[10] and the findings from previous research (Figure 14.1).

Model fit acceptability

The SEM analysis revealed that the model did not differ significantly from the observed data ($\chi^2 = 29.11$, df = 20, $P = 0.09$) and that the goodness of fit indices for this model were acceptable to good (CMIN/df = 1.46; IFI = 0.97; RMSEA = 0.07; SRMR = 0.07). The variables included within the model accounted for 44% of the variance of passive coping, 54% of OHRQoL, and 14% of HRQoL. The total significant effects, when combining direct and indirect pathways, are shown in Figure 14.2. Eight direct and 13 indirect significant pathways existed within the model (see Table 14.4).

Direct pathways

Eight direct significant pathways existed within the model (see Table 14.4). Individuals with more frequent sensations experienced higher levels of health anxiety, perceived their condition as more chronic, and felt they had less control over their condition than people with less frequent sensations. Those individuals with high levels of health anxiety were also more likely to use passive coping strategies and have negative emotional representations of their condition. Individuals who had negative emotional representations of the condition were found to use more passive coping strategies. Passive coping strategies, in turn, were associated with worse OHRQoL outcomes at follow-up. Worse OHRQoL was associated with worse perceived health state.

Table 14.2 **Participants baseline and follow-up scores for illness beliefs, health anxiety, pain-related coping strategies and quality of life outcomes**

	Number of items	Possible range	Baseline		Follow-up	
			Number	Mean (SD)	Number	Mean (SD)
OHRQoL total score[a]	34	34–238	101	134.4 (27.1)	99	127.5 (29.5)
Restriction[a]	4	4–28	101	19.4 (4.1)	99	18.5 (4.1)
Coping adaptation[a]	12	12–84	101	51.1 (11.4)	99	51.1 (12.4)
Social impact[a]	5	5–35	101	14.0 (5.8)	99	13.7 (5.7)
Emotional impact[a]	8	8–56	101	35.7 (8.1)	99	30.6 (8.6)
Identity[a]	5	5–35	101	14.1 (6.3)	99	13.6 (6.7)
HRQoL utility index	5 items (combined into index score)	0–1	101	0.9 (0.1)	99	0.9 (0.1)
Coping styles						
Passive[b]	11	11–55	101	21.0 (5.2)	99	21.0 (5.5)
Active[b]	7	7–35	101	20.9 (5.3)	99	21.0 (6.0)
Illness beliefs						
Time: cyclical[a]	4	4–20	101	11.9 (3.6)	100	12.3 (3.4)
Timeline: chronic[a]	6	6–30	101	22.7 (3.8)	100	22.9 (4.2)
Consequences[a]	6	6–30	101	11.3 (3.5)	100	11.3 (3.6)
Emotional aspects[a]	6	6–30	101	12.9 (4.4)	100	13.1 (4.8)
Personal control[c]	6	6–30	101	20.7 (3.8)	100	20.2 (4.1)
Treatment control[c]	5	5–25	101	15.6 (4.8)	100	14.8 (4.5)
Illness coherence[c]	5	5–25	101	15.6 (4.1)	100	16.5 (4.6)
Health anxiety	14	0–42	101	12.2 (4.9)	100	12.0 (5.8)

[a]Higher scores reflect increased negative consequences and cyclical/chronic timeline.
[b]Higher scores reflect increased use of coping style.
[c]Higher scores reflect increased control and coherence.

Table 14.3 Correlation coefficients between variables at baseline

Baseline variables	Duration of condition	Treatment control	Personal control	Time: cyclical	Time: acute/ chronic	Illness coherence	Emotional representations	Consequence	Health anxiety	Active coping	Passive coping	OHRQoL	HRQoL
Duration of condition	1												
Treatment control	−0.21*	1											
Personal control	−0.16	0.40**	1										
Time: cyclical	−0.41**	0.23*	0.13	1									
Time: acute/ chronic	0.51**	−0.25*	−0.31**	−0.35**	1								
Illness coherence	0.10	0.10	0.037**	−0.18	0.07	1							
Emotional representation	−0.07	0.00	−0.16	0.16	0.19	−0.18	1						
Consequence	−0.00	−0.07	−0.20	0.10	0.19	−0.17	0.78**	1					
Health anxiety	0.11	0.06	−0.08	0.06	0.24*	−0.04	0.38**	0.30**	1				
Active coping	−0.03	−0.06	0.20*	0.06	−0.04	0.15	−0.04	−0.04	0.09	1			
Passive coping	−0.08	0.11	−0.12	0.22*	0.05	−0.16	0.60**	0.54**	0.42**	0.07	1		
OHRQoL	0.12	−0.08	−0.17	0.00	0.28**	−0.29**	0.60**	0.58**	0.39**	0.10	0.55**	1	
HRQoL	−0.05	0.10	0.27**	0.04	−0.10	0.14	−0.25**	−0.18	−0.19	0.04	−0.27**	−0.32**	1

*$p < 0.05$ **$p < 0.01$ (2-tailed). $N = 101$.
Note: Spearman correlation coefficients calculated for categorical data.

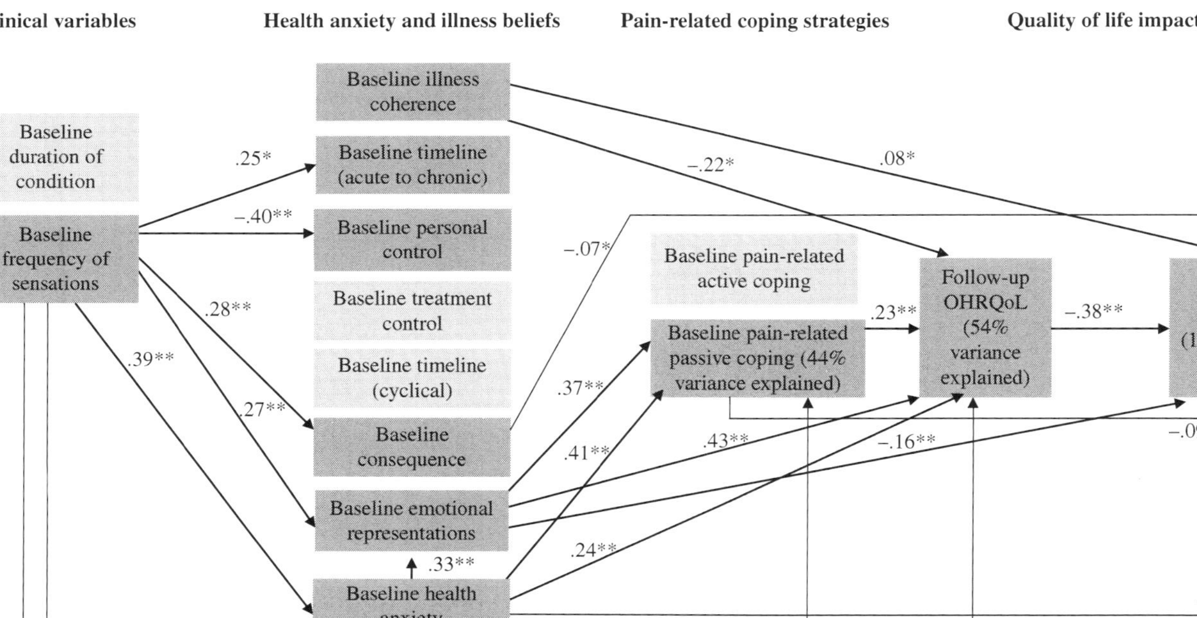

Figure 14.2 Significant total effects between baseline clinical variables, health anxiety, illness beliefs, pain-related coping and follow-up health-related outcomes proposed within model 1.

→ Significant "total" standardized pathways $^{*}p < 0.05$, $^{**}p < 0.01$.

Note: Variables in pale grey boxes not entered into final model because these were not significantly related to the primary outcome variable. Error terms not shown for simplicity.

Table 14.4 **Significant direct, indirect and total pathways between baseline variables and follow-up health-related outcomes proposed within the SEM**

Pathways	Direct pathways		Indirect pathways		Total pathways	
	β value	Bootstrap bias-corrected 95% CI	β value	Bootstrap bias-corrected 95% CI	β value	Bootstrap bias-corrected 95% CI
Clinical variables						
Frequency of sensations→health anxiety	0.39**	0.22 to 0.53	–	–	0.39**	0.22 to 0.53
Frequency of sensations→timeline (acute to chronic)	0.25*	0.05 to 0.42	–	–	0.25*	0.05 to 0.42
Frequency of sensations→consequence	0.20	−0.01 to 0.37	0.09*	0.01 to 0.19	0.28**	0.10 to 0.43
Frequency of sensations→illness coherence	−0.15	−0.34 to 0.06	–	–	−0.15	−0.34 to 0.06
Frequency of sensations→personal control	−0.40**	−0.56 to −0.19	–	–	−0.40**	−0.56 to −0.19
Frequency of sensations→emotional representations	0.14	−0.03 to 0.35	0.13**	0.05 to 0.22	0.27**	0.09 to 0.44
Frequency of sensations→passive coping	–	–	0.23**	0.09 to 0.36	0.23**	0.09 to 0.36
Frequency of sensations→follow-up OHRQoL	0.15	−0.06 to 0.32	0.20*	0.06 to 0.36	0.35**	0.13 to 0.52
Frequency of sensations→follow-up HRQoL	–	–	−0.13**	−0.26 to −0.04	−0.13**	−0.26 to −0.04

Illness beliefs						
Health anxiety→consequence	0.22	−0.02 to 0.42	–	–	0.22	−0.02 to 0.42
Health anxiety→emotional representations	0.33**	0.10 to 0.51	–	–	0.33**	0.10 to 0.51
Health anxiety→passive coping	0.25**	0.10 to 0.41	0.16*	0.05 to 0.29	0.41**	0.25 to 0.56
Health anxiety→follow-up OHRQoL	–	–	0.22**	0.09 to 0.38	0.24**	0.10 to 0.39
Health anxiety→follow-up HRQoL	–	–	−0.09**	−0.19 to −0.03	−0.09**	−0.19 to −0.03
Timeline (acute to chronic)→ passive coping	−0.12	−0.30 to 0.06	–	–	−0.12	−0.30 to 0.06
Timeline (acute to chronic)→ follow-up OHRQoL	0.04	−0.14 to 0.22	−0.03	−0.10 to 0.01	0.01	−0.18 to 0.18
Timeline (acute to chronic)→ follow-up HRQoL	–	–	−0.01	−0.06 to 0.07	−0.01	−0.06 to 0.07
Consequence→passive coping	0.18	−0.08 to 0.42	–	–	0.18	−0.08 to 0.42
Consequence→follow-up OHRQoL	0.15	−0.05 to 0.39	0.04	−0.01 to 0.16	0.20	−0.02 to 0.43
Consequence→follow-up HRQoL	–	–	−0.07*	−0.18 to −0.00	−0.07*	−0.18 to −0.00
Illness coherence→passive coping	−0.04	−0.25 to 0.13	–	–	−0.04	−0.25 to 0.13
Illness coherence→follow-up OHRQoL	−0.21	−0.43 to 0.00	−0.00	−0.06 to 0.03	−0.22*	−0.46 to −0.01
Illness coherence→follow-up HRQoL	–	–	0.08*	0.01 to 0.17	0.08*	0.01 to 0.17
Personal control→passive coping	−0.03	−0.20 to 0.18	–	–	−0.03	−0.20 to 0.18
Personal control→follow-up OHRQoL	0.08	−0.12 to 0.30	−0.00	−0.06 to 0.04	0.07	−0.13 to 0.28
Personal control→follow-up HRQoL	–	–	−0.03	−0.10 to 0.05	−0.02	−0.10 to 0.05

(*Continued*)

Table 14.4 **(Continued)**

Pathways	Direct pathways		Indirect pathways		Total pathways	
	β value	Bootstrap bias-corrected 95% CI	β value	Bootstrap bias-corrected 95% CI	β value	Bootstrap bias-corrected 95% CI
Emotional representations→passive coping	0.35**	0.06 to 0.58	–	–	0.37**	0.11 to 0.60
Emotional representations→follow-up OHRQoL	0.27	−0.02 to 0.52	0.09**	0.02 to 0.20	0.43**	0.16 to 0.65
Emotional representations→follow-up HRQoL	–	–	−0.16**	−0.33 to −0.05	−0.16**	−0.33 to −0.05
Coping						
Passive coping →follow-up OHRQoL	0.23**	0.05 to 0.41	–	–	0.23**	0.05 to 0.41
Passive coping→follow-up HRQoL	–	–	−0.09**	−0.19 to −0.02	−0.09**	−0.19 to −0.02
Oral health and health-related follow-up outcomes						
Follow-up OHRQoL→follow-up HRQoL	−0.38**	−0.55 to −0.20	–	–	−0.38**	−0.55 to −0.20

*$p < 0.05$, **$p < 0.01$.

Indirect pathways

Thirteen indirect significant pathways existed within the model (see Table 14.4). Frequent painful sensations were indirectly associated with negative emotional representations, negative consequences, and the use of passive coping strategies. These indirect effects were mediated mainly through increased levels of health anxiety. Frequent painful sensations were also indirectly related to worse OHRQoL and HRQoL, mainly through increased negative emotional representations. Negative emotional representations were indirectly associated with worse OHRQoL through increased use of passive coping strategies. An indirect relationship between higher levels of health anxiety and the increased use of passive coping strategies and worse OHRQoL was also revealed, and these relationships were mainly mediated by negative emotional representations. Negative emotional representations, low levels of illness coherence, and negative consequences were all associated with worse health status at follow-up through the mediating variable of OHRQoL.

Discussion

This research aimed to determine the clinical and psychological factors that influence how people cope and adjust to dentine hypersensitivity through examining direct and mediation pathways within an SRM of dentine hypersensitivity.

Clinical variables were directly associated with several illness beliefs. Individuals who experienced sensations on a more frequent basis had higher levels of health anxiety, more chronic timeline perceptions, believed there were more negative consequences as a result of their oral condition, had more negative emotional representations of it, and believed they had less personal control over their dentine hypersensitivity than individuals who experienced sensations in their teeth less frequently. Frequency of sensations was also positively associated with the use of passive coping strategies and worse OHRQoL and HRQoL at follow-up. This finding suggests that dentine hypersensitivity does have the potential to negatively influence people's daily lives. However, it is important to highlight that the relationship between frequency of sensations and OHRQoL was mediated through an individual's illness beliefs. This finding supports previous research that found illness beliefs to be strong mediators in the relationship between medical factors and oral health outcomes.[17]

Health anxiety was associated with negative emotional representations related to dentine hypersensitivity; however, surprisingly it was not associated with perceived consequences of the oral condition. Health anxiety and negative emotional representations were both associated with the use of more pain-related passive coping strategies. This supports previous research that has found health anxiety associated with the use of behavioral strategies of reassurance-seeking when confronted with illness information.[21] Personal control was associated with the use of active coping strategies, suggesting that people who felt more in control of their condition used coping strategies that helped distract them from their pain so that they could continue with their daily activities. This finding is consistent with previous research linking higher levels of perceived controllability with active coping styles.[16]

Health anxiety and illness beliefs were significant predictors of health outcomes through a combination of direct and indirect pathways. Individuals with low levels of illness coherence and high levels of health anxiety were more likely to report worse OHRQoL and HRQoL. This suggests that having a clear understanding of dentine hypersensitivity could enhance adjustment to this condition. Perceived negative consequences and negative emotional representations were associated with worse outcomes, thus supporting findings from previous research.[16,18,29] Controllability and timeline perceptions were not associated with individual health outcomes, which is somewhat surprising in light of previous findings that found a positive association between controllability and psychological well-being.[16] However, these findings are consistent with some research conducted within this field and could suggest that the amount of time you believe you will have to live with a condition is not predictive of the impact that condition has on your daily life.[18,36]

Pain-related passive coping was a direct predictor of worse health outcomes and played a mediating role in the relationship between illness representations and outcomes. These findings provide supporting evidence for the maladaptive role of passive coping in people's long-term adjustment to health conditions.[27,28,37] It has been proposed that passive strategies, such as reassurance-seeking, can be counterproductive because individuals can become dependent on this type of coping and can find it difficult to develop their own techniques to effectively manage the negative cognitions that could be contributing to their health problem.[28] However, it is not possible to determine causality between passive coping and health outcomes within the current study. It is plausible, for example, that individuals who experienced higher levels of negative health impacts because of their dentine hypersensitivity may have felt greater need to use passive coping to deal with these impacts.

It has been proposed that sophisticated analytic procedures need to be used in illness and coping research to test the causal processes hypothesized by the theoretical models used.[38] It is now increasingly recognized that only through testing direct and indirect interrelationships will we be able to gain an understanding of the complex mechanisms that underpin the impact of oral stressors in daily life.[39] In the current study, the theoretical underpinnings drove the longitudinal study design and statistical analysis used, strengthening the credibility of the research.

Although the current study took significant steps to understand the psychosocial impact of dentine hypersensitivity, it was not without some limitations. It should be recognized that individuals were recruited form one large university and, although they included individuals from all grades together with postgraduate students, such individuals may not necessarily be representative of the wider general population. Individuals who participated were "self-diagnosed" with dentine hypersensitivity; therefore, it is not possible to determine whether recession/enamel erosion was solely responsible for the sensations reported by participants. For example, it is possible that some sensations could have resulted from other dental conditions, such as caries or trauma. However, the inclusion criteria for the study ensured that only individuals who had attended routine dental appointments could participate, minimizing the levels of untreated dental disease and conditions in the sample.

Future research could use qualitative methods to explore the *development* and *maintenance* of maladaptive illness beliefs and health anxiety in individuals with dentine. People are thought to form illness beliefs through their interactions with significant others, previous social communications, and personal experiences with the condition or similar conditions.[16] Therefore, research could explore the way in which these pathways may be responsible for the individual's account of why he/she does not feel he/she has a good understanding of his/her condition (poor illness coherence). It is proposed that understanding the etiology of illness beliefs and the relationship between illness beliefs and adaptive health-related behaviors is key if interventions aimed at improving person outcomes are to be developed.[13,40,41]

It should be acknowledged that the HRQoL reported by the sample who participated in the study was comparable with the self-reported health of the general population, with people reporting very few general health difficulties (e.g., pain/ discomfort, anxiety, and depression). This suggests that the majority of participants with dentine hypersensitivity do not experience significant impacts on their general health because of this oral condition. However, the relationship between OHRQoL and HRQoL suggests that the impacts caused by this condition do have the potential to have more wide-reaching impacts on an individual's health state.

To conclude, the SRM of health and illness of Leventhal et al.[10] was able to explain just less than half of the variance in passive coping and just more than half of the variance in individual OHRQoL. This provides support for the usefulness of the SRM in aiding our understanding of the factors that influence coping and health outcomes in individuals with dentine hypersensitivity. Therefore, the findings from the present study provide evidence that an individual's understanding of the condition has the potential to influence how that individual manages pain and how much the dentine hypersensitivity impacts on the quality of life.

Acknowledgment

Our gratitude goes out to all the individuals who gave up their time to participate in the present study. This study would not have been possible without the rich information they so generously provided.

References

1. Cummins D. Dentin hypersensitivity: from diagnosis to a breakthrough therapy for everyday sensitivity relief. *J Clin Dent* 2009;**5**:1–9.
2. Gillam DG, Seo HS, Bulman JS, Newman HN. Perceptions of dentine hypersensitivity in a general practice population. *J Oral Rehabil* 1999;**26**:710–4.
3. Irwin CR, Mccusker P. Prevalence of dentine hypersensitivity in a general dental population. *J Ir Dent Assoc* 1997;**43**:7–9.

4. Ayad F, Ayad N, Zhang YP, Devizio W, Cummins D. Comparing the efficacy in redcuing dentin hypersensitivity of a new toothpaste containing 8/0% arginine, calcium carbonate, and 1450 ppm flouride to a commercial sensitive toothpaste containing 2% potassim ion: an eight-week clinical study on Canadian adults. *J Clin Dent* 2009;**9**:10–6.
5. Gibson B, Boiko OV, Baker SR, Robinson PG, Barlow APS, Player T , et al. The everyday impact of dentine sensitivity: personal and functional aspects. *Soc Sci Dent* 2010;**1**:11–21.
6. Leventhal H. Findings and theory in the study of fear communications. *Adv Exp Soc Psychol* 1970;**5**:119–86.
7. Maas M, Taal E, Van Der Linden S, Booneb A. A review of instruments to assess illness representations in patients with rheumatic diseases. *Annu Rheumatol Dis* 2009;**68**:305–9.
8. Diefenbah MA, Leventhal H. The common-sense model of illness representation: theoretical and practical considerations. *J Soc Distress Homel* 1996;**5**:11–38.
9. Donovan HS, Ward S. A representational approach to patient education. *J Nurs Scholarsh* 2001;**33**:211–6.
10. Leventhal H, Diefenbah MA, Leventhal EA. Illness cognition: using common sense to understand treatment adherence and affect cognition interactions. *Cognit Ther Res* 1992;**16**:143–63.
11. Meyer D, Leventhal H, Gutman M. Common-sense models of illness: the example of hypertension. *Health Psychol* 1985;**4**:115–35.
12. Weinman J, Petrie K, Moss-Morris R, Horne R. The illness perception questionnaire: a new method for assessing the cognitive representation of illness. *Psychol Health* 1996;**11**:431–45.
13. Petrie KJ, Cameron LD, Ellis CJ, Buick D, Weinman J. Changing illness perceptions after myocardial infarction: an early intervention randomised control trial. *Psychosom Med* 2002;**64**:580–6.
14. Horne R, Weinman J. Self-regulation and self-management in asthma: exploring the role of illness perceptions amd treatment beliefs in explaining non-adherence to preventer medication. *Psychol Health* 2002;**17**:17–33.
15. Cameron LD, Booth RJ, Schlatter M, Ziginskas D, Harman JE, Benson SRC. Cognitive and affective determinants of decisions to attend a group psychosocial support program for woman with breast cancer. *Psychosom Med* 2005;**67**:584–9.
16. Hagger MS, Orbell S. A meta-analytic review of the common-sense model of illness representations. *Pyschol Health* 2003;**18**:141–84.
17. Mccarthy SC, Lyons AC, Weinman J, Talbot R, Purnell D. Do expectations influence recovery from oral surgery? An illness representation approach. *Psychol Health* 2003;**18**:109–26.
18. Galli U, Ettlin DA, Palla S, Ehlert U, Gaab J. Do illness perceptions predict pain-related disability and mood in chronic orofacial patients? A 6-month follow-up study. *Eur J Pain* 2010;**14**:550–8.
19. Norman SA, Thompson R, Vedhara K. A prospective examination of illness beliefs and coping in patients with type 2 diabetes. *Br J Health Psychol* 2007;**12**:621–38.
20. Warwick HM, Salkovskis PM. Hypochondriasis. *Behav Res Ther* 1990;**28**:105–17.
21. Hadjisavropoulos HD, Craig KD, Hadjistavropoulos T. Cognitive and behavioral responses to illness information: the role of health anxiety. *Behav Res Ther* 1998;**36**:149–64.

22. Turner JA, Whitner C, Dworking SF, Massoth D, Wilson L. Do changes in patient beliefs and coping strategies predict temporomandibular disorder treatment outcomes? *Clin J Pain* 1995;**11**:177–88.
23. Affleck G, Tenner H, Keefe FJ, Lefebvre JC, Zuck SK, Wright K, et al. Everyday life with osteoarthritis or rheumatoid arthritis: independent effects of disease and gender on daily pain, mood, and coping. *Pain* 1999;**83**:601–9.
24. Tennen H, Affleck G, Zautra A. Depression history and coping with chronic pain: a daily process analysis. *Health Psychol* 2006;**25**:370–9.
25. Verbrugge LM. Health diaries. *Med Care* 1980;**18**:73–95.
26. Boiko OV, Baker SR, Gibson BJ, Locker D, Sufi F, Barlow APS, et al. Construction and validation of the quality of life measure for dentine hypersensitivuity (DHEQ). *J Clin Periodontol* 2010;**37**:973–80.
27. Brown GK, Nicassio PM. Development of a questionnaire for the assessment of active and passive coping strategies in chronic pain patients. *Pain* 1987;**31**:53–64.
28. Salkovskis PM, Rimes KA, Warwick HMC, Clark DM. The health anxiety inventory: development and validation of scales for the measurement of health anxiety and hypochondriasis. *Psychol Med* 2002;**32**:843–53.
29. Moss-Morris R, Weinman J, Petrie KJ, Horne R, Cameron LD, Buick D. The revised illness perceptions questionnaire (IPQ-R). *Psychol Health* 2002;**17**:1–16.
30. The Euroqol Group. Euroqol—a new facility for the measurement of health-related quality-of-life. *Health Policy* 1990;**16**:199–208.
31. Efron B, Tibshirani R. *An introduction to the bootstrap*. New York, NY: Chapman & Hall; 1993.
32. Hoyle RH, Panter AT. Writing about structural equation models. In: Hoyle RH , editor. *Structural equation modelling*. Thousand Oaks, CA: Sage; 1995.
33. Byrne BM. *A primer of LISREL: basic applications and programming for confirmatory factor analytic models*. New York, NY: Springer-Verlag; 1989.
34. Hu LT, Bentler PM. Cutoff criteria for fit indexes in covariance structure analysis: conventional criteria versus new alternatives. *Struct Equ Model* 1999;**6**:1–55.
35. Byrne BM. *Structural equation modeling with AMOS: basic concepts, applications, and programming*. Mahwah, NJ: Lawrence Erlbaum Associates; 2001.
36. Foster NE, Bishop A, Thomas E, Main C, Horne R, Weinman J, et al. Illness perceptions of low back pain patients in primary care: what are they, do they change and are they associated with outcome? *Pain* 2008;**136**:177–87.
37. Benson S, Hahn S, Tan S, Janssen OE, Schedlowski M, Elsenbruch S. Maladaptive coping with illness in women with polycystic ovary syndrome. *J Obstet Gynecol Neonatal Nurs* 2010;**39**:37–45.
38. Wallander JL, Varni JW. Effects of paediatric chronic physical disorders on child and family adjustment. *J Child Psychol Psychiatry* 1998;**39**:29–46.
39. Baker SR. Testing a conceptual model of oral health: a structural equation modeling approach. *J Dent Res* 2007;**86**:708–12.
40. Petrie KJ, Jago LA, Devcich DA. The role of illness perceptions in patients with medical conditions. *Curr Opin Psychiatry* 2007;**20**:163–7.
41. Philippot LN, D'hoore W, Bercy P. Improving patients' compliance with the treatment of periodontitis: a controlled study of behavioural intervention. *J Clin Periodontol* 2005;**32**:653–8.

Part Four

Dentine Hypersensitivity and the Construction of Meaning

The experience of health and illness: polycontextural meaning and accounts of illness

Barry J. Gibson and Olga V. Boiko
School of Clinical Dentistry, Claremont Crescent, University of Sheffield, Sheffield, UK

Introduction

> *Not only are we born into complex communal narratives, we also experience, understand, and order our lives as stories we are living out. Whatever human rationality consists in, it is certainly tied up with narrative structure and the quest for narrative unity.[1] (p. 264)*

Accounts of illness experiences are not only a central theme of attempts to deal with the complexity of illness in modern society but also part and parcel of sociology's attempt to grasp the importance of the changing nature of illness by making illness experience accessible and therefore more visible to society.[1–6] In doing so, sociology has sought to go beyond the surface of accounts of illness to explore the insider's perspective and promote a depth of understanding that explains what is "behind" these experiences.[1,7,8] In short, the literature on accounts of illness has enabled sociologists and those involved in health care to explore the nature of illness, embedded as it is in social and political relations.[1,8]

The discussion about accounts of illness has largely been preoccupied by the significance of these accounts and their emergence in modern society.[1,5,7,8] It also covers how they can serve various purposes for both individuals and sociology.[1,6,7,9,10] The literature therefore maps the consequences of illness for those suffering from various conditions and those undergoing complex treatments.[1,6,7,9–11] Much less discussed has been the structural aspects of illness accounts. Pierret[4] stated that social structure was the problem to be analyzed; she went on to say:

> *The challenge is to define a paradigm and methodology for handling the problems related to the social structure. This entails working out theories about the interrelations, reciprocal effects and feedback between subjectivity, cultural factors and social structure.[4] (p. 16)*

In particular, it seems that one of the major problems is disentangling the complex relationships that emerge in accounts of illness. This appears to involve being focused on the fact that what is said in these accounts is at the same time social and

Dentine Hypersensitivity. DOI: http://dx.doi.org/10.1016/B978-0-12-801631-2.00015-4

individual,[8,12] public, and private.[13] Therefore, whereas an important function of narrative is its unifying effect, frequently this unity is perhaps held at the expense of the complexity of relationships underpinning the meaning of illness. This is not to say that the structural aspects of illness accounts have not appeared.

The description by Bury[1] of "contingent narratives" is a useful starting point for this analysis. "Contingent narratives" focus on "the beliefs and knowledge about factors that influence the onset of disorder, its emerging symptoms, and its immediate or 'proximate' effects on the body, self, and others"[1] (p. 268). Such accounts are frequently dependent on the medical system as a source of information.[14] They are also cited to be dependent on important cultural and everyday sources for observations about illness.[1,4] Contingent narratives are said to have two purposes, to elicit the source of illness and to explain its relational impacts. In attempting to achieve these purposes, such narratives are frequently dependent on the medical and cultural context of illness. However, as Bury[1] indicates, these narratives quickly spill over into accounts about what to do in the face of illness. The contexts of narratives therefore often link closely to their purposes.[1] A common theme in the sociology of illness accounts is that the sources, purposes, and cause of narratives overlap in complex ways, generating many interrelations and interdependencies.

A good example of the overlap between concepts is how morality, culture, and society interchange as sources of narrative. For a long time, morality has been cited as an important source of narratives about illness.[2,15,16] Yet, morality is frequently seen as an aspect of the more enveloping concept of society[15] or culture.[1] The commentary by Bury[1] discusses the "theme of morality in chronic illness" that "is developed by locating patients' stories within a cultural framework that increasingly portrays health as a virtuous state" as described by William[15] (p. 275). On closer inspection, however, the bulk of the discussion by William[15] is, in fact, a discussion of the morality of illness. He sets out to look at the way "society is represented through individuals' discussions of the impact of chronic illness and disablement upon them"[15] (p. 93). There is, in fact, no explicit conceptualization of culture as the source of illness accounts, although of course morality and religion could be seen as a subset of culture.

"The social" also appears as a source for narratives about illness, and once more the discussion of "the social" quickly spills over into another discussion about values. For example, it is frequently cited that society is a source of negative attitudes or negative representations about old age.[14,17] We are told that we operate "within a definite 'social clock' which guides our expectations of events within the biographical context. Such expectations influence whether events are anticipated or unanticipated"[18] (p. 503). Very little is said about expectations beyond that they seem to be part of the taken-for-granted nature of everyday life. They seem to regulate whether an illness is to be seen as a problem or a normal part of everyday experience associated with aging.

In several studies it has been reported that "the social" is responsible for the fact that older people appeared to expect illness and when it occurred. It was seen as a normal part of aging.[14,18] Sanders et al.[17] discussed how there could be both normal and abnormal aspects to the experience of arthritis. Such experiences were described as complex and paradoxical. Therefore, expectations about age formed the background to a discussion about the degree to which they were disappointed, and society was described or attributed as the source or regulator of these expectations. The source of such narratives was described as "the social"; however, the discussion quickly focused on the realization of or disappointment with such expectations.

The social is also thought of as a source for the important distinction between public and private accounts of illness in the work of Jocelyn Cornwell.[13] Cornwell,[13] drawing on Douglas,[19] discussed the already recognizable nature of certain aspects of illness. These were defined as public accounts that were "sets of meanings in common social currency that reproduce and legitimate the assumptions people take for granted about the nature of social reality"[13] (p. 15). Public accounts obviously circulate in some way, providing a kind of security for people talking about many things, including illness; they conform to the "least common denominator morality." But there is more to it than this; they also relate to etiological accounts of illness. In this respect, public accounts were frequently derived from medical sources[13] (p. 148). These sources were used to legitimize claims being made in the accounts. In the study, Cornwell[13] stated that the overall concern of participants was with morality and responsibility for illness. Scientific concepts appeared to help absolve the individual from being responsible for the condition. In this complex account, we have an interaction between morality and medical knowledge operating through accounts of illness to provide a form of culturally conditioned medicalization.

After this discussion about "the social" as a source for accounts of illness, it is apparent that sociology has tended to focus on the fact that illness disturbs the normal aspects of everyday life that are taken for granted, i.e., has tended to focus on its consequences.[1,15,20,21] One of the principal effects of illness is having disappointing expectations about the normal flow of duties and responsibilities. Here, the discussion flows quickly from the source of illness accounts as "the social" toward the impacts of the illness on the everyday life and those around the person who is ill.[9,22,23] Both Navon[23] and MacRae[9] describe how stigma is produced and regulated by the social despite the well-meaning and best efforts to change such sources of stigma. However, the discussion moves from the source of these accounts and attempts to change them quickly to the effects of illness. The difficulty of focusing the discussion on the source of the accounts and their mechanisms seems to persist.

Much the same can be said of the concept of "culture" as a source for narrative accounts. In discussing the relationship between culture and illness accounts, Bury[1] uses the words of Gellner[24] to describe narrative as an infinite reservoir of meanings and comprehension. The "language of narrative" helps "to sustain and create the fabric of everyday life, they feature prominently in the repair and restoring of meanings when they are threatened"[1] (p. 264). Narrative combines "universal, cultural, and individual levels of human existence"[1] (p. 264). Narrative therefore draws on symbols that serve to connect different "levels of experience." The mechanism for how this happens is not directly discussed in the accounts of illness literature.

Bury[1] goes on to discuss the work of Kelly and Dickinson[25] and outlines how culture contributes to illness narratives in the form of genre and how such culturally derived structures can enable stories to be told. In short, culture, although a source of narrative *form*, simultaneously enables and constrains stories to be told. Culture is therefore analyzed in terms of its consequences for the individuals telling the stories.[1,3,25]

There is nothing inherently wrong with the way in which discussions about social structural factors associated with illness experience transform into discussions about the consequences of these factors for "individuals." Such discussions certainly have performed an important "function" for health care and have established an important niche for the sociology of health and illness.[2,4,14] This is certainly apparent when sociology draws on such sources. Take the example of Locker and Kaufert,[11] who explored how changes in the therapeutic context of the patient could dramatically affect their experience of care. Such accounts in many ways help improve the sensitivity of medicine to the "care" of patients.[11] The provision of such accounts therefore can serve the important function of advocating on behalf of illness sufferers, facilitate a kind of "empowerment ideology," and possibly help to limit the dominance of "biomedicine."[1,6,14] At this point, however, we are no longer talking about the structure of these accounts but rather their function and purpose.

Despite these problems, there remain many good discussions about the functions and purposes of narrative accounts. Such accounts suggest that narratives not only draw from a preexisting cultural or social "reservoir of meanings" but also serve the function of building new reservoirs of meanings for the future.[1] They help in the search for causes of illness and also serve to explain the relational impacts of illness and its consequences.[1,4,6,7] Finally, they help to connect disparate fields of meaning into a thread of individual experience and make those experiences accessible for society. Although people are the sources of narratives, in this sociology, they are also frequently acting "against" structures of illness and its relations.[7] It seems odd that despite the richness of the field and its major contributions to our understanding of health and illness, there have been very few accounts that have focused on making the social structural aspect of illness experience their focus.

It is almost as though the social is a "structure" acting as one source of illness accounts and, as such, social structure sometimes appears to be more or less taken for granted in the literature. It frequently appears briefly, only to quickly slip into the background of a discussion about consequences and purposes of narrative for individuals. Of course this sociology rightly serves the important function of placing illness accounts at the forefront of concerns about chronic illness.[1,4,15,20,26] When reading the literature, however, one cannot help wondering what more remains to be said. More remains to be said about narrative and its functions, and more remains to be said about the sources or social structural aspects of narrative.

It might be suggested that one of the problems is that current sociological accounts of illness experience do not adequately refer to the structural traditions in social theory. Anderton and Elfert [10] produced a comparative account of how the experiences of Chinese and white families in managing the care of their chronically ill children are shaped by social structural processes. These processes involve exploring how the ideology and moral discourse on normalization enables the categorization of families by professionals. The consequences of this process might eventually involve someone being refused treatment. The goal of their work is to expose the "complex historical, social, political, and economic nexus"[10] (p. 254) that underpins the differences in the accounts of illness between Chinese and white families.[10] They propose that "ideologies of normalization" operate to separate "compliant" from "noncompliant" families. Following Smith,[27] ideology originates from a "definite position of dominance in the society" and not the "neutral floating thing" called culture[10] (p. 255). Ideology is therefore something that originates from one's position and reflects relations of dominance and subordination more significantly. For Anderton and Elfert [10] ideology relates to one's economic position and it serves the dominant views of society. This is made clearer in their account of how normalization acts through nurses and health professionals to reinforce dominant, largely westernized, views of "normalization."[10] Normalization is not just a humanist discourse; it reinforces economic and political views of how to help people to maximize their potential. In their study, medical ideologies interacted with cultural and material circumstances to produce very different responses and accounts of the two chronic conditions. One family shared all of the same ideological resources as that of the health professionals, whereas the other was living under the conditions of severe deprivation. Further, the material circumstances produced their own materially patterned form of normalization.

The account itself is a very good example of the kind of structural analysis that could be produced as illness accounts. In some ways, a Marxist account is perhaps too reductionist. The goal of this chapter is to set, along with such structural accounts, the systems approach of Niklas Luhmann. To do this, we need a basic introduction to Luhmann's theory and we need to explore what he had to say about systemic knowledge (biomedical knowledge), morality, age, culture, medicine, and ideology.

Luhmann's social systems theory

Niklas Luhmann's systems theory touches on pivotal problems in understanding modern society through a description of the functioning of different social systems. Firmly embedded in European tradition of the humanities and social sciences (Husserl, Heidegger, Habermas, Derrida), systems theory also integrates the elements of theory of systems evolution (Maturana and Varela) and cybernetics (Spencer Brown, Heinz von Forester). For British sociology, one introduction to systems theory has been the publication of the English edition of *Social Systems*.[28] In this book, Luhmann outlined a way into a theory of social systems that was fit for purpose. Social systems introduces the problems that his proposed version of systems theory would need to consider, e.g., introducing the problem of the system and environment difference, structure and time, double contingency, meaning, and others. Instead one might find that the descriptions of various social systems, such as systems of law, politics, family, religion, art and mass media, risk, trust, and time, in his applied theory, are more accessible.[28–33]

The relationships between society, person, and culture in Luhmann's sociology are condensed in his imperative about social systems: "there are systems." Actorless, bodyless social systems are given ostentatious priority in his account. Social systems—political, legal, medical systems, the systems of religion and art—emerged and separated in the process of societal evolution, operating through relatively stable structures and specific functions. For Luhmann, there is a systemic functionalism, and this reigns over any other ontological premises about the social order. To him, the integral problem of modernity should be concerned with how systemic functional communications are organized and attributable to different systems rather than with the structure of social actions or cultural expectations.

Society is a product of communication in social systems theory. It is argued that "social systems communicate through communication"[27] or, more precisely, they consist of system-specific communications. In this respect, different social systems provide different functionally equivalent solutions to specific problems[27] (p. 15). This "functional structuralism" shifted the logic of the structural functionalism of Parsons (whom Luhmann studied in Harvard) to the point at which functions, supported by structures, serve to resolve different problems.[34]

Social systems communications emerge around attempts to solve stabilized problems through using various functionally differentiated media. These systems achieve this by communicating around various codes. Such codes have positive/negative sides, so that system-specific communication "oscillates" between them. For example, the political system, it was claimed, attempts to solve the problem of how to reach collectively binding decisions. This system therefore communicates around the distinction between power and no power. The system of law protects normative expectations through communications that turn on the distinction between legal and illegal. The system of art operates with the code beautiful/ugly; the economic system is concerned with payment or nonpayment. Finally, the system of medicine communicates around the distinction between sick and healthy.

As a consequence, concepts such as "culture," "society," and "the social" are not central to the differentiation of stable functional structures or stable distinctions within Luhmann's approach. Such concepts cannot serve as evolutionary principles for society. The notion of culture evoked a great deal of skepticism in Luhmann because of its endlessly divisible character. Culture as opposed to social systems only produces paradoxes without resolving them. It articulates ideology without necessarily engaging in practice[35] (p. 112). Luhmann found culture's inability to have boundaries "at the bottom"[36] (p. 101) to be liberating in all the wrong ways. However, he gave it more credibility in *Social Systems*; no matter how marginal the analysis remains, the functions of culture (for social systems) emerge in the linkages between meaning, social memory, semantics, and, finally, typifications in cultural forms. It would be useful for our purposes to briefly concentrate on these concepts.

Meaning, in Luhmann's terms, refers to the actualizations of the certain forms of thinking and communication, selected out of the horizon of other possibilities. It has three dimensions: the factual, the social, and the temporal. In the factual dimension, meaning gives form to the world of objects. In the temporal dimension, it serves to connect past and future events. In the social dimension, meaning makes interactions plausible, but never predetermined, as far as understanding between humans is concerned. Contradictory to the optimism of early symbolic interactionists, Luhmann stipulated that reduplication of interpretative possibilities occurs regardless of subjects and the minutiae of interaction. It is believed in social systems theory that the minds of individuals, termed psychic systems in his theory, during interaction (Ego and Alter) are opaque and therefore cannot be responsible for mutually shared meaning.

Given the central position that culture maintains in exiting accounts of illness, it is necessary to explore this concept in more detail and explore what it might mean within the framework of social systems theory. For Luhmann, culture as a term was prominent at the end of the eighteenth century and was used to observe how others observe. In summary, it enables the observation of society in comparison with society. He also argues that "typified forms, or cultural elements, are usually used within communication as always already known by all participating psychic systems"[37] (p. 69). In other words, culture functions to condense meanings and, by doing so, it acts as a memory function for society. Memory is important because social systems operate through chains of recursively connected communications where one utterance connects to another. Communications are not just sound or noise, but rather are something understood. Each new communication links to what went before and what comes after in a constantly shifting horizon of meaning. Social systems are described by Laermans[37] as "constantly 'enchained' communications" (p. 70). Each new communication is tested with reference to the previous communication.

What helps to preserve commonalities in meaning is semantics—the complexes of forms or schemes used in communication. These forms (or distinctions in his original idea) are held by social systems to restrict and stabilize observations and communications. Semantics is defined as the social stock of standard forms or rules for the normal treatment of meaning[27] (p. 163). Laermans[37] comments on the role of typical forms in systems theory by emphasizing semantics as those typical, abstract generalizations that society uses to specify, retain, and stabilize meaning. Communication progresses by reproducing "a supply of possible themes that is available for quick and readily understandable reception in concrete communicative processes. We would like to call this supply of themes culture, and if it is reserved specifically for the purposes of communication, semantics"[27] (p. 193). In other words, if we follow Laermans[37] interpretation of Luhmann, then culture is the result of a specific way of looking at the typified forms that belong to memory of social systems[27] (p. 47).

Again, culture as a term helps us to compare social systems. The typical cultural forms and the relationships between them are dispersed over different social systems.[38] (p. 88). This is because different social systems enforce particular meanings, themes, and boundaries on everyday communication. Communications as they are uttered emerge as polysemantic or polyphonic. By this, we mean that each communication can have many participating systems and many participating horizons. This concept enables us to explore the complexity of systemic communications around health and illness. In addressing the complexity of communications in modern social systems, King[39] develops the Luhmannian point about nonanthropomorphic organization of communication:

> *One way to achieve this is to start not by looking at people and their actions, but by asking what different the forms of communication are produced by different systems and how these different forms of communication filter the external world.[38] (p. 153)*

In a similar vein, Andersen[40] introduces the notion of polyphonism in relation to communication in organizations. He views communications as polyphonic because of the different semantics that are brought into organizations. Organizational polyphony has emerged as functional systems have "exploded" and exceeded their original organization forms. For example, welfare organizations use legal, political, economic, and moral communications in tackling the range of issues applicable to them in attempting to address social contracts with the clients of the welfare system.[39] Such logic is also at work when understanding communication about particular organizational forms; Andersen[41] specifically argues for polyphonism in partnerships.

From the perspective of systems theory, one always accounts for the different semantics of social systems that may be manifest in different forms of communication. From this perspective, health communications are polyphonic and they involve the participation of the semantic forms of many social systems. It is expected that what initially appears to be mundane "bodily" experiences of health are narrated through the voices and semantics of different systems, such as medical communications, economic communications, and family communications, to name but a few. We return to these themes in our section on data analysis.

The study

The current (secondary) study is based on the qualitative study that, in turn, is a part of three-stage project aimed at constructing and validating quality of life measure for the people with dentine hypersensitivity. The overall project involved individual interviews, focus groups, and questionnaire validation. It was supported by GlaxoSmithKline and received research ethics approval from the University of Sheffield. As far as qualitative interviews were concerned, participants were purposively recruited from the general population with the goal of providing a range of experiences and views about the everyday impact of dentine sensitivity.[42]

Participants were initially identified through the research team's contacts and through a process of snowball sampling. After this, we recruited from the general population using a recruitment agency and also from e-mails sent to staff and students of the University of Sheffield. Potential participants were phoned and/or e-mailed and invited to take part in the study. They were asked about their sensitive teeth, their age, and availability for the study. The goal of the study was described in general terms as being "to explore their experiences of living with sensitive teeth." After this initial approach, participants were sent a letter containing an information sheet and consent form. They were then called and an interview date was arranged at a time and place suitable to them. Full written consent was obtained on the day of the interview. When interviews were arranged in the homes of participants, safety arrangements for the researcher were made such as having a mobile phone nearby and leaving a message about where and what time the interview was to be held. Interviews lasted from 20 to 40 min and were transcribed as quickly as possible after the interviews and the recordings were deleted after a short time. To preserve anonymity, all identifying information was avoided during the interview and if any identifying information did emerge it was subsequently removed from the transcripts.

The interviews were guided by a few questions to cover all aspects of the individual experience of sensitivity. Yet, interviewers attempted to be as open as possible to the participants' narratives and flexible in switching between the interview topics, which included screening the dentine sensitivity (discomfort with hot/cold, physical pressures on teeth), pain profile, and experience of the onset and illness carrier. Specifically, we were interested in functional limitations and impact on everyday activities such as eating, drinking, talking, toothbrushing, and social interaction. Then, we focused on individual reactions to sensitivity episode such as emotional burden, adaptation, and coping strategies. Finally, the questions distinguished the relations between individual experiences with sensitivity and health identity.

Analytical strategies of Luhmann's social systems theory

Luhmann's social systems theory avoids any anthropocentrism. Any analytical strategy that uses this approach faithfully will not be focusing on the consequences of the phenomena of illness for individuals. Such an approach will instead focus on communication and meaning as a social event. For the traditional sociology of health and illness, this might be a problem. However, as we have seen, there are already a plethora of articles examining the consequences of illness for the individual.

For Luhmann, communication is composed of a tripartite distinction between information, utterance, and understanding. Each of the concepts of communication has specific meaning within Luhmann's systems theory. Information is "an event that selects system states"[27] (p. 67). In other words, communication selects one possibility from a range of possibilities. Information also changes the state of the system. It is a momentary actualization that marks a change in the system from that point forward. For example, once someone has a bad experience at the dentist, all future experiences can be tainted and there is the expectation that such bad experiences can recur. Socially, however, it is certainly not new that people have had bad experiences at the dentist. In some respects, such experiences are "expected." In this way, not every communication carries information that counts as "information" for both psychic and social systems.[27]

Communication also involves utterances. Utterances seek to leave information outside of the communication but reformulate it appropriately "for example, by providing it with a linguistic (eventually an acoustic, written, etc.) form"[27] (p. 142). Utterances are therefore contingent on information. One can present one's experience of illness, but we are always confronted with the problem that everything said could also be very different, and we cannot be sure that what we think is information is also information for the other. Likewise, what we have said about our illness places us as part of "the meaning world in which information is true or false, is relevant"[27] (p. 141). In other words, in stories about illness, we must interpret ourselves as "part of what can be known about the world"[27] (p. 141). The information we give refers back to us. The result is that we are forced to control ourselves as self-referring systems and as part of what we are presenting.

If communication includes information and utterance, then it also includes understanding. Understanding has a specific meaning in Luhmann's systems theory. Understanding happens only if "one projects the experience of meaning or of meaningful action onto other systems"[27] (p. 73). In this respect, understanding differs from experience. We can understand that people have had experiences with illness, but we can never grasp fully what that experience means "*for them*." Our understanding is always "*our*" understanding and not theirs. The space between the unity of information, utterance, and understanding as tripartite selections frequently produces misunderstandings and, in this respect, communication continues generating its own autopoiesis.

It should be clear that communication is a complex phenomenon in Luhmann's theory. However, the important point we introduce is that his theory is designed to keep us focused on systems and communication. We have already introduced the tripartite distinction between the factual, social, and temporal dimensions of meaning. Luhmann defined meaning as "the continual actualization of potentialities"[27] (p. 65). The definition of meaning according to Luhmann[27] is close to the phenomenological approach because it focuses on the world as it appears, not on psychological approaches.

As we have discussed previously, his approach represents a very broad revision of sociology and as a consequence there are a number of analytical strategies that can be used. Some of the most important have been provided by Andersen[43] and are summarized in Table 15.1.

Table 15.1 Luhmann's analytical strategies

Analytical strategy	General question	Examples
Form analysis	What is the unity of the distinction? And which paradox does it establish?	In what way are organizations systems that communicate through the form "decision"? Which paradoxes does this form establish?
Systems analysis	How does a system come into being in a distinction between system and environment? How is the systems boundary of meaning and autopoiesis defined?	In what way does the politicization of the organization become apparent in the internal construction by the organizations of their environment so that they not only construct the environment as market but also as political public?
Differentiation analysis	How are systems differentiated? What is the similarity in the dissimilarities of the systems? What are the conditions, therefore, of the formation of new systems of communication?	In what way does the politicization of the organization challenge their internal form of differentiation and force them to institutionalize internal reflections of themselves as closed communication (e.g., through a so-called ethics officers)?
Semantic analysis	How is meaning condensed and how does it produce a pool of forms, that is, stable and partially general distinctions available to the systems of communication?	How is meaning condensed with respect to environment, human rights, ethics, animal welfare, health and prevention, into the concept of the "socially responsible corporation," and bring about new conditions for corporate communication?

(*Continued*)

Table 15.1 (Continued)

Analytical strategy	General question	Examples
Media analysis	How are media shaped? How do they suggest a specific potential for formation?	In what way does the politicization of the organization mean that the organization is no longer only supposed to form the medium of money, but it is also expected to form a number of other communicative media such as power, information, and morals? In what way does this change the conditions of the organization from homophony to polyphony?

Source: Reproduced with kind permission from Andersen (2003, p. 93).

Form and semantic analysis

The two analytical strategies we focus on in our study of communication regarding dentine sensitivity are those of form and semantic analysis. Form analysis was typically used by Luhmann when he was beginning a set of communications for observation[43]. It involves identifying the "unity of a difference" within communications. The best example from his work is the distinction between risk and danger. According to Luhmann[32], communication about risk cannot help but carry with it the problem of danger. Risks are losses or damage that are attributed to decisions made, whereas dangers are losses or damage that result from the environment. It should be clear that a risk for one person can present itself as a danger to someone else. Someone taking a risk by smoking a cigarette can present as dangerous to someone who would not take that risk. Risk communications are bound to the form of the difference between risk and danger; they cannot help processing this difference and the distinction is biased in social communication toward risk. As a result, risk communication converts dangers into risks.

Our interest in form analysis was to explore the unity of distinctions within "everyday" communication about dentine sensitivity. This analysis is principally achieved by reading the transcripts line-by-line and asking what distinctions are being used in these communications. Knowing this, we subsequently went on to ask "what must necessarily follow the shaping of these distinctions?" regarding communications about dentine sensitivity. Finally, we asked how these distinctions restrict communications about dentine sensitivity. Each of these strategies are outlined and explained in more detail in Andersen.[43] For the purposes of simplicity, we focus on just a few of the distinctions that emerged in our data.

A secondary element to our analysis was to suggest how these forms of communication may have emerged as social semantics of dentine sensitivity. This involves establishing how forms of meaning may have "condensed" around guiding differences or forms.

> *Semantics are characterised as the accumulated amount of generalised forms of differences (for example, concepts ideas, images and symbols) available for the selection of meaning within the systems of communication. In other words semantics are condensed and repeatable forms of meaning, which are at our disposal for communication.[42] (p. 87)*

Although this analysis in application to dentine sensitivity is the subject of further work, it involves exploring other forms of social communication about dentine sensitivity to establish how the guiding distinction of communications on dentine sensitivity may have been established. We see that, in some respects, some form of semantic analysis is needed when it comes to understanding the emergence of forms and how they become established. These analyses involve looking at how the factual, social, and temporal aspects of communication about dentine sensitivity have taken shape. What themes and objects are present in communications about dentine sensitivity (the fact dimension)? How do communications about dentine sensitivity explore the theme of ego and alter ego (social dimension)? How does communication about dentine sensitivity explore the distinction between before and after the condition (temporal element of communications)?

Our purpose in this chapter is to illustrate each of these elements of communications about dentine sensitivity. In so doing, we hope that our analysis can illustrate the polyphonic nature of communications about the condition. We also hope it can establish how narrative can be seen as a polyphonic unity.

The imperative of dentine sensitivity

A total of 23 interviews were conducted; 15 participants were female and 8 were males. Participants included those who were currently experiencing sensitivity in their teeth and who were adults of all ages (18–65 years) and backgrounds. In what follows, pseudonyms for each participant are used. A central feature of accounts of dentine sensitivity was their referral to what we call "the imperative of sensitivity." This imperative masked a deep paradox in communications about the condition. The imperative was that dentine sensitivity was a problem to be suffered independently and, more often than not, privately. Dentine sensitivity meant having a watchful sensitivity to a battery of stimuli that appeared, somewhat unpredictably, in everyday life. Dentine sensitivity, for our participants, involved learning to control one's exposure to these stimuli and, when mistakes or failures did occur, to avoid disturbing others. Dentine sensitivity involved being sensitive to the stimuli and sensitive to others, but paradoxically insensitive to one-self. This paradox ran through the core of the accounts of dentine sensitivity. Participants were both sensitive and insensitive at the same time. Accounts of dentine sensitivity turned on the subtle nature of this distinction.

The following excerpts demonstrate this imperative. Each participant communicates doing this in subtly different ways. In the first example, Jana relates how she was sensitive to a range of things in the immediate environment. The sensitivity is, however, unpredictable. The only apparent condition is how she responds to her sensitivity—by keeping it from others, including her partner. She shares her sensitivity with her son who has the same condition. She had just described the sensations as a kind of "brainfreeze," a common term for jarring sensations of pain common to the experience of the condition.

> *My eldest son and I we share with one another because we make a joke out of it erm.*
>
> *Brainfreeze?*
>
> *Brainfreeze yeah brainfreeze emm don't tend to say anything to my husband because he wouldn't understand in any case it's a bit of a 'don't be soft situation. Umm and no probably that's more a bi-product of up-bringing, you sort of get on with things more than show if anythings the matter so that's more a bi-product of um sort of the way we've been taught to get on with things. Ermm and not make a fuss.*
>
> *Umm is it possible to predict the pain? Anticipate, expect? [pause] Do you predict?*
>
> *That's a difficult question [pause] I suppose I predict, if its going to be really cold stuff, I know before I put it in my mouth that its going to be painful hence the reason why I leave it again. Or leave it full stop. Umm sometimes I can end up with sensitivity with stuff that I've least expected so occasionally no its not predictable that I don't expect, that's triggers it. Umm I try to think of something. Marmite for instance its something that you wouldn't expect to trigger of sensitivity but it does and somethings that you've tended. And its hugely focused around eating apart from as I say during the winter when its cold.*
>
> *(Jana, Interview 5 [S1.5], page 6)*

It is clear from this account that dentine sensitivity is unpredictable and can have a range of impacts in everyday life. From the perspective adopted in our analysis, we can see that the account has factual, social, and temporal dimensions. Factually, it thematizes a range of objects that can cause sensitivity: the winter and marmite. The condition can be triggered at any moment by hot and cold stimuli. The communication also has a temporal dimension; the stimuli can appear at both predictable and unpredictable moments in everyday life. It is, however, in the social dimension that we see the paradox of sensitivity acting to limit the impact of the condition. Jana can communicate about her sensitivity to another sufferer, her son, but feels she should not to speak to her partner. In this respect, the paradox of sensitivity acts to limit what can and cannot be communicated or, more importantly, with whom one can communicate about the condition. Jana went on to say:

> *I'm just aware its painful occasionally if I'm in company I'll shut off from the conversation so I'll miss sometimes what people are talking about. Because it's painful for a few seconds and its controlling yup. So I shut down because you don't want to make a fuss in front of other people.*
>
> *(Jana, Interview 5 [S1.5], page 6)*

The paradox at the heart of sensitivity controls Jana's account and effectively isolates her from communicating about her sensitivity to others. The next account from Alicia describes the paradox of communication about dentine sensitivity. The interviewer (OB) had been asking about the severity of the condition.

> *Do you reckon its, well will you tell me, does it fall into the category of a minor nuisance or a real problem.*
>
> *It's a nuisance when it happens so for example if my daughter hands me the half eaten apple then I think straight away 'don't throw it away it's a waste' and then I eat and of course my teeth hurts because I'm eating an apple so that and I just think 'why have I got this'. But I wouldn't say it's, you know like I say sometimes I might not eat something because it's too cold erm so you know you don't want to offend people but sometimes its...but I don't know.*
>
> *(Alicia, Interview 7 [s1.7], page 9)*

The condition is a condition and it is communicated as a condition rather than a true illness. It therefore occupies an ambiguous position with respect to health and illness. Throughout the interviews, participants found it difficult to decide if it was something that warranted serious attention or if it was simply a minor health condition. This account clearly limits dentine sensitivity to the status of a condition to be dealt with in isolation. For example, Alicia communicates experiencing pain from eating the apple that her daughter was going to "waste." So, we have the experience of pain and the problem of wasting food.

In the next example, Kerry had already communicated her sensitivity as a minor thing. She hated ice in her drink and would frequently scoop it out to avoid the intensity of the cold.

> *But umm, still can you bring back. I know it's a minor thing. When it happened last time, maybe with the cold drink and you couldn't avoid drinking so you drank and you felt 'oh I forgot'?*
>
> *Yeah like if I go out with friends.*
>
> *Exactly.*
>
> *You know say if I go out for a meal with my friends from work and they bring me a drink from the bar they always put ice in and I hate ice. If nobody's looking I scoop it out and put it in, well it used to be an ash tray but there aren't any now are there! Yeah I do not like drinks with ice in.*
>
> *(Kerry, Interview 14 [s1.14], page 7)*

Again, the necessity to suffer the condition as a condition and not as an illness limits Kerry's ability to communicate about it to others. This suggests that communication about dentine sensitivity is conditioned by the fact that it is not considered a proper illness but rather a health state and something one can expect as part of oral health on a normal basis. Communications about dentine sensitivity therefore establish that one should experience a range of impacts from various objects over an unpredictable temporal horizon. The ability to talk about these impacts, however, seems to be conditioned by the social dimension of meaning. We found the same thing time and time again in accounts of dentine sensitivity. There was an inability to talk to others about the condition and the condition was very much a private problem to be discussed and experienced privately.

The consistency with which this occurred suggests that communications about dentine sensitivity are "polyphonic." We suggest that they are polyphonic in two ways. First, they contain factual, social, and temporal dimensions. Second, they are also "polyphonic," because semantics of multiple systems are affecting what is said. Simply put, the social dimension conditions when someone should or should not talk about dentine sensitivity. It also limits the condition of sensitivity to the status of a health condition and not an illness. How this happened may well be related to our next distinction, the nonproblem problem of dentine sensitivity.

The nonproblem problem of dentine sensitivity

There was another aspect to the polyphonic character of the accounts. Underlying the form of communication about dentine sensitivity seemed to be the participation of the health system. In Luhmann's social systems theory, health systems operate through the code of sick/healthy and there is a clear emphasis or bias on sickness rather than health. The emphasis has consequences for what can be communicated in medical settings. In this respect, communication tends to thematize sickness and the problems people have, rather than their health. We found significant evidence of a similar paradox in the accounts of dentine sensitivity, although there was some ambiguity in this respect. In the following account, Annie talked about consulting her dentist at the beginning of her interview. She was talking about the location of her sensitivity:

> *So it's just one side, the lower jaw?*
>
> *Yeah, it's lower and when I first got it I thought the tooth needed filling I visited the dentist, and she said 'no, the problem you've got there is that part of your tooth is exposed to dentine' and there's very little we can do about that, other than, the only thing she's asked me to do is to brush with like a pro-enamel toothpaste for a 12-week period and then it will be reviewed at the end of that 12 weeks.*
>
> *(Annie, Interview 10 [s1.10], page 1)*

Here, the problem itself is conditioned as a problem for Annie to take care of through the use of specialty toothpaste, in this case "Pronamel." Bill was already using toothpaste (Sensodyne) for his sensitivity and as a result was not going to talk about his sensitivity to the dentist.

> *Umm has it (Bill's sensitivity) been diagnosed by a dentist.*
>
> *No, its never been that bad. I got to the stage where I thought oh it's a little bit sensitive. I'll get some Sensodyne, it seems to work therefore I don't worry about it. So I wouldn't say to my dentist oh I've had a pain because by the time I get to the dentist which is what every six months or every year or something. I've had the problem and I've dealt with it.*
>
> *(Bill, Interview 11 [s1.11], page 1)*

Here, dentine sensitivity as a problem comes and goes; it is a minor problem that is generally dealt with through the use of specialty toothpastes. This is not always the case, however, and some ambiguity in communication still exists. In some cases, dentists will use Duraphat to put a protective layer over the problem.

> *Well I just basically I was doing some routine observation of my teeth and I noticed that there was a little groove at the base of one of my top teeth, incisor and it, it didn't bother me at that point it was just something strange thing I haven't noticed at that point and then I noticed that the groove was getting deeper and deeper and basically after that I started feeling uncomfortable when I was eating some sweet stuff like honey as I say of syrup.*
>
> *And noticeable basically you can see that, what can you see exactly?*
>
> *Well you can't see at the moment because I do have a coating, a special coating the dentist put on to protect it but yeah there was a greyish sort of groove.*
>
> *(Marisha, Interview 3 [1.3], page 2)*[1]

Such solutions are typically temporary. In these accounts, dentine sensitivity was simultaneously a problem and not a problem. It was clear that such communications operated through a code that primarily indicated that sensitivity was more or less a problem for the person and not the dentist, although some nuanced variations did emerge and care could be given. The main outcome was that people should treat their problem themselves by using the appropriate toothpaste. Annie's account indicates that she had a temporal period through which she was to use the toothpaste and then she could communicate with her dentist about it again. Her problem was not the dentist's problem, but rather her problem. Once more, the social dimension of the communication predominated as the conditioning factor. Annie cannot seek help until after a specified period of time.

Bill's self-care was working to the point where he did not need to "bother" his dentist. His problem was his problem and not a problem for the dentist. The social

[1] In the second quotation Marisha is clearly able to go to her dentist for a solution to her sensitivity, and this might indicate that there is a degree of trust between herself and her dentist. Our problem is that this is a secondary analysis of these data and that, sadly, we do not have enough data to be able to elaborate on this further.

dimension of the paradox of sensitivity once more operated to condition whether he should make his sensitivity the subject of communication with his dentist. His problem was a problem; it was just not a problem for the dentist. We suggest that semantics of the medical system were operating in his communications concerning whether he was sick or healthy. Finally, Marisha had clearly brought the subject of her sensitivity to her dentist and made it the theme of their discussions. Her problem was also a problem for her dentist. So even though a predominant suggestion was that sensitivity was simultaneously a problem and not a problem, it can be made a theme for communications in a health care organization such as a dental clinic.

The polyphonic nature of these communications indicates the continual interplay of the factual, social, and temporal dimensions. In other words, there are many participating horizons in each communication. It also indicates how systems such as the health system may well be participating in these communications to condition them and determine whether dentine sensitivity is to be considered a problem or not a problem for health care. As we have seen, more often than not the problem was conditioned as a nonproblem in this respect. This brings us to a semantic history of dentine sensitivity.

The emerging semantics of dentine sensitivity

We don't have enough space here to go into detail with respect to the semantics of dentine sensitivity. As a consequence, the analysis remains necessarily brief. It is important to notice that the interview data are not enough to help us disentangle the nature of the relationship between the accounts of dentine sensitivity and their sources. Verbal communication alone is not enough to explain how communications about dentine sensitivity may have become conditioned in this way. The forms of meaning we found in talks about dentine sensitivity, as with all forms of communication, have a semantic history. This history can be found in several places; for now, we focus on scientific communication about the problem.

The nature of dentine sensitivity has been the subject of some debate in dental science. The initial scientific problem with dentine sensitivity was to separate the symptoms of dentine sensitivity from pulpal disease. Such research found almost no correlation between pulpal pathology and the symptoms of dentine sensitivity. A consequence of this was that the term was frequently described as somewhat of "an enigma"[44] (p. 342). This description also resulted in some confusion over what it should be called, the terms "dentine hypersentivity" or "dentine sensitivity" seemed equally appropriate. Although clearly the term "hypersenstivity" seems to indicate that there is something over and above, or something unusual in individuals with the condition. Likewise, dental science has traditionally distrusted patients' perspectives. This distrust was also manifest in the science of dentine sensitivity, where pain measurement was described as "subjective" and, as such, this made "an accurate assessment of the extent of the problem difficult."[44]

It eventually took a consensus conference to agree a definition. This definition states that dentine sensitivity is "characterised by short, sharp pain arising from exposed dentine in response to stimuli typically thermal, evaporative, tactile, osmotic, or chemical and which cannot be ascribed to any other form of dental defect or pathology."[45,47] Dentine sensitivity therefore requires a diagnosis that eliminates other conditions first, before the diagnosis can be made. The reasons for this are that the symptoms for the condition are very similar to other conditions such as dental decay. As a consequence, the condition is diagnosed as a last resort.

Although several theories have been proposed to explain the underlying mechanism for dentine sensitivity, the evidence seems to support a "hydrodynamic mechanism." In this theory, it is postulated that rapid fluid movement through small "tubules" that run through exposed enamel results in a stimulus to the pulp that produces a pain response.[44–48] In addition, there are seemingly numerous etiological factors. Attrition, erosion, and abrasion, alone or in combination, may be responsible for exposing dentinal tubules.[45,46] Abrasion from toothbrushing seems to be insufficient to produce sensitivity on its own.[49] As a result, it has been suggested that it takes a combination of these factors, most notably erosion and abrasion, to result in sufficient tissue loss for sensitivity to occur.[45,46] In general, it is agreed that the occurrence of gum recession is one of the most important aspects of dentine sensitivity. There is much more that can be said here, but our purpose is not to provide a definitive analysis of the semantic history of dentine sensitivity; that is the subject of ongoing work. We simply want to illustrate that there is a scientific history and medical history that explain the conditioning behind these accounts.

This complex history may well be a major determinant in why the condition has the character it has. In particular, it gets us to think about how different systems are "structurally coupled" to each other. Structural coupling in Luhmann's theory has been very nicely defined in Moeller.[50] as "a state in which two systems shape the environment of the other in such a way that both depend on the other for conditioning their autopoiesis and increasing their structural complexity" (p. 19). In other words, systems that are structurally coupled help each other develop complexity. The relationship is one of mutual coevolution to the point where a change in one can provoke changes in the other. This could be observed in our data on dentine sensitivity. In Annie's narrative, the dentist explained that her problem was one of exposed dentine. The dentist could only have explained it to her in this way because the medical system drawing on the complexity of the scientific system in some way has conditioned the way dentine sensitivity is discussed, diagnosed, and managed. Several systems are participating in her account. The scientific system will have participated in the diagnosis because the dentist would have been aware of the etiology of the condition as defined by this system. In this respect, dentists are structurally coupled to the science of dentistry; they draw on its complexity. In the same way, the science of dentistry would not become a reality for this patient and the dentist if the knowledge was not communicated in the interaction. Dentists are structurally coupled to the medical system, which decides what Annie should do about her condition. Annie's account is therefore a complex account that is derived from her recollection of an encounter that was the result of several structural bindings between the dentist and dental science, the dentist and the medical system, and, finally, Annie's encounter with the dentist. Annie's account is therefore "polyphonic" in the sense that her encounter resulted from multiple systems participating in and conditioning her original encounter, and by way of recollection her account of dentine sensitivity. In Luhmann's terms, the medical and scientific systems have made their own complexity available to Annie, via her encounter with her dentist, to help her make sense of her condition.[39] What we know from the science concerning the etiology of dentine sensitivity is that it can occur as a result of not brushing one's teeth or, somewhat ironically, from brushing them too much, in combination with other factors such as acid erosion from drinking carbonated beverages. This has obvious consequences for the moral character of the condition.

Morality and dentine sensitivity

A common theme in accounts of health and illness is moral character.[8,13,15] It should not be surprising that accounts of dentine sensitivity are no exception to this. Moral communication operates with the distinction between what is good and bad about something. During our analysis, such distinctions emerged relatively frequently. What we were interested in was how morality conditioned and restricted communication about dentine sensitivity. Cora's sensitivity was described as a consequence of brushing too hard:

> *Both sides, upper not my lowers, no not my lowers just my uppers erm but I've got quite a lot of recession on my gums and I was told that by my dentist quite a few years ago, because I went to my university dentist and he told me that I'd got quite like recession and he told me to stop brushing my teeth as harsh. And then I changed back to my local dentist and he's never raised it as an issue he's never raised it as a problem so he said 'yeah carry on everything is fine, everything is perfect' type of thing so I just carried on brushing my teeth the way I've always brushed them erm and then quite recently within the last few months a spoke to somebody who is a dentist and erm, for a different problem actually, I was asking for a bit of advice about a different problem and they said I'd got really bad recession. Like severe recession because of the abrasion from the tooth brushing erm and then after that I was more aware of it and I was more conscious and I think I have had sensitivity up there before obviously but because I'm so aware now that actually there's quite a big area where I've not got any enamel on my tooth.*
>
> *(Cora, Interview 4 [s1.4], Page 1)*

Obviously brushing one's teeth is considered a good thing. However, this problem had a double edge to it for Cora, who later stated:

> *I think sometimes erm admitting, I don't know admitting problems with your teeth erm opens yourself up to judgements and I don't think I would share it. Because I think people automatically assume with unless, had Cora got bad teeth then does she not look after her teeth then, does she not brush her teeth properly. And umm people might not think like that but that would be my fear so I wouldn't really like people other than my close friends and family to know otherwise, in case people might make judgements about me.*
>
> *(Cora, Interview 4 [s1.4], page 8)*

"Bad" recession and having "problems" (factual dimension of meaning) subsequently (temporal dimension) may open Cora to "judgments" (social dimension), and this possibility limits who she will and won't tell about her condition. From the perspective of this chapter, we can see how moral communication through the factual, temporal, and social dimensions of meaning communicate that Cora should restrict her toothbrushing and brush less harshly. Morality conditions the communication in an absolute sense. In the following communication, Adam describes how he feels about his teeth and that he and doctors/dentists are to blame for not treating them very well:

> *I know my teeth are not very good, I feel that they are not very good, I feel that they have not been good treated by me or by doctors as well. You know but there are not much things to do about it...*
>
> *(Adam, Interview 22 [s1.22], page 7)*

This communication once more has factual, social, and temporal dimensions. What is interesting, however, is that the order is slightly different; the social dimension places Adam and the doctor/dentist as coconspirators in damaging his teeth and on the temporal dimensions that nothing remains to be done. Likewise, what is "good" is associated with "good" oral "care":

> *So oral hygiene and self-care that's really important.*
>
> *Yeah especially being a smoker as well its even more important but yeah I'm very, I'm in a good routine with that. Brush minimum morning and night, I always floss at night and use a mouth wash and sometimes. Because at the minute I'm actually pending having a surgical procedure which I've got next Tuesday ironically so I actually brush at work as well during the day just because I'm conscious of keeping my mouth as clean as possible.*
>
> *(Annie, Interview 10 [s.1.10], page 5)*

In this communication, the "bad" of smoking makes "good" oral hygiene practices more significant. What is important is to note how the good practice continues despite the ever-shifting horizon of settings. Annie has an upcoming surgical procedure and she communicates that she is making sure she cleans her teeth even at work. Once more, morality helps Annie codify her efforts to maintain good oral hygiene as "good." It was clear that morality could be bound with emotional reactions. In the following excerpt, Cathy is responding to a question regarding how the sensitivity makes her "feel":

> *Oh that's interesting, I um, that really an interesting question, um I guess a little bit scared because I don't want my teeth to be sensitive, I think I might actually feel a little bit guilty like somehow I contributed to this problem because my gums are receding and maybe that is an indicator that I didn't take good enough care of my teeth, even though, my dentist tells me this is common, this has to do with the age process and blah blah blah and so I think that's what's going on I think to myself oh when was the last time I used that fluoride toothpaste you know the one that I was prescribed to me and have I been using that and have I been flossing oh dear you I think that that's what's going on in my head, what can I, you know, oh theres that pain again have I not been careful about this particular issue recently yeah.*
>
> *(Cathy, Interview 1 [s1.1], page 4)*

Here, morality codes the "goods" and "bads" of Cathy's experience of sensitivity and enables her to express what she feels is bad about her treatment of her teeth. The "bads" are linked closely with emotional reactions, in particular, fear and guilt. The account she gives shifts through morality and emotion in a fluid fashion, connecting the changing horizons of meaning. These horizons combine with the factual, social, and temporal dimensions of her communication and with morality for Cathy to evaluate her own performance with her oral health.

It is perhaps not surprising to find that morality codifies dentine sensitivity as a bad thing. Dentine sensitivity can be caused by both being overly vigilant and being neglectful. We can brush our teeth too hard or we might not brush them enough. In addition, when morality codifies communication on sensitivity, it seems to condition communications toward an exploration of the degree of care and the rights and wrongs of personal hygiene regimes. It also serves to further reinforce the isolation associated with dentine sensitivity by limiting who can and cannot be talked to about the condition.

"My teeth," "the teeth," and sensitivity

In the literature on chronic illness, it is common to see accounts of the body as both a subject and an object. In *The Body Silent*, Robert Murphy described how he gradually became alienated from his body as his condition advanced (Murphy, 1990).

> *Ok. What do you think causes it?*
>
> *Absolutely no idea. I don't know if its some thinning of the enamel on the ends of the teeth, I don't know. I mean I'm just guessing I honestly don't know what causes it.*
>
> *(Andrew, Interview 20 (s1.20), page 2)*

Here, Andrew responds to the question by directly drawing on the knowledge of the dental system. He talks of "the teeth." Both Bury[1] and Ballard[14] indicate that medical knowledge can act as a source for accounts of illness. Bury[1] tells us that such narratives have two purposes, to elicit the source of the illness and to describe its relational consequences. As we have seen repeatedly, these two aspects can be related to Luhmann's tripartite distinction between the factual, social, and temporal aspects of communication. In fact as Bury[1] has already noted, such accounts frequently spill over into the temporal dimension, i.e., respondents start to consider what to do about the illness. In Cathy's interview, both subject and object positions appeared in the same section:

> *That's interesting I think I'm more concerned about gum receding versus the sensitivity of my teeth even though the two are related and my dentist would say it's the gum recession that's causing the sensitivity to the teeth I'm not so concerned about teeth sensitivity on a day to day basis so much as gums receding.*
>
> *(Cathy, Interview 1 [s1.1], page 4)*

Although it is tempting to see this as evidence of Cathy adopting both a subject position and an object position with her condition, in more extreme forms of this argument, Cathy may be said to be experiencing a form of alienation from her teeth because they are talked about as though they are an object. However, it seems, to us at least, that we are *not* seeing the same problem experienced by Murphy.[51] Cathy's biggest problem was her gum recession and not the sensitivity, which she said she could deal with. From the perspective of Luhmann,[28] we understand such communications to be drawing on the complexity of the health system to communicate about her sensitivity. In this respect, the system *enables* her to communicate about what is happening.

The polyphonic unity of accounts of illness

So what can we say about the structural aspects of accounts of illness? We describe an interpretation influenced by Luhmann's social systems theory along with existing approaches. Taking the concept of Bury[1] regarding contingent narratives, we add the concept of "polyphony." The accounts of dentine sensitivity in this chapter have a degree of polyphonic unity. They are polyphonic in the sense that they capture three different horizons of meaning (the factual, social, and temporal) simultaneously. More crucially, they are polyphonic in the sense that they draw on the complexity of various social systems. In this case, accounts of dentine sensitivity draw on the scientific and medical systems and they also contain evaluations of personal esteem, i.e., morality. It must be kept in mind that the data are secondary and a more direct study of polyphonic communication in dentine sensitivity is warranted. Nonetheless, we hope that we have been able to illustrate how these multiple horizons of meaning appear through the unity of the accounts.

We support the ideas of Bury[1] regarding the contingent narrative but feel that it is worthy of further exploration. The theory of social systems would certainly agree that accounts of illness contain a search for the causes of illness. Bury[1] indicates that such accounts are dependent on medical and everyday sources. However, we feel that the theory of social systems may also potentially allow us more precision in exploring *how* this happens. What we would suggest is that, in dentine sensitivity at least, accounts of health and illness are drawing on the complexity of dental science, medicine, and other general codifications such as morality. The relationship is one of interpenetration, where systems, in this case science and medicine, make their complexity available to other systems, in this case dentists and their patients. Science and medicine would not exist if not for psychic systems; likewise, the participants in the study reported here would not be able to communicate in the way they do without recourse to science and medicine. They also would not be able to evaluate what is good and bad about their situations if not for morality. Sure, they might communicate about illness in a different way if they lived in a different society, but then they would not be living in a functionally differentiated society.

Society is not an interchangeable concept with morality and culture, as it seems to be in the illness accounts literature. For Luhmann, we exist in the environment of a functionally differentiated society, a society where multiple systems communicate and produce multiple different realities.[39] The accounts in this chapter highlight how three different horizons of meaning simultaneously emerge in communications about dentine sensitivity. In addition to this, we have suggested that two different social systems (the medical and morality) are used in the accounts to explain the fact that dentine sensitivity is not really a problem for dentists and that it results from excess care and from neglect of teeth, and this produces particular communicative problems for people with the condition. Rather than a reservoir of meanings being made available, we have a series of systemic distinctions that condition the meaning of dentine sensitivity. In summary, society's semantics serve to condense and stabilize meaning so that we know what to expect under certain conditions.

This brings us to the problem of culture as a source of these accounts. If we accept the analysis of Laermans,[37] then our task becomes more complex. To what extent can we suggest that the conundrum of dentine sensitivity is as much a cultural problem as opposed to a problem of science, medicine, and morality? There is a considerable degree of attraction in using the term culture to help us unpick how the accounts of dentine sensitivity are conditioned. In many ways, the accounts presented here are very similar to the public accounts of Cornwell.[13] As we saw in her analysis, public accounts made what was already recognizable available in communications about health and illness. This is similar to what Laermans[37] claims for culture from a systems theory perspective. The problem is, however, that the resemblance would end there. Cornwell[13] clearly demonstrates through a very careful analysis of her data that the medical system influences public accounts of illness, especially when it comes to etiology and supporting the more moral private claims in the accounts of illness in her study.

If we agree with Laermans,[37] then we could argue that the accounts of dentine sensitivity presented here do in some ways reference readily recognizable responses to the condition that could be termed "cultural." Participants did not like to bother people during meals and when out for drinks. They did not like to be seen as making an issue out of such a short-term problem, despite its many unpredictable and significant impacts. This may be interpreted to refer to some kind of "cultural etiquette." These did not, at least in these data, appear to be dominant. It seems from these preliminary analyses that the accounts of dentine sensitivity were conditioned by science and medicine. The truth is, however, that without more direct data and analyses derived primarily from a social systems approach, we cannot say more. Associated with this is the need to undertake a much more detailed literature review of how science has handled dentine sensitivity. However, our purpose in this chapter was to introduce the problem of the structural aspects of illness accounts, not to provide a definitive account of dentine sensitivity from a systems theory perspective. That is the subject of ongoing work.

In conclusion, the imperative of dentine sensitivity forces people to suffer from the condition in isolation from others. In some respects, this problem is related to the fact that the medical system has decided that sensitivity is not really a true illness but rather an everyday health problem. In some senses, then, the status of dentine sensitivity is conditioned by science and medicine, but perhaps also to some extent by culture. The interesting thing about all of these accounts is that they seem to be pushing people toward individualized solutions. In this respect, dentine sensitivity was a normal part of having a healthy mouth. In summary, the imperative of dentine sensitivity seems to be forcing people into a socially conditioned solution, i.e., the use of specifically produced toothpastes. Therefore, it effectively causes sufferers of the condition into the arms of consumer capital. The rule is that if you have sensitivity, you should buy toothpaste and use it as an isolated consumer. It is "good" (sic) to do this for this problem.

Here is precisely where we hit on a number of limitations of the analysis presented in this chapter. From the beginning, we have sought to argue that the sociology of chronic illness needs to begin to analyze the origins and sources of illness accounts. To this end, we have argued that form analysis is one way to unpick such sources. Yet, from what we can see, this analysis does not provide the full picture. What does this chapter achieve? It provides us with a map of the core distinctions and paradoxes of accounts of dentine sensitivity and, in some way, the analysis shows how these paradoxes are unraveled in communication. What the analysis cannot tell us is where these paradoxes and distinctions can be located historically. Nor can it tell us if there is a determining or core distinction that tends to predominate when it comes to dealing with the problem. We tend to think the answer lies in the predominance of the nonproblem problem nature of dentine sensitivity. However, we feel that only a semantic history could be a serious contender to unpick the emergence of this paradox. Then, if Luhmann's approach is to have any value in disentangling the roots of accounts of illness, it may well be necessary to combine analytic strategies, in this case form analysis with semantic analysis. In addition, the analysis suggested here has, in fact, taken us to the possibility of an alternative interpretation for the emergence of the paradox of dentine sensitivity. It is actually quite possible that the determinate force in the development of the accounts presented here have their origins in the commodification of sensitive pain rather than the development of the knowledge of the condition. It is only with ongoing historical analysis that we will be able to disentangle the relative merits of such competing explanations.

Acknowledgment

This chapter started as a discussion with Prof. David Locker about the nature of the meaning of illness. He was looking forward to engaging in the arguments as we developed subsequent drafts of the chapter but, sadly, he passed away before it could be completed. We therefore dedicate this work to his memory. In addition, we acknowledge the very kind support of GlaxoSmithKline, which funded the original study of the everyday impact of dentine sensitivity. This chapter is based on a secondary analysis of that data and is also related to the many debates and discussions we have had with members of that organization over the past 2 years.

References

1. Bury M. Illness narratives: fact or fiction? *Sociol Health Illn* 2001;**23**(3):263–85.
2. Bury M. Chronic illness as biographical disruption. *Sociol Health Illn* 1982;**4**(2):167–82.
3. Bury M. The sociology of chronic illness: a review of research and prospects. *Sociol Health Illn* 1991;**13**(4):451–68.
4. Pierret J. The illness experience: state of knowledge and perspectives for research. *Sociol Health Illn* 2003;**25**(3):4–22.
5. Hyden L-C. Illness and narrative. *Sociol Health Illn* 1997;**19**(1):48–69.
6. Crossley M. "Sick role" or "empowerment"? The ambiguities of life with an HIV positive diagnosis. *Sociol Health Illn* 1998;**20**(4):507–31.
7. Pinder R. Bringing back the body without the blame? The experience of ill and disabled people at work. *Sociol Health Illn* 1995;**17**(5):605–31.
8. Radley A, Billig M. Accounts of health and illness: dilemmas and representations. *Sociol Health Illn* 1996;**18**(2):220–40.
9. MacRae H. Managing courtesy stigma: the case of Alzheimer's disease. *Sociol Health Illn* 1999;**21**(1):54–70.
10. Anderton J, Elfert H, Lau M. Ideology in the clinical context: chronic illness, ethnicity and the discourse on normalisation. *Sociol Health Illn* 1989;**11**(3):253–78.
11. Locker D, Kaufert J. The breath of life: medical technology and the careers of people with post respiratory poliomyelitis. *Sociol Health Illn* 1988;**10**(1):23–40.
12. Herzlich C. *Health and illness*. London: Academic Press; 1973.
13. Cornwell J. *Hard-earned lives: accounts of health and illness from East London*. London: Tavistock Publications; 1984.

14. Ballard K, Kuh D, Wadsworth M. The role of the menopause in women's experiences of the "change of life". *Sociol Health Illn* 2001;**23**(4):397–424.
15. Williams G. Chronic illness and the pursuit of virtue in everyday life. In: Radley A, editor. *Worlds of illness: biographical and cultural perspectives on health and disease*. London: Routledge; 1993.
16. Riessman C. Strategic use of narrative in the presentation of self and illness: a research note. *Sociol Health Illn* 1990;**30**(11):1195–200.
17. Sanders C, Donovan J, Dieppe P. The significance and consequences of having painful and disabled joints in older age: co-existing accounts of normal and disrupted biographies. *Sociol Health Illn* 2002;**24**(2):227–53.
18. Pound P, Gompertz P, Ebrahim S. Illness in the context of older age: the case of stroke. *Sociol Health Illn* 1998;**20**(4):489–506.
19. Douglas M. *Natural symbols: explorations in cosmology*. London: Routledge; 1970.
20. Locker D. *Disability and disadvantage: the consequences of chronic illness*. New York, NY: Tavistock Publications; 1983.
21. Williams SJ. Chronic illness as biographical disruption or biographical disruption as chronic illness? Reflections on a core concept. *Sociol Health Illn* 2000;**22**(1):40–67.
22. Shilling C. Culture, the "sick role" and the consumption of health. *Br J Sociol* 2002;**53** (4):621–38.
23. Navon L. Beyond constructionism and pessimism: theoretical implications of leprosy destigmatisation campaigns in Thailand. *Sociol Health Illn* 1996;**18**(2):258–76.
24. Gellner E. *Reason and culture*. Oxford: Blackwell; 1992.
25. Kelly MP, Dickinson H. The narrative self in autobiographical accounts of illness. *Sociol Rev* 1997;**45**(2):254–78.
26. Williams S, Calnan M. The limits of medicalisation? Modern medicine and the lay populace. *Soc Sci Med* 1996;**42**:1609–20.
27. Smith D. An analysis of ideological structures and how women are excluded: considerations for academic women. *Can Rev Sociol Anthropol* 1975;**12**(1):353–69.
28. Luhmann N. *Social systems*. Stanford, CA: Stanford University Press; 1995.
29. Luhmann N. *Love as passion: the codification of intimacy*. Stanford: Stanford University Press; 1986.
30. Luhmann N. *Ecological communication*. Cambridge: Polity Press; 1989.
31. Luhmann N. *Political theory in the welfare state*. Berlin, NY: Walter de Gruyter; 1991.
32. Luhmann N. *Risk: a sociological theory*. New York, NY: Aldine De Gruyter; 1993.
33. Luhmann N. *The reality of the mass media*. Cambridge: Polity Press; 2000.
34. King M. What's the use of Luhmann's theory? In: King M, Thornhill C, editors. *Luhmann on law and politics: critical appraisals and applications*. Oxford: Hart Publishing; 2006.
35. Gershon I. Seeing like a system: Luhmann for anthropologists. *Anthropol Theory* 2005;**5** (2):99–116.
36. Luhmann N. *Observations on modernity*. Stanford, CA: Stanford University Press; 1998.
37. Laermans R. Theorizing culture, or reading Luhmann against Luhmann. *Cybern Human Knowing* 2007;**14**(2–3):67–83.
38. Martens W. The distinctions within organizations: Luhmann from a cultural perspective. *Organization* 2006;**13**:83–108.
39. King M. *Systems, not people, make society happen*. Holcombe Publishing; 2009.
40. Åkerstrøm-Andersen N. Creating the client who can create himself and his own fate—the tragedy of the citizen's contract. *Qual Sociol Rev* 2007;**3**(2):119–43.

41. Åkerstrøm-Andersen N. *Partnerships: machines of possibility*. Bristol: The Policy Press; 2008.
42. Sandelowski M. Words that should be seen but not written. *Res Nurs Health* 2007;**30** (2):129–30.
43. Åkerstrøm-Andersen N. *Discrusive analytical strategies: understanding Foucault, Laclau and Luhmann*. Bristol: Policy Press; 2003.
44. Dowell P, Addy M. Dentine hypersensitivity—a review: aetiology, symptoms and theories of pain production. *J Clin Periodontol* 1983;**10**:341–50.
45. Orchardson R, Gillam DG. Managing dentin hypersensitivity. *J Am Dent Assoc* 2006;**137**(7):990–8.
46. Dababneh R, Khouri AT, Addy M. Dentine hypersensitivity – and enigma? a review of terminology, epidemiology, mechanisms, aetiology and management. *Bri. Dental J.* 1999;**187**:606–11.
47. Hypersensitivity CABoD. Consensus-based recommendations for the diagnosis and management of dentin hypersensitivity. *J Can Dent Assoc* 2003;**69**(4):221–6.
48. Holland G, Narhi M, Addy M, Gangarosa L, Orchardson R. Guidelines for the design and conduct of clinical trials on dentine hypersensitivity. *J. Clin. Periodontol.* 1997;**24** (11):808–13.
49. Absi E, Addy M, Adams D. Dentine Hypersensitivity - the effect of toothbrushing and dietary compounds on dentine in vitro: an SEM study. *J. Oral Rehabi.* 1992;**19**:101–10.
50. Moeller H-G. *Luhmann explained: from souls to systems*. Illinois: Open Court; 2006.
51. Murphy RF. *The body silent*. London: WW. Norton; 1990.

Differentiation and displacement: unpicking the relationship between accounts of illness and social structure*

16

Barry J. Gibson and Ninu R. Paul
School of Clinical Dentistry, Claremont Crescent, University of Sheffield, Sheffield, UK

Introduction

Exploring accounts of illness as a social and political phenomenon has been a prominent preoccupation of the sociology of health and illness. Sociologists have helped to establish the significance of these accounts in a changing society.[1–5] In addition, the literature has focused on how understanding accounts of illness can help patients and physicians[6–8] and how they can enable society to observe the consequences of different illnesses and their treatment.[3,5–7,9] Janine Pierret, writing in 2003, stated that one of the key remaining challenges for those working on the problem of illness experience was to:

> *define a paradigm and methodology for handling the problems related to the social structure. This entails working out theories about the interrelations, reciprocal effects and feedback between subjectivity, cultural factors and social structure.[10] (p. 16)*

Although the literature at that time did have some examples of attempts to address this aspect of illness experience, it was clearly felt that this subject required more sustained exploration. Attempts to explain the relationship between "subjectivity, cultural factors, and social structure" have involved the use of numerous concepts. Some have used the concept of "contingent narratives"[3] and others have used concepts such as morality, "the social," culture, and ideology.[2,3,10–18] In what follows, we briefly explore how each of these concepts has been used to help disentangle the relationship between social structure and illness experience before outlining the specific aim of this chapter.

Bury[3] is most notable for developing the concept of "contingent narrative" to articulate the focus individuals have on their beliefs about the causes of a disorder. Common themes of such discussions are that the medical system, culture, and

*Previously published as Gibson B, Paul NR. Differentiation and displacement: Unpicking the relationship between accounts of illness and social structure. *Social Theory & Health* advance online publication 11 June 2014; doi: 10.1057/sth.2014.6

Dentine Hypersensitivity. DOI: http://dx.doi.org/10.1016/B978-0-12-801631-2.00016-6

everyday life are significant sources of information for such narratives.[3,10,17] Yet, despite the identification of these as sources, *how* this actually might work is not explained. A similar thing happens when morality is discussed as a significant source on which accounts of illness rest.[1,3,11–13] Although it is widely regarded as a "source" for accounts of illness, there is little explicit discussion of *how* this happens. This is important because there is a tendency within the literature for many of the concepts to overlap and for some to be seen as a subset of others. For example, Bury[3] in a commentary on the subject of illness narratives explores how Williams[14] uses the concept of morality in "chronic illness" and how this is developed to locate patients' stories "within a cultural framework"[3] (p. 275). Is morality a part of culture? In some respects we might see it this way, but then is culture itself also a horizon of meaning? Does it subsequently encapsulate other aspects of meaning?

Bury[3] provides some solutions to the problem, although this is not the direct goal of his review. He states, while drawing on the work of Gellner,[19] that narrative can be seen as an important resource, a way of constituting an infinite "reservoir of meaning" with the goal of making sense of illness and its impacts[3] (p. 264). He shows, drawing on the work of Kelly and Dickinson,[15] that culture can contribute to illness narratives by providing a series of frameworks (genres) that can then act as a "cultural resource" on which people can draw when making sense of illness. In doing so, culture is both a restraint and an enabling resource.[3,15,19]

The problem of the relationship between society and the experience of illness has also been explored within the Marxist tradition. Anderton et al.[6] uncovered how the experiences of white middle class families contrasted with those from Chinese families living under conditions of deprivation. The concept of "normalization" was mobilized as an ideological resource to obtain better outcomes for children from white middle class backgrounds. In this sense the relationship between the experience of illness and the accounts given to understand, explain, and act on this experience is both social and political.[6,11,20]

More recently, Gibson and Boiko[18] have explored how the work of Niklas Luhmann might contribute to the wider understanding of the relationship between society and accounts of illness; they argued that experiences of illness are narrated through "the voices and semantics of different systems"[18] (p. 166). Their study explored what these semantics might be in relation to the everyday impact of dentine hypersensitivity. Of the key distinctions that they uncovered, the most important seemed to be that it was a "non-problem problem." This peculiar distinction referred to the fact that although dentine hypersensitivity was problematic and had extensive and often difficult everyday impacts, it was not a problem in general. It was not a serious problem for the dentist and it was something that was not a major problem in everyday life. It was simultaneously a problem and not a problem. Gibson and Boiko's[18] interest was in the sociological significance of this distinction. They stated, in contrast to Bury,[3] that:

> *Rather than a reservoir of meanings being made available we have a series of systemic distinctions that condition the meaning of dentine sensitivity. In short, society's semantics serve to condense and stabilise meaning so that we know what to expect under certain conditions.[18] (p. 183)*

They went on to speculate whether the distinctions they had uncovered bore some relationship to particular "social systems" such as science or medicine, but were unable to tell if this was the case from their interview data. Their data only directed attention to the content of the pool of meaning and could not tell us *how these distinctions may have developed.* This brings us to the theoretical background for this chapter.

Theoretical background: sense-making, narratives of illness, semantic displacement, and social structure

As we have seen, this chapter shows, as its starting point, that narratives of illness are attempts at "making sense" of illness. In some respects this is not much different than what can be readily be determined from a cursory glance at this field of research.[1–3,10,14,21] However, the approach we propose goes further. We argue that by adopting the tools of Luhmann's[22] social systems theory, this "sense-making" process should be seen to have a relationship with more general processes of "sense-making" that happen in a virtual plane.[22,23] What this means is that the sense of what is being said in a communication about illness is not contained directly in narratives of illness but is displaced somewhere else. This other place is referred to as the virtual plane. Luhmann's focus has been on the abstract processes that shape this virtual plane.

The virtual plane of sense-making is shaped by communications that happen over time. But what is communication? For Luhmann,[22] communication is the unity of three things: information, utterance, and understanding. It is not the transmission of a message or content from a sender to a receiver and, as a consequence, does not travel through space and time. It is an event that occurs and as soon as it happens it disappears.[24] The key event in communication is how understanding is selected. For example, someone might grin or wince when someone else talks of their illness experience. What they do in response has meaning for that experience as it is communicated. Communication "happens retroactively, in the moment of understanding"[23] (p. 7). It cannot be reduced to information, utterance, or understanding but rather it combines these different elements.

Adopting this approach is similar to what Bury[3] and others have been proposing. Here, however, we propose a set of concepts and a framework with which to understand sense-making, not only in communications about illness but also in how the virtual plane of sense-making developed. Sense-making, both in relation to an illness communication and in relation to the background "virtual plane," is therefore something that is actualized out of a surplus of references to other possibilities of experience and action[22] (p. 60).

> *Sense thus opens up a virtual world for communication, a world that does not encompass a collection of pre-existing things, but is rather an unlimited and unpredictable reservoir of lines of communication.*[23] *(p. 8)*

Narratives of illness are communications that actualize this medium in relation to illness and disease; however, they also simultaneously direct our attention to this virtual plane. In some senses they "displace" the problems that are being referenced within the narratives to this virtual plane. Yet, the "sense-making" processes that build this virtual plane in the first place often remain hidden from view. But how does such a virtual plane get constituted?

In systems theory, the study of how society develops such pools of meaning is the study of social semantics.[25–28] Åkerstrøm-Andersen[29] states that semantics are:

> *characterised as the accumulated amount of generalised forms of differences (for example, concepts ideas, images and symbols) available for the selection of meaning within the systems of communication. In other words semantics are condensed and repeatable forms of meaning, which are at our disposal for communication.[29] (p. 87)*

Semantics are established in society over long periods of time. For these to develop, communications must link to each other. This linkage entails a fourth selection that must take place in communication. This refers to either the acceptance or the rejection of the sense of communication.[22] What matters is how previous communications are linked to what follows.[30] Mapping how this happens is termed semantic analysis. Before we move on to what this entails, a few further points are necessary.

Communication linkages tend to differentiate along different types of reference problem. If, for example, communications link around the problem of scarcity, then these communications become economized. When they link around making collectively binding decisions, they become politicized. When they link around if something is an illness or a health condition, they are medicalized. These problems of reference cannot be deduced theoretically and they are not ahistorical. They are contingent historical products that develop from concrete situations in everyday life.[23,31] These virtual linkages and problems are never resolved; they come and go as reference problems develop and disappear.

Virtual reference problems are biased. They have a positive and negative code. Luhmann's historical studies sought to uncover genealogies of how highly specific problems of reference developed as ways of processing specific sense-making problems. In this perspective, the reference problems that are evident in narratives of illness (the social, morality, medicine, and culture) are not simply amorphous pools of meaning. Rather, they should be seen as "forms of difference" composed of tensions and problems that can facilitate sense-making processes. We would suggest that narratives of illness point towards these pools of meaning and their underlying reference problems.

A final distinction relevant to the study of social semantics is the idea of concept and counter-concept. As we have seen, semantics are generalized forms of differences that society has decided to preserve, which semantics achieve through the use of concepts. Concepts condense multiple meanings and expectations. The effect of this is to produce a horizon of meaning associated with the concept. These multiple meanings are always locked into the concept through its counter-concept. So, for example, a conceptual pair might be dentist/patient. The dentist might evoke a series of expectations such as "small business owner," "health care professional," "sadist," "someone who fills teeth," to "someone who whitens teeth." The concept is linked to a horizon of meanings that are never completely obvious but that have developed over time in association with the concept. The idea of a patient might also evoke various different expectations. Changes to the expectations of the patient such as "demanding," "lacking trust," "unwilling," or "nervous" will have an effect on the concept of the dentist. There is often a battle over such concepts and change in one concept can change the other.

As we have seen, Gibson and Boiko[18] uncovered evidence that dentine hypersensitivity was commonly experienced and understood as a "nonproblem problem." Understanding how this became central to the accounts of illness they were studying involves establishing if such sense-making processes are in some way related to a virtual plane of sense-making along with its underlying reference problems. Therefore, the aim of this study was to go beyond Gibson and Boiko's[18] analysis to explore how such a sense-making process might have developed in relation to dentine hypersensitivity, including diagnosing if there were any underlying reference problems. We report on a study exploring the semantics of dentine hypersensitivity.

Methodology

When it comes to the study of social semantics, Luhmann was heavily influenced by *Begriffsgeschihte* or conceptual history.[22,27,32–40] This approach is inspired by Saussure's observation that there is a distinction between the synchronic and diachronic aspects of language. Although a language might change over time, at any point in time it has a definite structure. The study of conceptual history tends to alternate between synchronic and diachronic analysis of a particular semantic field.[34] Another important principle of conceptual history is not to identify the concept undergoing study with any single word or words but to realize that there will often be a range of terms designating the same underlying concept.

In the present study, this involved the use of three strategies deployed in two overlapping stages of analysis. First, we studied the semantic field *in the current literature*, which meant analyzing the range of terms (synonyms and antonyms) that form part of the broad vocabulary associated with the current understanding of dentine hypersensitivity. Second, we drew on the techniques of onomasiology and semasiology to establish the range of terms and references within the pool of meaning. Onomasiology is the "study of different terms available for designating the same or similar thing or concept"[34] (p. 2). Semasiology "seeks to discover all of the different meanings of a given term"[34] (p. 2). Third, because an important aspect of communication and sense-making is the fourth selection of meaning, we explored how communications linked to each other and what this meant for the meaning of the concepts being studied. Our focus was on diagnosing how the chain of sense-making within the literature became constituted over time. We describe how these techniques were deployed in the two overlapping stages of synchronic and diachronic analysis.

Synchronic analysis

In the beginning, we used onomasiology and semasiology to explore the current professional literature on dentine hypersensitivity. Our initial analysis focused on establishing the synchronic meaning of dentine hypersensitivity through a careful analysis of definitions and discussions of the terms used to refer to the concept. We began by searching standard online databases, including Web of Knowledge, PubMed, and OvidSP, using the terms "dentin* hypersensitivity," and this search revealed a wide range of citations. The search results revealed a total of 749 citations in the Web of Knowledge, 905 in PubMed, and 151 in Ovid Medline. Obviously, not all of these articles were relevant for this study.

It is important to bear in mind that we were interested in the sense-making process related to the changing meaning of the concept within the professional literature. Although a great number of studies use the language and concepts associated with "dentine hypersensitivity," very few discuss the meaning of the concept. The term is often used to report the findings of *in vivo* or *in vitro* studies, but nothing is said about the condition beyond that. In addition, such studies often did not add to the changing pool of meaning other than by reporting if a particular desensitizing agent was useful or not. As a result, we narrowed the initial search by looking for those articles with "dentine hypersensitivity" and "review" in the title. This reduced our search to a detailed study of 75 articles. By scanning through these articles, we found that many were not central to the definition and specification of the underlying problem or concept, and they often took for granted its meaning. They were also based on a smaller subset of key articles. These articles therefore were much more central to the current meaning of the condition. They were also more widely cited and, as a result, we could be sure they were central reference points for the current meaning of the concept. This reduction left us with 17 articles that became central to the *synchronic analysis* (see Table 16.1). The work of Dowell and Addy,[41] Pashley,[42,43], and Absi et al.[44] are central nodes in the literature; they are most widely cited and are central to the contemporary meaning of the term. Of particular note is another article that provided a detailed history of the diagnosis and treatments for the condition.[45] This article was particularly useful because it reviewed other work going back to the nineteenth century and therefore provided us with a way into the historical literature.

When it comes to the current meaning of dentine hypersensitivity, the work of Dowell and Addy[41] in particular highlighted the central problem of using the concept. They began by stating:

> *The pain arising from exposed dentine, typically in response to chemical, thermal, tactile and osmotic stimuli, is varied in both frequency and severity. For most patients the pain, which typically follows instantaneously upon application of the offending stimulus, is short-lived and usually resolves immediately after withdrawal of the stimulus. diagnostic difficulties are created since the symptoms do not differ from those which may be reported with dental caries and its associated pulpal changes. However, in the absence of other dental pathology, when such symptoms*

Table 16.1 Synchronic analysis: starting references

Date	Paper details	Citations
1976	Harris R and Curtin JH (1976) Dentine hypersensitivity. Australian Dental Journal, 21(2), 165–169	**5**
1983	Dowell P and Addy M (1983) Dentine hypersensitivity – A review: Clinical and in vitro evaluation of treatment agents. Journal of Clinical Periodontology, 10, 351–363.	**172**
1985	Berman LH (1985) Dentinal sensation and hypersensitivity. A review of mechanisms and treatment alternatives. Journal of Periodontology, 56(4), 216–222.	**18**
1986	Pashley DH (1986) Dentine permeability, dentine sensitivity and treatment through tubule occlusion. Journal of Endodontics, 12 (10), 465–474.	**154**
1987	Absi EG, Addy M and Adams D (1987) Dentine hypersensitivity – a study of the patency of dentinal tubules in sensitive and non-sensitive cervical dentine. Journal of Clinical Periodontology, 14, 280–284.	**137**
1987	Absi EG, Addy M. and Adams D (1987) Dentine hypersensitivity – the development and evaluation of a replica technique to study sensitive and non-sensitive cervical dentine. Journal of Clinical Periodontology, 16, 190–195.	**37**
1990	Rosenthal W (1990), Historic review of the management of tooth hypersensitivity. The Dental Clinics of North America. 34(3); 403–427.	**14**
1990	Pashley DH (1990) Mechanism of dentine sensitivity. The Dental Clinics Of North America, 34(3), 449–469.	**129**
1990	Addy M (1990) Etiology and clinical implications of dentine hypersensitivity. The Dental Clinics of North America, 34(3), 503–514.	**65**
1994	Addy M and West N (1994), Etiology, mechanisms, and management of dentine hypersensitivity. Current Opinion in Periodontology, 71–77.	**9**
2000	Addy M (2000) Dentine hypersensitivity: definition, prevalence, distribution and etiology. In Addy M et al. (eds) Tooth wear and sensitivity- clinical advances in restorative dentistry. London, Martin Dunitz.	**53**[¥]
2002	Addy M (2002) Dentine hypersensitivity: new perspectives on an old problem. International Dental Journal, 52(5), 367–375.	**51**
2003	Canadian Advisory Board for Dentine Hypersensitivity (2003), Consensus Based recommendations for the diagnosis and management of dentine hypersensitivity. Journal of Canadian Dental Association, 69(4), 221–26.	6[¥]
2005	Addy M (2005) Tooth brushing, tooth wear and dentine hypersensitivity – are they associated? International Dental Journal, 55(4), 261–267.	37

(*Continued*)

Table 16.1 (Continued)

Date	Paper details	Citations
2005	Walters PA (2005) Dentinal hypersensitivity: a review. The Journal of Contemporary Dental Practice, 6(2),107–117	35
2006	Bartold PM (2006). Dentinal hypersensitivity: a review. Australian Dental Journal, 51(3), 212–218.	20
2006	Orchardson, R and Gillam DG (2006). Managing dentin hypersensitivity. Journal of the American Dental Association, 137 (7), 990–998.	64

arise from the dentine exposed to the oral environment, the term 'dentine hypersensitivity' is used to describe the condition.[41] *(p. 342)*

They went on to describe the difficulty of distinguishing dentine hypersensitivity from dentine sensitivity:

To date no attempt has been made to clearly define the term 'dentine hypersensitivity'. In fact, since for the majority of sufferers, pain from exposed dentine only occurs on the application of the stimulus, the term 'dentine sensitivity' could equally well be applied.... This situation has arisen for several reasons. There is a dearth of information concerning the pulp changes, if any, associated with dentine hypersensitivity. Most studies correlating clinical signs and symptoms with pulpal pathology have been concerned with dental caries and its sequelae.... Dentine hypersensitivity is thus, perhaps, more a symptom complex than a true disease and the severity of the pain or the patient's interpretation of this, appears to determine whether treatment is sought.[41] *(pp. 342–343).*

It was clear at that time that dentine hypersensitivity suffered from problems of definition and that the concept seemed to be either confused or associated with a related problem, that of dentine sensitivity. At this point, we were wondering if we had an example of a concept and its counter-concept. *Diagnosing how these terms became retroactively connected to each other, how they came to be central to the pool of meaning, including how they provided a surplus of references for the concept of dentine hypersensitivity, became the focus of the second stage of analysis.*

Diachronic analysis

We had established that there was a link between two closely related concepts, dentine sensitivity and dentine hypersensitivity, and sought to trace the emergence of these concepts over time. We did this by mapping all the associated terms that had been deduced from the onomasiological and semasiological analyses in the literature.[34] In doing so, we traced the earliest reference we could find to words and terms we knew had been associated with the concepts of dentine hypersensitivity and dentine sensitivity (e.g., "sensibility of dentin," "sensitiveness of dentin," "dental neuralgia"). This was achieved by paying close attention to the texts and looking back through the citations that formed part of the synchronic analysis and reading those articles, too. So, for example, we uncovered a citation to the work of Gysi,[46] who referred to earlier sources and concepts. *Our approach was to identify key points in time when the underlying concepts and distinctions in dentine hypersensitivity changed.* In this respect, we follow Åkerstrøm-Andersen's[40] approach in tracing several different movements in the meaning of dentine hypersensitivity. This included looking to establish how these two concepts became entangled in the sense-making process and looking at changes in the underlying reference problems.[22,40] We present the findings of this diachronic analysis. We do this because a presentation of the synchronic analysis is less sociologically interesting and there is a need to conserve space.

Differentiation and displacement: the emergence of dentine hypersensitivity

Nettleton,[47] following Foucault,[48,49] argues that the mouth became separated from the body *through its creation as an object of knowledge and governance.*[47] This process involves opening up the mouth to the clinical gaze, and understanding its colors, textures, and cell arrangements. It also involves understanding how to manage these through the techniques of surveillance and governance.[47,48] In this study what we uncovered is that embedded within this general "sense-making" process are important processes of *differentiation.*[31,50] This process involves distinguishing different qualities of sensation and pain in thinking feeling subjects. We describe the historical process of how the *differentiation* of "sensibility" from "sensitivity" occurred *and* how this is different in reference problems for dentistry. We do this by first going through the changing semantics of concepts associated with sensitive teeth before making some general remarks.

Concerns about the underlying problem of sensitive teeth date back thousands of years, although the earliest reference we could find to anything specific about the sensitivity of teeth was by Perry in 1827 discussing "sensibility of the tooth" as a consequence of the loss of enamel in response to acid attacks. This loss of enamel was said to occur in the early stages of disease.[51] The discipline continued to distinguish and differentiate the underlying mechanisms of this "sensibility".[52] For example, it was proposed that small holes called "tubules" ran from the circumference of the tooth to the center and these were filled with fluid secreted from the tooth pulp. It was suggested that the "sensibility of the tooth" was a consequence of water moving in and out of these "tubules" and this was conceptualized as a "hydrostatic pressure." This movement could also happen as a result of various stimuli such as cold water, rubbing or scraping, and contact with certain substances.[53,54] But the differentiation process really began to take shape when, in 1900, a distinction began to develop between "sensibility" and "sensitiveness." The sensitiveness of dentine:

> *...is only of a secondary nature, and is not physiologic. The physiological sensibility of dentine is sufficiently provided for by the pulpa and the periosteum, so that the supply of nerves in dentine would be superfluous. If the dentinal canalicules contained nerves, the progress of caries would be painful, which is not the case as long as the pulp is not attacked.*[46] *(pp. 865–866)*

These different concepts, the "sensitiveness of dentine" and the "physiological sensibility of dentine," concerned *differentiating* biological mechanisms and, therefore, producing *different* sensations of pain.[54] The differentiation process was therefore producing different reference points for communication about sensations in teeth as a result of different underlying pathologies. It was also differentiating the sensations that occurred during *dental treatment*. It is obvious that sensitivity was getting in the way of the work of dentists. For example, the principal source of "sensibility" of dentine was found to be the stimulation of the pulp as a consequence of the movement of fluid through tubes that run through dentine. These tubes, called "canalicules," were filled with an "aquous content," so that:

> *A pressure or drawing exercised upon the aquous content of a dentinal canalicule that opens into a carious cavity is directly transmitted to the other end of the dentinal canalicule where it is loosely closed by the odontoblasts and then the odontoblasts which are abundantly interwoven with nerves feel the pressure or drawing as a sensation of pain.*[46] *(p. 866)*

The "pressure" being referred to are the sensations that occur in response to the "drawing" or excavation of dental caries (disease) *in dental treatment*.[46] The quality of these sensations perhaps became *the* central reference problem for dentistry marking the point at which the "sensitiveness of dentine" was *differentiated*[50] from the "sensibility" of teeth. This differentiation process is important. An early sign of dental disease was understood to be enamel loss, and this became understood as the main cause of the *sensibility* of dentine.[51] By the turn of the twentieth century, a key cause of "sensitivity of dentine" was dental treatment.[46] As we can see from Table 16.2, before this point "sensitivity" *included* incipient disease, dental treatment, and chemical action.[53,54] The *differentiation* of "sensibility" from "sensitivity" resulted in different reference problems. One reference was to the clinical management of disease (sensitivity) and the other was to a pain experience in the absence of disease (sensibility). Even though these were *differentiations* in the pool of meaning, each of the various concepts remained tied to each other *because they needed to be distinguished from each other* and acted on in different ways.

The practical work of the clinic generated different forms of sensitivity referring to different underlying reference points for problems of diagnosis and management. These findings are very similar to the work of Mol[67] in "The body multiple," which

Table 16.2 **The differentiation of dentine sensitivity and dentine hypersensitivity**

Date	Reference	Conceptual implications		
1827	Perry (1827)[51]	'Sensibility of the tooth' – recognised as a quality associated with the progression of acid attack and an early sign of disease	↓	**Undifferentiated communication.** These publications do not link with each other. It isn't until Gysi (1900) that the process of differentiation starts to take hold.
1851	**Blandy, A (1851)[52]**	**Sensibility explanation suggested to be due to fluid moving in and out of 'tubules' in the dentine**		
1856	White, JD. (1856); Discussion at American Dental Convention (1856)[53]	Sensibility explained in terms of the exposure of dentine as a result of destruction of enamel and gum recession		
1857	Cartwright, S (1857)	Sensitivity could vary dramatically even on different surfaces of the same teeth		
1860–1865	Barker (1860), Atkinson, WH (1864), Francis, CE (1865);	Debate about causes of sensitivity along with various claims about who discovered the causes first		
1871–1875	Yingling (1871); Harriman, GB (1875)	New treatments recommended and further arguments about physiological properties of dentine		
1878	Harris, CA (1885)[54]	'Sensitive dentine' described as a painful condition as a result of the 'disturbance of the soft fibrils radiating from the pulp into the tubules of the tooth' or 'to the conduction of the shock of the instrument to the pulp' or to 'local chemical action and a pathological condition of the general system.' Used the term and used illustrations to demonstrate how tubules in dentine may be the cause.		

(*Continued*)

Table 16.2 (Continued)

Date	Reference	Conceptual implications		
1891–1889	Harris, CA (1891); Smale and Colyer, (1893); Harris, CA (1897); Eames, CF (1899); Burchard (1998)	Treatments had expanded and it was noted that sensitivity during cavity preparation was not uncommon. Harris (1897) talked about "the sensitiveness of dentine", long list of treatments.	↓	
1900	**Gysi (1900)[46]**	**Distinguishes 'sensitivity of dentine' from the 'sensibility of teeth', concept linked with dental treatment**		
1900–1915	Guerini (1909); **Burchard (1915)[55]**; Prinz, H (1916)	**First mention of the concept of 'Dentine Hypersensitivity'** – an over-reaction to *dental treatment* and distinct from 'dentine sensitivity'.		**Instability, controversy and technology.** During this time there was considerable debate concerning the causes of dentine hypersensitivity. It became known as 'an enigma'. Specialist toothpastes enter the market (Thermodent and Sensodyne).
1916–1950	Prinz, H (1922); Hopewell Smith, A (1924)[56]; Council on Dental Therapeutics (1932); Grossman, L (1935); McGehee, WHO., (1936); Prinz, H and Rickert, UG (1938); Lukomsky, EH (1941); Hoyt, WH and Bibby, BG (1943)	Beginning of Organised dentistry in the United States results in the formation of the Council on Dental Therapeutics who reject Sensitex as suitable to treat dentine sensitivity and hypersensitivity. Controversy over the mechanism for dentine hypersensitivity, do the tubules have nerve fibres or not?		
1951–1955	Kramer, IRH (1955); **Pawlowska, J and WJ Znaczenie (1956)[57]**	**Discovery of strontium chloride for use in home treatments**		
1956–1960	**Fitzgerald, G (1956)[58]**; Abel, I (1958); Scott, D (1960); Zelman, H and Hillyer, CE (1963)	**Development of 'Thermodent' and later 'Sensodyne' as home remedies.**		
1961–1965	**Ross, MR (1961); Brännström, M (1961;1962a; 1962b;1963a;1963b;1965)[59–64]**; Scott, D and Tempel, (1963); Skurnik, H (1963); Yamada, M (1963); Zelman, H and Hillyer, CE (1963); **Rosenthal, MW (1964)**; Smith, BA and Ash, MM (1964)	**A series of studies by Brännström are critical in rejecting the hypothesis that dentine is enervated. Support is given to the 'hydrodynamic theory'.** **Patent granted for Sensodyne.**		

1966–1982	Everett, FG et al., (1966); Anderson, D and Matthews, B (1966); Brännström, M (1967); Carrasco, HP (1971); Edwall, L and Scott, D (1971); Horiuchi, H and Matthews, B (1973); Johnson, G and Brännström, M (1974)[65]; **Harris, R and Curtin, JH (1976);** Green, et al. (1977); Gangarosa, LP (1978); Brough, K et al. (1979); Greenbill, J and Pashley, D (1982); Johnson, R B Zulqar-Nain, et al. (1982)	Continued study of mechanisms but also the emergence of clinical studies including the discovery of the psychosocial in dentine hypersensitivity. Dentine Hypersensitivity described as 'an enigma'.		
1983–2003	**Dowell, P and Addy, M (1983)[41];** Berman, LH (1984); Collins and Perkins (1984); Collins et al., (1984); Berman, LH (1985); Brough, K et al., (1985); Clark et al., (1985); Flynn et al., (1985); Pashley, DH (1986)[42]; Absi, EG et al. (1987)[44]; Addy et al. (1987); Absi, EG, Addy, M, and Adams, D, (1987)[44]; Clark et al., (1987); Harris, CA (1987); Rosenthal, W (1990)[45]; Pashley, DH (1990)[43]; Addy, M(1990); Mazor et al. (1991); Collaert, B and Fischer, C (1991); Addy, M (1992); Scharman, A and Jacobsen, PL (1992); Addy, M and West, N (1994); Maguire, K (1996); Dababneh, RH et al., (1999); Addy, M (2000); Addy, M (2002); Rees, JS and Addy, M (2002); Suge, T et al., (2002); Brandt, PD., and de Wet, FA (2002); Canadian Advisory Board for Dentine Hypersensitivity (2003)[66]	**Dentine sensitivity and dentine hypersensitivity become differentiated as different qualities of sensations that can be experienced in teeth.**	↓	**Differentiation:** The condition is given a definition that involves a 'differential diagnosis as: "a short, sharp pain arising from exposed dentin in response to stimuli typically thermal, evaporative, tactile, osmotic or chemical and which cannot be ascribed to any other form of dental defect or disease"

draws on a material semiotics and actor network theory. In this work the disease, atherosclerosis, is "enacted" through different practices in different parts of the hospital. In one place, atherosclerosis is the thickening of the intima visible under a microscope. In another, it becomes an inability to walk a certain distance without pain. Mol[67] indicates several key things. First, diagnosis is the practical management of the body; something is done with the body. It involves poking, touching, dicing, and manipulating the body. Second, diagnosis involves material relationships, and this enables particular aspects of the body to become intelligible. Third, multiple practices enact disease in multiple forms that, in turn, are called multiple realities. In this study it was primarily processes of *differentiation*,[22] which led to different concepts of sensitivity, sensitiveness, and sensibility, which were nonetheless in some way all related to each other. This process of *differentiation* contributed to the production of different lines of observation and communication about sensitivity in teeth. They also contributed to the pool of meaning that was developing around sensitivity.

These differentiation processes were critical because they prepared the way for the eventual emergence of the concept of dentine hypersensitivity. At this time "sensitive dentin" was distinguished from "hypersensitive dentin":

> *...term sensitive dentin applied to this condition is a misnomer; all vital dentin is sensitive, and its degree of sensitivity differs markedly in individuals; it is only when hypersensitivity is observed that the condition becomes pathological. Hypersensitivity of dentin may be defined as such a degree of sensitiveness as interferes with the proper excavation and shaping of a carious cavity; or which, in the absence of dental ministrations, causes painful symptoms, as a rule reflected about neighbouring parts.*[55] *(p. 393)*

In this quotation, "sensitive dentin" is considered "normal," and hypersensitivity of dentin is considered "pathological." However, "hypersensitivity of dentin" is pathological *because it interferes with dental treatment.* In this instance one concept, sensitive dentin, becomes a "normal" problem to be managed. This concept is supported by another concept, hypersensitivity, which became an "abnormal" over-reaction. Dentine sensitivity became a normal problem to be managed as part of the practical management of teeth in the clinic, "dentin hypersensitivity" became an unmanageable condition and a pathological over-reaction. We would suggest that dentine hypersensitivity became a counter-concept to dentine sensitivity.

As we have already noted, these ideas relate to challenges coming from the emerging sociology of disease. In this approach, the challenge has been for sociology to develop a nuanced understanding of the physiological diversity of different conditions.[67] Following this approach, Gardner et al.[68] have proposed that specific "diseases involve specific and often multiple disease diagnostic practices, which, in turn, can generate particular experiences of the body"[68] (p. 849). Here, we find that sensitivity of teeth developed very different meanings, and these differences in meaning refer to very different problems. First, the classification of sensitivity into different categories happened because of the practical problem of making sense of and managing sensitivity during the work of the dental clinicians. Second, the classification of different sensations as sensitive or hypersensitive enabled dentistry to focus on what would be its central concern: the management of sensitivity during dental treatment. This emphasis had an impact on the status of dentine hypersensitivity, which remained a highly unstable and indeed controversial category:

> *...there are no nerves in dentine...consequently, no sensation in dentine. There is no hypersensitivity of dentine.*[56] *(p. 76).*

The body is a site for multiple lines of investigation and these jostle for position, push up against one another, produce different emphases, and generate diverging plans of action. They change the pool of meaning by enabling a wider repertoire for making sense of various sensations in teeth. Even though dentine hypersensitivity was an unstable category, efforts continued to try to understand the condition. There was, for example, a series of communications in the literature seeking out the underlying mechanisms of dentine hypersensitivity (see Table 16.2). Several mechanisms were proposed for the condition. It was claimed that the nerves in dentine were acting as pain receptors; however, osmotic changes in the fluid-filled dentinal tubules caused activation of pain receptors in the pulp and, finally, the tubules themselves were said to be part of the sensory mechanism.[69] No individual hypothesis was refuted and this left dentine hypersensitivity more or less with the appearance of a medically "unexplained symptom"[a].[70–73]

A critical stage in the process of differentiating dentine hypersensitivity from dentine sensitivity occurred in a series of studies by Martin Brännström (see Table 16.2) involving experiments examining the condition in relation to various stimuli.[59–65,74] As a consequence, the "hydrodynamic theory" was confirmed as the best explanation for the continued presence of sensitivity in the absence of other causes.[62] These developments eventually culminated in a consensus statement in Canada that dentine hypersensitivity was "a short, sharp pain arising from exposed dentin in response to stimuli typically thermal, evaporative, tactile, osmotic or chemical and which cannot be ascribed to any other form of dental defect or disease".[66] Yet, despite this, dentine hypersensitivity remained conceptually bound to dentine sensitivity. It could only be diagnosed *after dental pathology had been ruled out.* This "differential diagnosis" has been shown to be common in other "medically unexplained" conditions such as neuralgia.[71–73] However, unlike these conditions, in which there is a lack of legitimacy in diagnosis, dentine hypersensitivity as a category has become relatively well-established. But how was this achieved? So far, in our analysis, we have uncovered how dentine hypersensitivity became established *as a line of communication in dental clinics*. In the next section we explore how this line of communication also involved exploring practical solutions to the problem.

[a]We acknowledge one of the anonymous reviewers for suggesting that we consider this concept.

The market and dentine hypersensitivity

So far, our analysis has been focused on diagnosing the effects of the *differentiation* of separate points of reference for dentine sensitivity and hypersensitivity. The differentiation of these reference points had consequences for the solutions that were developed for each problem. The solutions to the problem of sensitivity and hypersensitivity took different forms because they were thought of as different problems. These differences acted as reference points for communication about the various problems as well as the eventual solutions. Dentine sensitivity remained a central problem for dentistry because it interfered with dental treatment. It would eventually be managed through the use of local and general anesthetics. Dentine hypersensitivity became something to be managed at home through the use of various "dentifrices."

As stated previously, the treatment of sensitivity goes back centuries; however, for our purposes, the earliest references we could find to dentine sensitivity were in 1825.[75] By the 1920s, commercial preparations such as Sensitex had become available. Rosenthal[45] recorded approximately 104 different substances that were used as solutions to the problem of dentine sensitivity. These included asbestos, anodyne cement, carbolic acid, morphine, oleate of cocaine, and quinine sulphate, among other things[45] (p. 406). Rosenthal[45] also explained how a report by Pawlowska[57] detailed that strontium chloride could produce a good effect on dentin hypersensitivity.[57] Based on this, Sensodyne toothpaste was "formulated with strontium chloride hexahydrate"[45] (p. 417). Eventually "Sensodyne," a toothpaste for dentine hypersensitivity, was introduced in the United States in 1961. Sensodyne is not the only dentifrice that has been developed for dentine hypersensitivity, however. Emoform toothpaste existed in Switzerland in the late 1940s to the early 1950s and contained formaldehyde, calcium carbonate, magnesium carbonate, and a mineralizing salt composed of sodium bicarbonate, sodium chloride, potassium sulphate, and sodium sulphate.[45] This substance was eventually licensed to Thomas Leeming Co. and sold in the United States as Thermodent toothpaste. This toothpaste was the leading brand until the early 1960s, when Sensodyne eventually surpassed it in terms of sales. During this time, research into the effectiveness of these toothpastes was being sponsored.[58]

We suggest that these processes had several implications. First, the production of various remedies produced different solutions that reflected how the underlying problems were envisioned. These differentiations produced different lines of communication and action that then enabled dentists and patients to investigate and manage sensitivity in different ways. The differentiation of dentine sensitivity and dentine hypersensitivity not produced the underlying diagnostic categories but also had an effect on the solutions that were foreseen. Although Mol[67] observed the same condition having multiple realities, here the differentiation of the same sensations produced different conditions and solutions. Dentine hypersensitivity can only be established through a differential diagnosis and, as such, it remains dependent on the elimination of dentine sensitivity as a condition. This had consequences for the experience of dentine hypersensitivity. It became a problem that dentists tended to avoid. On one level it became a "nonproblem problem"[18] because it was not a problem for the dentist.

However, the production of commercial solutions to the problem of dentine hypersensitivity also *produced* it as a condition to be managed *at home* through the use of specialist dentifrices. In so doing, dentine hypersensitivity became a nonproblem problem, precisely because it could be easily managed. The suggestion is that although considerable legitimacy emerges from being defined as a medical problem, in keeping with the work of Nettleton[71] and others,[72,73] legitimacy can also be derived from the market. This source of legitimacy however, appeared to generate a different horizon for the meaning of the condition. It means that when we talk of dentine hypersensitivity, we talk of a health condition rather than a disease, something that is trivial rather than problematic. In contrast to medically unexplained symptoms,[71–73] patients have a label that tells them how they can speak about their condition and, more importantly, what they can do about it.

Discussion

The starting point for this chapter was that there is a "pool of meaning"[3] that acts as a resource from which accounts of dentine hypersensitivity can be drawn. What it suggests is that with respect to dentine hypersensitivity, this pool of meaning is not simply a surplus of references; it is composed of a series of different "lines of communication".[22] The range of concepts and lines of communication developed over the past 186 years to enable society not only to make sense of sensitivity in teeth but also to manage this sensitivity in different ways. When communication about dentine sensitivity happens, it refers in some way to this pool of meaning in *an attempt to make* sense of the underlying sensation. This attempt to make sense seeks to tease out whether the sensations are a consequence of pathology or if they are relatively "trivial." This "differentiation" process then resulted in the same underlying symptom being understood through different meanings and solutions. Dentine sensitivity became sensations that resulted from dental disease and required management during dental treatment. Dentine hypersensitivity is an abnormal reaction to various stimuli that could be "remedied" through the use of different toothpastes. The narratives of dentine hypersensitivity in Gibson and Boiko's[18] study refer to this pool of meaning and operate through these distinctions.

By adopting the approach of conceptual history, we have sought to determine how one "conversation in mankind"[76] has distinguished different qualities of sensitivity in teeth. Throughout we have suggested that the development of dentine hypersensitivity as a concept involved the constant displacement of various underlying reference problems. For example, dentine hypersensitivity, when it first emerged, was developed to articulate abnormal reactions to dental treatment. By focusing on normal sensitivity, dentists left these over-reactions to the patient to manage. This displacement to outside of the clinic had consequences for hypersensitivity. It remained an unstable concept and lacked legitimacy. Despite this displacement, it remained a persistent problem that required an explanation and a solution. This is why we can see other communication that "scientized" the underlying reality and others that "marketized" the solution in the form of sensitive toothpastes. It might be suggested that behind these developments are the effects of functional differentiation. Dentine hypersensitivity was simultaneously medicalized, scientized, and economized.[22] It suggests that what Gibson and Boiko[18] uncovered in their narratives was not a distinction that had its roots in the medical system, but rather a distinction that got its meaning from the varied effects of this broad social process.

The story we present here is the story *behind* the way society has developed a specific language for communicating about sensitivity in teeth. It is the task of conceptual history to pay attention to which groups take charge of concepts and the biases this produces.[76] In this story, dentistry and dentists took charge of sensitive teeth, and this became their domain. In doing so, they set boundaries around what was normal and what was abnormal. The discipline focused on controlling, studying, and treating the causes of "normal," "pathological" sensitivity, leaving dentine hypersensitivity to become a peculiar abnormality or an irritation to the workers in the clinic. However, this story is not simply about how a particular clinical gaze operated. Interest in dentine hypersensitivity did not stop. Over the period of 186 years, remedies were developed for the problem of dentine sensitivity and dentine hypersensitivity. These developments had an effect on the meaning of dentine hypersensitivity. It became a problem with a relatively "simple" solution: use toothpaste. The problem, it seems, was defined, displaced, trivialized, and transformed into a nonproblem problem.

Acknowledgments

We thank the members of the Unit of Dental Public Health at the University of Sheffield for their kind feedback and support in reviewing earlier versions of this manuscript. In particular, we thank Sarah Baker, Olga Boiko, and Peter G. Robinson for all their help. Finally, although this research was not directly funded by GlaxoSmithKline, we still acknowledge the stimulating debates and discussions we have had with the team concerning the nature of dentine hypersensitivity. We also thank them for enabling this manuscript to be made available as an open access publication.

References

1. Bury M. Chronic illness as biographical disruption. *Sociol Health Illn* 1982;**4**(2):167–82.
2. Bury M. The sociology of chronic illness: a review of research and prospects. *Sociol Health Illn* 1991;**13**(4):451–68.
3. Bury M. Illness narratives: fact or fiction? *Sociol Health Illn* 2001;**23**(3):263–85.
4. Hyden L-C. Illness and narrative. *Sociol Health Illn* 1997;**19**(1):48–69.
5. Crossley M. 'Sick role' or 'empowerment'? The ambiguities of life with an HIV positive diagnosis. *Sociol Health Illn* 1998;**20**(4):507–31.

6. Anderton J, Elfert H, Lau M. Ideology in the clinical context: chronic illness, ethnicity and the discourse on normalisation. *Sociol Health Illn* 1989;**11**(3):253–78.
7. Pinder R. Bringing back the body without the blame?: the experience of ill and disabled people at work. *Sociol Health Illn* 1995;**17**(5):605–31.
8. MacRae H. Managing courtesy stigma: the case of Alzheimer's disease. *Sociol Health Illn* 1999;**21**(1):54–70.
9. Locker D, Kaufert J. The breath of life: medical technology and the careers of people with post-respiratory poliomyelitis. *Sociol Health Illn* 1988;**10**(1):23–40.
10. Pierret J. The illness experience: state of knowledge and perspectives for research. *Sociol Health Illn* 2003;**25**(3):4–22.
11. Herzlich C. *Health and illness*. London: Academic Press; 1973.
12. Cornwell J. *Hard-earned lives: accounts of health and illness from East London*. London: Tavistock Publications; 1984.
13. Riessman C. Strategic use of narrative in the presentation of self and illness: a research note. *Soc Sci Med* 1990;**30**(11):1195–200.
14. Williams G. Chronic illness and the pursuit of virtue in everyday life. In: Radley A, editor. *Worlds of illness: biographical and cultural perspectives on health and disease*. London: Routledge; 1993.
15. Kelly MP, Dickinson H. The narrative self in autobiographical accounts of illness. *Sociol Rev* 1997;**45**(2):254–78.
16. Pound P, Gompertz P, Ebrahim S. Illness in the context of older age: the case of stroke. *Sociol Health Illn* 1998;**20**(4):489–506.
17. Ballard K, Kuh D, Wadsworth M. The role of the menopause in women's experiences of the 'change of life'. *Sociol Health Illn* 2001;**23**(4):397–424.
18. Gibson B, Boiko O. The experience of health and illness: polycontextural meaning and accounts of illness. *Soc Theory Health* 2012;**10**:156–87.
19. Gellner E. *Reason and culture*. Oxford: Blackwell; 1992.
20. Herzlich C, Pierret J. The social construction of the patient—patients and illness in other ages. *Soc Sci Med* 1985;**20**(2):145–51.
21. Williams SJ. Chronic illness as biographical disruption or biographical disruption as chronic illness? Reflections on a core concept. *Sociol Health Illn* 2000;**22**(1):40–67.
22. Luhmann N. *Social systems*. Stanford, CA: Stanford University Press; 1995.
23. Farias I. Virtual attractors, actual assemblages: how Luhmann's theory of communication complements actor network theory. *Eur J Soc Theory* 2013;1–18.
24. Clam J. System's sole constituent, the operation: clarifying a central concept of Luhmannian theory. *Acta Sociol* 2000;**43**(1):63–79.
25. Luhmann N. *Love as passion: the codification of intimacy*. Stanford, CA: Stanford University Press; 1986.
26. Luhmann N. *Risk: a sociological theory*. New York, NY: Aldine De Gruyter; 1993.
27. Luhmann N. The paradox of form. In: Baecker D, editor. *Problems of form*. Standford, CA: Standford University Press; 1999. p. 15–26.
28. Luhmann N. *The reality of the mass media*. Cambridge, MA: Polity Press; 2000.
29. Åkerstrøm-Andersen N. *Discrusive analytical strategies: understanding Foucault, Laclau and Luhmann*. Bristol: Policy Press; 2003.
30. Schneider W. The sequential production of social acts in conversation. *Hum Stud* 2000;**23**(2):123–44.
31. Luhmann N. *Essays on self-reference*. New York, NY: Columbia University Press; 1990.
32. Luhmann N. *Observations on modernity*. Stanford, CA: Stanford University Press; 1998.

33. den Boer P. The historiography of German Begriffsgeschichte and the Dutch project of conceptual history. In: Hampsher-Monk I, Tilmans K, van Vree F, editors. *History of concepts: comparative perspectives*. Amsterdam: Amsterdam University Press; 1998.
34. Hampsher-Monk I, Tilmans K, van Vree F. *History of concepts: comparative perspectives*. Amsterdam: Amsterdam University Press; 1998.
35. Hampsher-Monk I. Speech acts, languages or conceptual history? In: Hampsher-Monk I, Tilmans K, van Vree F, editors. *History of concepts: comparative perspectives*. Amsterdam: Amsterdam University Press; 1998.
36. Koselleck R. Social history and Begriffsgeschichte. In: Hampsher-Monk I, Tilmans K, van Vree F, editors. *History of concepts: comparative perspectives*. Amsterdam: Amsterdam University Press; 1998.
37. Koselleck R. *The practice of conceptual history- timing history, spacing concepts*. Stanford, CA: Stanford University Press; 2002.
38. Koselleck R. *Futures past—on the semantics of time*. New York, Colombia: University Press; 2004.
39. Åkerstrøm-Andersen N. Polyphonic organisations. In: Bakken T, Hernes T, editors. *Autopoietic organization theory*. Copenhagen: Copenhagen Business School Press; 2003. p. 151–82.
40. Åkerstrøm-Andersen N. Conceptual history and the diagnostics of the present. *Organ Hist* 2011;**6**(3):248–67.
41. Dowell P, Addy M. Dentin hypersensitivity—a review. 1. Etiology, symptoms and theories of pain production. *J Clin Periodontol* 1983;**10**(4):341–50.
42. Pashley D. Dentine permiability, dentine sensitivity and treatment through tubule occlusion. *J Endod* 1986;**12**(10):465–74.
43. Pashley D. Mechanism of dentine sensitivity. *Dent Clin North Am* 1990;**34**(3):449–69.
44. Absi EG, Addy M, Adams D. Dentin hypersensitivity—a study of the patency of dentinal tubules in sensitive and non-sensitive cervical dentin. *J Clin Periodontol* 1987;**14**(5):280–4.
45. Rosenthal W. Historic review of the management of tooth hypersensitivity. *Dent Clin North Am* 1990;**34**(3):403–27.
46. Gysi A. An attempt to explain the sensitiveness of dentine. *Br J Dent Sci* 1900;**43**:865–8.
47. Thorogood N. 'London dentist in HIV scare': HIV and dentistry in popular discourse. In: Bunton R, Nettleton S, Burrows R, editors. *The sociology of health promotion*. London: Routledge; 1995.
48. Foucault M. *The birth of the clinic: an archaeology of medical perception*. London: Routledge Classics; 1963.
49. Foucault M. *The history of sexuality: the will to knowledge*. Middlesex: Penguin Publishers; 1998.
50. Luhmann N. *The differentiation of society*. New York, NY: Columbia University Press; 1982.
51. Perry. *Perry's treatise on the prevention and cure of the tooth-ache*. London: Messers. Butler; 1827.
52. Blandy A. On the sensibility of teeth. *Am J Dent Sci* 1851;**1**:22–8.
53. Discussion at American Dental Convention. Pathological condition of diseased dentine and its treatment. *Am J Dent Sci* 1856;**VI**(2 series):507–33.
54. Harris C. *The principles and practice of dentistry- including anatomy, physiology, pathology, therapeutics, dental surgery and mechanism*. 11th ed. Philadelphia, PA: WM. L. Fell and Co.; 1885.

55. Burchard H. *A textbook of dental pathology and therapeutics for students and practitioners*. Philadelphia, NY and New York, NY: Lea and Febiger; 1915.
56. Hopewell Smith A. The non-innervation of dentine. *Proc R Soc Med* 1924;**17**:63–79.
57. Pawlowska J, Jego Znaczenie W. Zeboleznictwie I profilaktyce. *Czas Stomatol* 1956;**9** (7):353–61.
58. Fitzgerald G. A clinical evaluation of a new agent for the relief of hypersensitive dentin. *Dent Dig* 1956;**62**:494–7.
59. Brännström M. Dentinal and pulpal response. V. Application of pressure to exposed dentin. *J Dent Res* 1961;**40**:960–70.
60. Brännström M. The elicitation of pain in human dentine and pulp by chemical stimuli. *Arch Oral Biol* 1962;**7**(1):59–62.
61. Brännström M. Dentinal and pulpal response: VI. Some experiments with heat and pressure illustrating the movement of odontoblasts into the dentinal tubules. *Oral Surg Oral Med Oral Pathol* 1962;**15**(2):203–12.
62. Brännström M. A hydrodynamic mechanism in the transmission of pain-producing stimuli through the dentine. In: Anderson D, editor. *Sensory mechanisms in dentine*. Oxford: Pergamon Press; 1963. p. 73–106.
63. Brännström M. Dentine sensitivity and aspiration of odontoblasts. *J Am Dent Assoc* 1963;**66**:366.
64. Brannstrom M. The surface of sensitive dentine. An experimental study using replication. *Odontol Revy* 1965;**16**(4):293–9.
65. Johnson G, Brännström M. The sensitivity of dentine. *Acta Odontol Scand* 1974;**32**:29–38.
66. Canadian Advisory Board for Dentine Hypersensitivity. Consensus-based recommendations for the diagnosis and management of dentine hypersensitivity. *J Can Dent Assoc* 2003;**69**(4):221–6.
67. Mol A. *The body multiple: ontology in medical practice*. Durham and London: Duke University Press; 2002.
68. Gardner J, Dewb K, Stubbe M, Dowell T, Macdonald L. Patchwork diagnoses: the production of coherence, uncertainty, and manageable bodies. *Soc Sci Med* 2011;**73** (6):843–50.
69. Ambrose J. The innervation of dentine. *J Dent Res* 1943;**22**:13–25.
70. Undeland M, Malterud K. Diagnostic work in general practice: more than naming a disease. *Scand J Prim Health Care* 2002;**20**(3):145–50.
71. Nettleton S. 'I just want permission to be ill': towards a sociology of medically unexplained symptoms. *Soc Sci Med* 2006;**62**(5):1167–78.
72. Jutel A. Medically unexplained symptoms and the disease label. *Soc Theory Health* 2010;**8**(3):229–45.
73. Greco M. The classification and nomenclature of 'medically unexplained symptoms': conflict, performativity and critique. *Soc Sci Med* 2012;**75**(12):2362–9.
74. Brännström M. The hydrodynamic theory of dentinal pain: sensation in preparations, caries and the dentinal crack. *J Endod* 1986;**12**(10):453–7.
75. The Examiner Classifieds. *Tooth ache and ear ache—Perry's essence*. The Examiner; 1825, Monday, January 3rd.
76. Scholz B. Conceptual history in context: reconstructing the terminology of an academic discipline. In: Hampsher-Monk I, Tilmans K, van Vree F, editors. *History of concepts comparative perspectives*. Amsterdam: Amsterdam University Press; 1998. p. 87–101.

Consumer advertising and the meaning of dentine hypersensitivity

Barry J. Gibson and Melanie Hall
School of Clinical Dentistry, Claremont Crescent, University of Sheffield, Sheffield, UK

Introduction

Previously, we investigated lay experiences of dentine hypersensitivity (DH) by, for example, explaining how individuals describe their pain, their behaviors in relation to the condition, and how it can become accepted as a relatively unproblematic aspect of everyday life.[1,2] This suggested that the management of DH could be integrated into normal everyday health care practices that, although articulated as a problem, were not a major problem for individuals. The impacts associated with DH could be considered something we might normally expect. Gibson and Boiko went on to identify some of the key forms that appeared to underlie the way people communicated about DH.[1] The claim was that different social systems, for example, the scientific and medical systems, helped to structure people's accounts of their condition.

Furthermore, Gibson and Boiko highlighted the potential for social systems theory as a means to explore how accounts of DH were structured and organized. Further analysis of their data concluded that DH was experienced in isolation from others and was not generally regarded as an illness, but rather as an everyday experience.[1] In other words, it was a "nonproblem problem." They pointed out that dentists did not see any real reason to take responsibility for caring for this issue. They (Chapter 15) argued that there was something about the forms of communication regarding DH that seemed to direct the patient away from the dental clinic, forcing "sufferers of the condition into the arms of consumer capital"[1]. But what does this mean and how can it be studied further? Gibson and Boiko indicated that there were several options. One was to study the historical emergence of the problem through a semantic history.[1] The other was to explore, in more detail, processes of "commodification" and how these might be related to experiences of DH.

Dentine Hypersensitivity. DOI: http://dx.doi.org/10.1016/B978-0-12-801631-2.00017-8

Commodification and DH

Commodification has its origins in Marxist sociology. It refers to the process whereby objects (including toothpastes) produced by human labor gain independence from that labor.[3] They are then traded and develop a life of their own. This concept is closely related to two other concepts within Marxism, reification and fetishism. Reification refers to the process whereby our labor becomes transformed into an object and seen as a thing in itself. The commodity is an object that in some respect satisfies human desires or wants. Fetishism is the process by which the commodity becomes such an object.[3,4] In this case, fetishism is the process by which an inanimate object such as toothpaste might become animated with meaning to come to serve our desires and needs.[5–8] The traditional Marxist perspective on concepts such as commodification, reification, and fetishism is that they enable us to see how capitalist society transforms inanimate objects such as toothpastes into animated forms with new meaning. This process has traditionally been seen as inherently negative, resulting in a kind of false consciousness. Silva has recently countered this negative view by arguing that reification and fetishism are productive.[5] They enable action and the transformation of meaning and therefore do not necessarily result in "disengagement" and apathy[9] (p. 94). As Silva claims, "Reifacts-fetishes" enable us to think about things in different ways; therefore, they enable us to act in different ways.

In this chapter we are interested in how sensations of sensitivity in teeth become animated in the commodity[9] of toothpaste. This involves what Parry calls a deal of "positioning or re-representation" of the items "that must occur before they are 'qualified' as commodities."[10–12] Gibson and Boiko used form analysis to explore how everyday accounts of DH were communicated and speculated that there may be multiple sources from which these communications are drawn (Chapter 15). One such system might be the mass media and advertising. In this chapter, we aim to explore how the meaning of DH is constructed in advertising.

It should be clear from this brief introduction that sociology sees toothpaste as more than an inanimate object. Toothpastes communicate, and in doing so they animate that which is inanimate. This chapter goes on to explore how toothpastes designed to treat hypersensitivity are animated. We do so by going beyond a negative view of Marxism to explore the relationship between the structure and form of advertisements about DH in exactly the same way that Gibson and Boiko read everyday communications earlier.[1,13] Rather than reviewing the theoretical background of this method, we refer the reader to Chapter 15.[1]

Materials and methods

Because the aim of this chapter is to ascertain how the meaning of DH is constructed in advertising, we explore the relationship between the structure and form of communication in advertisements about DH. Our goal is to identify how advertisements seek to organize the meanings we associate with the condition.[1] As we have seen, one of the findings of this work related to the role of products for the condition. It was suggested that sensitivity toothpaste could become integrated as part of the "normal" response for the condition and that these toothpastes, in some way, communicate a solution to the pain people experience with DH. How might this happen and what are the implications for the production of toothpaste?

This work involved a detailed visual analysis of a sample of contemporary broadcast and print advertisements. The research centered on an analysis of Sensodyne advertising campaigns between 2005 and 2011 and includes television and print advertisements. The analysis was guided by the principles of semiotic and form analysis. This process is outlined here.

Analysis of advertising campaigns

The research team contacted an advertising agency responsible for an archive of print and television commercials. The agency made 11 print advertisements and 27 television commercials available. These were initially for *Sensodyne* and *Pronamel* products. For the purpose of this study, only *Sensodyne* advertisements were analyzed. This left us with 6 print and 18 television advertisements. Recent advertisements for *Colgate Sensitive* were also collected as a useful comparison.

Data analysis

Because our aim was to explore how the meaning of DH is constructed in advertising, we explored the "forms of meaning" that we could see in communications about DH toothpastes. Forms of meaning are basically distinctions that are used to explain the meaning of a phenomenon. Our secondary goal was to explore how these forms might be related to the forms of meaning in narratives about DH. The content of the advertisements was analyzed from a range of perspectives incorporating form analysis and semiotic analysis.

Form analysis

The work presented here involved establishing the distinctions being used in advertisements about DH. This method has been covered in some detail by both Boiko and colleagues and by Gibson and Boiko.[1,14] Figure 17.1 illustrates the process of analysis and the implications of the underlying theory. When communication occurs, the observation that is being communicated "draws a mark"; that is, it marks out what is included in communication and what is to be left out. It indicates what is valued and what is not valued. All communications after the basic distinction is drawn will "connect" on the basis of this value. A simple object is indicated when it is called and everything else that is outside of the basic distinction is left unobserved (see Figure 17.1). In the first distinction in Figure 17.1, we have indicated a chair and what is outside of this is not defined. We might simply call this

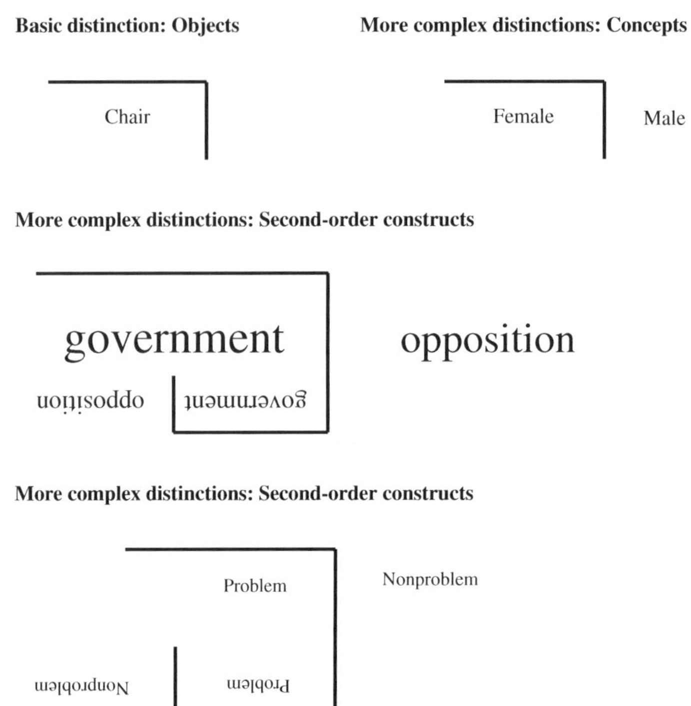

Figure 17.1 The use of distinctions in Luhmann's social systems theory.

the world. The world is composed of a virtually infinite number of objects that can be made the subject of communication. However, there are more complex distinctions, which Luhmann called "concepts." Previous work has demonstrated this in some detail with respect to communication in dental clinics. In this case, the analysis seeks to uncover the distinctions that guide communications about DH.[13]

Concepts occur when the other side of the distinction is called or indicated. In Figure 17.1, the concept of gender is composed of female and male. These two indications subsequently restrict communication on the theme of gender. Communication about gender then becomes thematized around the competition and distinction between male and female. In Figure 17.1, female is valued over male. This kind of value distinction can be readily seen in feminist communication. There are, however, more complex "second-order" distinctions. The classic case in Luhmann's systems theory is the distinction between government and opposition. Second-order distinctions such as these fold back on themselves and produce a re-entry of the distinction into itself. So, whereas every government has an opposition that is outside itself, it can also have an opposition within itself. The ruling elite within government are opposed by others who are nevertheless in the same government. The evolution of these complex forms of communication gives society its particular character. We suggest that the basic distinction between DH as a nonproblem problem operates through all communication about the condition. However, it is deployed very differently by different perspectives. These differences expose the biases of various systems and observers who are seeking to communicate about the condition. Consumers, organizations such as multinational companies, and the dental system have different ways of observing the condition.

Previous work on communication about DH emphasized the *problem* side of the distinction at particular points, such as when people are in pain,[1] so much so that they consult medical advice. However, for dentists the *nonproblem* aspect of the distinction is more significant because DH doesn't require treatment. As a consequence, after talking with the dentist and when the diagnosis of DH is confirmed, the patient is directed toward the market to buy products to remedy this problem. The producers of toothpaste and desensitizing products place emphasis on the *problem* posed by sensitivity, but this problem can be transformed into a *nonproblem* through the use of their technology. In this sequence, the problem of DH, something that was originally a problem for the medical system, is now a problem for the individual consumer. The dental system is displaced and the problem becomes solvable in the home.

Semiotic analysis

Semiotic analysis is the process by which media may be examined in close detail in terms of signs and their meanings. Media, such as advertisements, present fertile ground for exploring words, objects, images, sounds, and gestures.[15] Social semiotics is a research tool that can reveal insights into images that are not immediately apparent. For Kress and van Leeuwen, each image is representative of the social world.[16] Images can be analyzed for the following:

a. *Representational meaning*: in which people, places, or objects that are depicted in images convey representational meaning. For instance, esthetic aspects, such as the color and shape of a toothpaste tube, may play a role here.
b. *Interactive meaning*: referring to the relations between viewers and the image, distance, contact, and point of view.
c. *Compositional meaning*: consisting of informational value. For example, the position of elements in advertisements known as "framing" refers to whether elements of a composition are separated or connected in saliency and modality.

In addition to these basic techniques, we drew on social systems theory and, in particular, followed the analytical strategy associated with form and semantic analysis.[17,18] Through this analysis, it was possible to explore the messages that were communicated and how this was achieved. For example, we observed how DH was described, how it should be treated, and who communicated these messages (that is, whether advertisements featured professionals or lay people). Furthermore, the analysis explored how these might be interpreted, for example, what distinctions were used.

Process

Establishing a proforma

The advertisements were scanned and watched by the research team. A proforma (Table 17.1) was completed for each advertisement.

Table 17.1 Proforma for data analysis

- Summary of advertisement: This involved writing a short description of the advertisement to aid familiarity with the data. This description focused on what the advertisement was about whilst highlighting the main features of the advertisement. Features might include dentists, 'real' people, descriptions of sensitivity, locations, illustrations and food.
- Transcription of the advertisement: This entailed recording details the script of the advert for detailed exploration.
- Type: The 'type' of advertisement was recorded as a 'customer testimonial', 'dentist' or 'other health professional'. This enabled the team to ascertain whether there was an emphasis on particular 'types'.
- Themes: The research team interrogated the transcripts for emergent themes such as, the focus on pain, visiting the dentist, ease of use of the product. This enabled the research team to obtain an understanding of the main themes explored in the advertisements. These themes were then explored for the underlying distinctions[18] they were using.
- Images: Any images were studied to identify the individuals or objects in the advertisements, personal objects, dental instruments, the products. This facilitated an understanding of the context of the product; for example, whether located in the home or in a shopping mall.
- Distinctions employed: the themes could be organised into distinctions, in line with Gibson and Boiko's[1,14] research and systems theory[17]. Examples included pain/not pain and take action/not action. These were recorded in order to identify how dentine hypersensitivity is constructed and to identify any differences between the distinctions employed by lay people, as identified by Gibson and Boiko[1] and those in the advertisements.

Results

Broadcast advertisements

Broadcast advertisements consisted of lay testimonials (11 advertisements), dentist testimonials (2 advertisements), and "other experts" (2 advertisements). The lay testimonials were typically accounts of sensitivity from the perspective of "real" people. Delivered in their homes by young adults, these accounts described how sensitivity became a problem, what it felt like, and what prompted it. The narrative in these advertisements often involved a description of how the individual experienced pain, how the individual visited the dentist, how the individual began to use a specific toothpaste, and how the sensitivity went away. Two advertisements featured several individuals stating that they could not eat ice cream or drink cold beverages because of their sensitivity. As a result of using *Sensodyne*, they were able to consume these again. Each advertisement lasted approximately 40 s. Lay testimonials span all years of the sample (2005–2011).

Another group of advertisements involved either oral health professionals or dentists. Dentist testimonials tended to be given by dentists in a dental clinic; they described what sensitivity feels like and indicated that it is something many of their

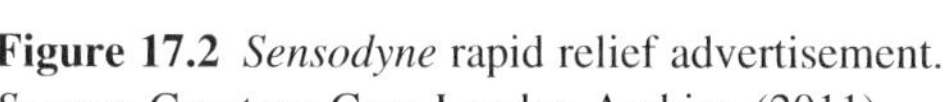
Figure 17.2 *Sensodyne* rapid relief advertisement.
Source: Courtesy Grey London Archive (2011).

patients experience and that using special toothpaste can remedy the sensitivity. Images used in these advertisements include dental instruments, photographs of families, and the product. Again, each lasted approximately 30 s. These testimonials were used sporadically (2005, 2010, 2011) and form the current advertising strategy (2013). Two other testimonials featured "nondental" experts. For example, a "director of oral health care" and an "ideation director" (2009). These testimonials described the sensation of DH and how the product works. They were accompanied by illustrations and animations of how this happened. Scripts of some of these advertisements are provided below.

Print advertisements

The research team analyzed six print advertisements for *Sensodyne* products. In three, the images depicted foods that may trigger DH, indicating that this could be remedied quickly through the use of toothpaste (Figure 17.2). A caption at the top of the page states "Love hot chocolate again in 60 seconds." An image of a mug of hot chocolate in the middle of the page becomes heart-shaped with the mirror image of the shadow of the handle on the side of the mug. The drink in the mug is topped with cream and chocolate in an appealing way. Below this mug is a picture of the product, followed by a text explanation of how it works.

Another advertisement described the product in simple text, explained how it worked, and mentioned that it was "clinically proven." A picture of the product was included. Two were described as "technical." These presented an image of a tooth and illustrated how the toothpaste worked to remedy the sensitivity. This image was accompanied by a description of DH and of the product.

The following paragraphs explore the central distinctions observed in all advertisements in one form or another.

Before/after

A striking feature of the advertisements was their plot and narrative structure, which were particularly vivid in the customer testimonials.[15,16] For example, the following advertisement presents a young woman, Amanda, in her living room, describing her experience with DH:

> *Caption*: The problem with sensitive teeth
>
> Female, "Amanda, Crystal Palace" sitting in her living room, surrounded by CDs
>
> *Amanda*: I have sensitive teeth. I'd always have a hot drink in the morning. I started to feel a bit of a twinge, I chose to ignore it, it'll go away. It just got continually worse. I thought, I've got to go to a dentist.
>
> *Caption*: What did her dentist recommend?
>
> *Amanda*: He said to use a toothpaste for sensitive teeth, and I thought, well that's quite easy. I might as well give it a go. I use Sensodyne two times a day. You can really notice a difference. You know, you hear these things and you think, no not a toothpaste, but it has, it really has. So I just wish that I'd used it before. Ice cream, I had some ice cream.
>
> *Source*: Courtesy Grey London Archive (2011).

Such advertisements portraying "real" people sharing their experiences of DH typically begin with text such as "Amanda's story." The person appears on the screen, accompanied by a caption stating their name and town. The individual describes the feeling of sensitivity, e.g., a "twinge," and the context during which it occurred, e.g., eating ice cream or having a cold drink. This is then followed with a caption, "What did he/she do?," which is then followed by a description of the interaction with the dentist who recommended *Sensodyne*. The person describes how it is used, e.g., as regular toothpaste, twice per day, and that as a result, the person no longer experiences the sensitivity and is able to resume activities such as eating particular foods without discomfort. A caption is displayed advising people to consult their dentist about sensitivity.

The plot suggests that dental sensitivity is experienced suddenly, perhaps when consuming pleasurable food or drinks. In this respect, consumers are being told that they cannot know when to expect it. But here there is an interesting paradox, because they know that it may appear after a cold drink. So although it is sudden, it is also known. Moreover, not all cold drinks might cause sensitivity, but some might. This predictable unpredictability is what makes the condition so tricky for the consumer. These findings mirror those uncovered in the narratives of individuals with the condition (Chapters 15 & 1).[1,2] In some respects, the advertisements animate the toothpaste, which becomes associated with the sensations experienced with the condition. However, the consumption of a particular product can help an individual prepare for this uncertain certainty, and perhaps prevent it. The uncertainty of DH becomes embedded within a particular program of action that happens over time.

There is an order to the action plan at the heart of the advertisements. The stage denoted by the caption "What did he/she do?" indicates that action *should* be taken to address this pain. The action is a consultation with a dentist for diagnosis, which brings sensitivity into the professional domain. However, after diagnosis, a "simple" solution is offered that takes sensitivity out of the professional sphere. This is achieved through the consumption of a specific product that can be readily purchased at a pharmacy or supermarket and consumed within the confines of an individual's own bathroom. Customers then incorporate the product into their oral health routine, which results in a happy ending for the consumers; they are no longer afflicted by DH. The advertisements communicate in time but are also located within a particular scheme as the solution and end result of sensations of sensitivity. The advertisements, then, we would argue, prepare expectations by suggesting patterns of appropriate action in response to feelings of sensitivity. They also articulate DH as predictable and unpredictable, and this establishes the necessity of having the product integrated into everyday life. The inanimate toothpaste makes the unpredictable predictable. In this sense, the product becomes animated, not only with the properties of sensitivity but also with the notion that all unwanted sensations may be banished.

Another interesting feature was how the advertisements animate the toothpaste within a particular space, the home. Each testimonial is situated in the home, in a living room, dining room, or kitchen. This might be interpreted as the experience of sensitivity being a "normal" occurrence and *Sensodyne* as a "normal" product, because homes, living rooms, dining rooms, and kitchens are each ordinary environments. During the testimonial the camera focuses on the person, then moves to personal items, and subsequently moves to the product in the bathroom. This pattern places the product in someone's bathroom as part of that person's lifestyle. Conveying the message that this is where the product belongs, i.e., in a familiar context. The product becomes subsumed into the home space and is transposed into a normal, mundane, everyday thing. It is not presented as something special; it does not belong in the medicine cabinet, it belongs on your bathroom countertop. Significantly, it substitutes the previous toothpaste, because it does the same job *and more*.

Already we can see how advertisements about DH constitute complex communications, articulating a number of distinctions at once. In doing so, the advertisements communicate several changes in meaning and do so simultaneously. They transform a technology for the removal of sensitivity in everyday life into a technology that is embedded within everyday life. Luhmann stated that the particular problem that the mass media attempts to solve is to tell us all we need to know about the world.[19] These advertisements animate the inanimate and tell us what to do when we have sensitivity. They program a pattern of action and shape our expectations about what will happen when we use the technology in the form of toothpaste. In doing this, the technology, which is presented as a medical solution to our everyday woes with our teeth, becomes transformed from a medical product to a home-based remedy. The advertisements therefore use the distinction of before and after to communicate what to do about DH, and at the same time they transform the meaning of the technology, moving it from an external space and social system, into the home and eventually the body.

This section has explored the journey from sensitivity to treatment. This temporal ordering has implications for place and space. The advertisements account for how people experience DH and how they overcome it. Part of this entails the use of space; advertisements feature consumers in their homes with the product in their bathroom, i.e., personal space. These advertisements animate the product through time and space from the dental clinic into the home, into the family. The advertisements do not simply animate the toothpaste, they do so in a particular way. This involves movements around various distinctions that the advertisements use. The toothpaste changes from being a scientific solution to an everyday problem to becoming an everyday thing embedded within the home.

Normal/not normal

Another important distinction indicated that sensitivity was both abnormal and normal. It is abnormal in the sense that the sensations are uncomfortable and shouldn't be ignored. Consumers are urged to take action because it is not normal to be prevented from enjoying ice cream or cold drinks; these are cast as everyday activities and they should be resumed. Dental sensitivity is clearly problematic from this perspective. Take Natalie's "story" as an example. This advertisement gives the audience a young woman's account of DH, which she tells while at home, and recounts the limitations her sensitivity placed on her.

Caption: Natalie's story

Natalie's account begins with her sitting in her kitchen, discussing her experience:

Natalie: I can remember going for dinner with friends and it being very uncomfortable to have quite a sweet drink, having Sangria, to the point where I'd say 'that's it, I can't have anymore', my dentist diagnosed that I did have sensitivity. I felt like I hadn't been looking after my teeth properly.

There is another caption: *What did she do?*

Natalie's story continues:

Natalie: I started using *Sensodyne*. There's the whitening version, that's my favourite. It sits in my bathroom all the time. It ticks all the boxes really. It keeps my teeth clean and white. I'm not sure how it does it, but it seems to work.

A caption accompanies the image of Natalie: Helps restore your teeth's natural whiteness.

Source: Grey London Archive

This person's everyday experience of DH portrays her as a young woman with a busy social life. DH had impacted on this by limiting her consumption of sangria with her friends. This narrative is consistent with the findings from our study of the impact of DH in everyday life,[1] because it reflects an *affects/stop* form that relates to the previous section on *before/after*. This knowledge is again problematic, because there is not only physical impact but also social impact, which is not normal. The narrative goes on to describe how Natalie addressed her sensitivity and resumed her normal behavior. This pattern of communication is not simply restricted to consumer narratives; it can also be discerned in the scripts presented by dental health professionals in other advertisements. Here, the presentation is of the condition as normal and abnormal as well as common and uncommon.

Caption: A dentist talks about sensitive teeth

Dr Cliff Rush: It's a very common problem. It's increasingly common in 16-24 year olds. Patients usually complain of a very sharp pain caused by something cold or sweet. Chocolate is extremely painful if it gets next to that sensitive part of the tooth.

Caption: What does he recommend?

Dentist: Patients who are suffering from sensitivity are delighted to hear they can treat their problem with something as simple as changing their toothpaste. There are a range available if used like any normal toothpaste will reduce the sensitivity and in many cases, cut it out altogether.

Caption: Ask your dentist about sensitivity

Source: Courtesy Grey London Archive (2011).

Note how it is described as common; the implication here is that it is "normal" because *lots* of people experience it, yet it is also a *problem*. Problems are not normal, but common problems are. Patients complain about this common problem and seek out treatment.

Once again the advertisement follows similar distinctions as those uncovered in our study of the impact of DH on everyday life. The experience of sensitivity as *normal* and *not normal*, *common* and *uncommon*, was uncovered within everyday narratives of DH[1] (see Chapter 15). What is unclear, however, is which side of the underlying distinction is being valued. Advertisements use distinctions to communicate about a product. The way the distinctions are unraveled reveals the meaning of the product. In this example, DH starts out as a problem; the implication is that if you have this problem, then you should visit your dentist. The problem becomes a problem for the dentist. But the dentist suggests using *Sensodyne*. The problem becomes a nonproblem through the consumption of the toothpaste. Furthermore, the advertisements emphasize taking action in terms of managing DH. This contributes to the idea that DH is *not normal* and something should be done to stop it. However, the "treatment" is, in fact, normal and easy. These findings contrast with the testimonials of lay people. Those advertisements urge individuals to "consult their dentist;" however, when they do, they are subsequently instructed to purchase the product. This places patients on a kind of "merry-go-round" of booking an appointment, perhaps having to pay for it, taking time out of their lives, occupying the precious time of a medical professional, and going through the anxiety of maybe having a "real" problem, when they could simply visit a pharmacist.

Although it might seem that the advertisements are narrating an unnecessary complication, we should look more closely. Some of the DH advertisements were about *Sensodyne*. These advertisements indicated that the symptoms of DH might mask underlying disease and that it was best to attend the dentist in case there was a more serious problem. These advertisements clearly articulate the history of DH. As we have seen throughout this volume, and in the dental literature, DH is a diagnosis of exclusion. Therefore, it might be argued that it would be unethical for a product that claims to remedy the problem to not include the dentist in the narrative.[20–22] The "merry-go-round" thus reflects the historical position of DH as something that was not a "real" disease and therefore wasn't something dentists should worry about. It is clear from the history of DH that it represented an "enigma" in the clinic.[23] Nonetheless, it is important to make sure that there is no other underlying pathology.

The combination of the distinction between before and after and the distinction between sensitivity as a problem and a nonproblem also indicate something else. *Sensodyne* is simultaneously a scientific, medical, and consumer product. The scientific, medical, and economic systems participate to produce "the reality" of DH, with all using these distinctions in different ways and unraveling them to expose different aspects of the meaning of the condition. The advertisement presents this "reality" in the form of a series of distinctions, communicating that behind this simple tube of toothpaste are these different "realities": a medical reality concerning the diagnosis of the condition, a scientific reality exploring and establishing whether the toothpaste "works," and an economic reality that if it works, then you should use it.

How these distinctions are being used speaks to changes in how health is understood in society. Sociologists have argued that modern forms of power are productive.[24] Nettleton[25–28] has shown how dentistry now operates through patients as psycho-emotional entities who take care of their own oral health by disciplining their bodies through toothpaste consumption. The dental "gaze" thus became incorporated into the everyday lives of individuals as they control, monitor, and evaluate their own health[25–29]:

> *Health status becomes something that consumers must continually monitor and evaluate, so as to be aware of their needs and wants to take the appropriate steps to satisfy them. This discourse draws upon the rhetoric of marketing: 'clients' or 'consumers' are provided with health promotional services, just as the ill are provided with biomedical services.*[23] *(p. 75)*

Lupton goes on to add that "There is no coercion involved: consumers are 'free' to make their own choices on the information provided to them, and change because they 'want to'"[29] (p. 75). However, consumers cannot "know" what to do when they have sensitivity unless these advertisements support very particular forms of meaning. The advertisements analyzed for this chapter reflect mechanisms for the preservation and stabilization of meaning. In so doing, they support expectations as "stabilized assumptions."[17] These expectations, along with the associated forms of meaning are worth preserving. In some respects, this is what these advertisements help achieve.

The purpose of the advertisements, as with all commercials, is to bring the attention of the customers to the availability of the product. As we have seen, they also indicate how the product should be located with respect to dentistry. Consumer testimonials place dentists as experts in the plot. The dentist is listed as the first port of call. In this respect the product is subordinated to the discipline of dentistry, but it also gains authority from this subordinate position. DH is something that can be "treated" in the domestic sphere. On diagnosis, DH becomes a nonproblem problem and is displaced from the clinic to the home. The clinic becomes transposed into the home and the toothpaste has the authority of the discipline with it.

The location of *Sensodyne* with respect to dentistry can be illustrated by a change in the advertising strategy. Between 2005 and 2010, the focus was largely on the experiences of lay people, with customer testimonials persuading individuals to switch to a product targeting DH. More recently, a change in advertising standards has allowed advertisers to use professional testimonials. A 2011 advertisement with a dentist in a clinic talking about *Sensodyne Repair and Protect* toothpaste begins with the following caption:

Caption: I think it's going to be very beneficial.

The camera focuses on Dr Mark Hughes, who states: I think it's going to be very beneficial. This new toothpaste, Sensodyne Repair and Protect can actually repair the sensitive part of the tooth, there are tiny little tubes in the dentine of the tooth [illustration of holes] are responsible for the sensitivity [animation] using an advanced novomen technology, *Sensodyne Repair and Protect* will seek out and repair these areas using the same building blocks the tooth is made from.

[camera cuts back to dentist, picture of product, caption 'Ask your dentist about sensitive teeth', caption, website]

Dr Mark Hughes: I would recommend *Sensodyne Repair and Protect* to my patients, it's not just short term relief, it's long term protection'.

Source: Courtesy Grey London Archive (2011).

Here, professional ideas explain the condition to lay people, therefore locating the product within the discipline (recommended by a dentist). At the same time the product is displaced from the responsibility of the dentist. Although sensitivity requires diagnosis by a dentist, dentists do not have to be present to treat it. Instead the clinical "gaze" is transposed into the toothpaste and the practice of toothbrushing. Furthermore, this advertisement introduces a new distinction, repair/protect. This new *Sensodyne*, we are told, both repairs and protects the tooth from further sensitivity. It is as though reducing sensitivity is not enough; toothpaste must now also repair the tooth and protect it. This advertisement intensifies meaning in the sense that the toothpaste needs to do more than simply reduce sensitivity. The background to this intensification may have been a competitor toothpaste appearing on the market at this time, *Colgate Sensitive*. The *Colgate Sensitive Pro-Relief* advertisements reveal something very interesting.

Caption: Colgate Sensitive Pro Relief

Presenter: "Welcome to the *Colgate Sensitive Pro-Relief* challenge. We're coming to a town near you during April to give away free samples and to do the challenge live. Let's see what people are saying about *Colgate Sensitive Pro-Relief*."

Young woman starts talking in a Shopping Mall. "I've got really sensitive teeth and I happened to bump into the *Colgate* people. Who introduced me to the toothpaste. Put some on my tooth that was like the most sensitive one. After that I took iced water and my teeth are absolutely fine and now I am having a Strawberry Sorbet. I haven't tried *Colgate Sensitive Pro-Relief* before but I definitely will now."

Another young woman. "I have sensitive teeth and I just took the *Colgate Sensitive Pro-Relief* Challenge and it really took away the sharpness and the intensity of the cold or hot drink so yeah it definitely made a difference."

Another young woman "I have sensitive teeth and I didn't think it was that serious until I did drink some very cold water then. But I realised that only after I tried the toothpaste how amazing it was. I didn't think the effects would be immediate but it's like an analgesic effect that I didn't feel anything after that. So it did make a difference."

Young man "I've got sensitive teeth and in particular one sensitive tooth. And I took the *Colgate* challenge with *Colgate Sensitive Pro-Relief* and it eased all the pain. It will make me enjoy things a lot more now I reckon."

Voice over "Prove it for yourself. Go online and take the *Colgate Sensitive Pro-Relief* challenge. You won't find more effective pain relief for sensitive teeth or your money back."

This advertisement makes radically new references. Its talk of instant relief and challenges others to test a very different set of distinctions than that proposed by Sensodyne. The product is located in a shopping mall. Although there is a scientific background, a striking feature is the absence of references to the discipline of dentistry in any of these advertisements. In this respect, *Colgate Sensitive Pro-Relief* operates with an instant/relief distinction and is placed in the marketplace and not the home. Likewise, the advertisement has at its core a very deliberate challenge presented as something everyone can participate in. The advertisers have added a participate/not participate distinction and have gone global with their challenge. Everyone can have sensitivity and everyone can use *Colgate Sensitive Pro-Relief.*

Discussion

Gibson and Boiko suggested that the imperative of DH is conditioned by science and medicine (Chapter 15).[1] They also suggested that culture may also play a role. Accounts of DH in these advertisements, whether from professionals or lay people, have one solution, to use specifically designed toothpaste. Bury argued that accounts of illness draw on a pool of meaning, but he did not explain how this might happen.[30] Nor did he discuss how such a pool of meaning was sustained. In this chapter, we interpret Luhmann's social systems theory to suggest that the pool of meaning regarding DH is sustained through multiple social systems. One participating system is that of the mass media.

A consequence of this analysis is that we have seen advertising as a process whereby inanimate toothpastes become animated with various distinctions.[5] These distinctions subsequently enable communication about DH to follow along certain pathways. The pathways being presented in the advertisements include movements from science, to dentistry, and, finally, the home. The product moves through several semantic spaces, becoming imbued with meaning from each and finally being placed in the bathroom of participants. The animation process also changes the status of DH itself. DH moves from being a problem to being a nonproblem, from abnormal to normal, all through the consumption of the toothpaste!

Finally, the advertisements produce an order for communication and, thus, behavior. One toothpaste communicates a process whereby the protagonist has pain and seeks out the dentist and then uses *Sensodyne*. In the other, the toothpaste produces a magical transformation that abolishes DH without dentists. This product is placed within the supermarket. Behind such communications other social systems are watching. For example, the legal system observes if what is being claimed is legal. The industry is also self-regulated; therefore, any claim made in advertising can be challenged. Even here there are references that are beyond the presentation of meaning that is visible to us.

According to Nettleton, power is a productive process.[25–28] We know that more than 60% of the variance in oral health can be accounted for through social factors, but there are very few studies exploring what these are and even fewer still examining what they mean.[31,32] This chapter explains how this might be happening. Advertisements generate distinctions that enable inanimate substances to develop meaning and become incorporated into the everyday lives of individuals. We have shown how advertisements support particular forms of meaning. They reflect the specific mechanism for the preservation and stabilization of meaning. By doing so, they support expectations as “stabilized assumptions.”[17] These assumptions are worthy of preservation because people experience sensitivity in their teeth and they need to know what to do about it. In this sense, then, the toothpaste becomes the solution to a need that has been constructed and identified for them. As we have already said, the purpose of advertisements, as with all commercials, is to bring the attention of the customers to the availability of the product. They also indicate how the product should be located with respect to science, dentistry, and everyday life. The clinic becomes transposed into the home, with the toothpaste bringing with it the authority of the discipline into the home.

References

1. Gibson B, Boiko O. The experience of health and illness: polycontextural meaning and accounts of illness. *Soc Theor Health* 2012;**10**:156–87.
2. Gibson B, Boiko O, Baker S, Robinson P, Barlow A, Player T, et al. The everyday impact of dentine sensitivity: personal and functional aspects. *Soc Sci Dent* 2010;**1**: 11–20.
3. Marx K. *Capital*, vol. 1. London: Penguin Books with the New Left Review; 1976.
4. Hoeyer K. Person, patent and property: a critique of the commodification hypothesis. *BioSocieties* 2007;**2**:327–48.

5. Silva S. Reification and fetishism: processes of transformation. *Theor Cult Soc* 2013;**30**:79–98.
6. Berger PL, Luckmann T. *The social construction of reality: a treatise in the sociology of knowledge*. London: Harmondsworth; 1971.
7. Ellen R. Fetishism. *Man* 1988;**23**:213–35.
8. Berger P, Pullberg S. Reification and the sociological critique of consciousness. *Hist Theor* 1965;**4**:196–211.
9. Martens L, Southerton D, Scott S. Bringing children (and parents) into the sociology of consumption: towards a theoretical and emprirical agenda. *J Consum Cult* 2004;**4**: 155–81.
10. Parry B. Entangled exchange: reconceptualising the characterisation and practice of bodily commodification. *Geoforum* 2007;**39**:1133–44.
11. Radin M. *Contested commodities*. Cambridge, MA: Harvard University Press; 1996.
12. Appadurai A. *The social life of things: commodities in a cultural perspective*. Cambridge, MA: Cambridge University Press; 1986.
13. Gibson B, Boiko O. Luhmann's social systems theory, health and illness. In: Scambler G, Scambler S, editors. *Contemporary theorists for medical sociology*. London: Routledge; 2012.
14. Boiko O, Ward P, Robinson P, Gibson B. Form and semantic of communication in dental encounters: oral health, probability and time. *Sociol Health Illn* 2011;**33**:16–32.
15. Van Leeuwen T, Jewitt C. *Handbook of visual analysis*. London: Sage Publications; 2002.
16. Kress G, van Leeuwen T. *Reading images: the grammar of visual design*. London: Routledge; 1996.
17. Luhmann N. *Social systems*. Stanford, CA: Stanford University Press; 1995.
18. Andersen N. *Discrusive analytical strategies: understanding Foucault, Laclau and Luhmann*. Bristol: Policy Press; 2003.
19. Luhmann N. *The reality of the mass media*. Cambridge, MA: Polity Press; 2000.
20. Dowell P, Addy M. Dentin hypersensitivity—a review.1. Etiology, symptoms and theories of pain production. *J Clin Periodontol* 1983;**10**:341–50.
21. Dowell P, Addy M, Dummer P. Dentine hypersensitivity: aetiology, differential diagnosis and management. *Br Dent J* 1985;**158**:92–6.
22. Addy M, Urquhart E. Dentine hypersensitivity: its prevalence, aetiology and clinical management. *Dent Update* 1992;**19**:407–8 410–2.
23. Dababneh RH, Khouri AT, Addy M. Dentine hypersensitivity—an enigma? a review of terminology, epidemiology, mechanisms, aetiology and management. *Br Dent J* 1999;**187**: 606–11.
24. Foucault M. *The history of sexuality: the will to knowledge*. Middlesex: Penguin Publishers; 1998.
25. Nettleton S. Protecting a vulnerable margin—towards an analysis of how the mouth came to be separated from the body. *Sociol Health Illn* 1988;**10**:156–69.
26. Nettleton S. Power and the location of pain and fear in dentistry and the creation of a dental subject. *Soc Sci Med* 1989;**29**:1183–90.
27. Nettleton S. Wisdom, diligence and teeth: discursive practices and the creation of mothers. *Sociol Health Illn* 1991;**13**:98–111.
28. Nettleton S. *Power, pain and dentistry*. Buckingham: Open University Press; 1992.
29. Lupton D. *The imperative of health: public health and the regulated body*. London: Sage; 1995.
30. Bury M. Illness narratives: fact or fiction? *Sociol Health Illn* 2001;**23**(3):263–85.

31. Nadanovsky P, Sheiham A. The relative contribution of dental services to the changes and geographical variations in caries status of 5- and 12-year-old children. *Community Dent Health* 1994;**11**:215–23.
32. Nadanovsky P, Sheiham A. Relative contribution of dental services to the changes in caries levels of 12-year old children in 18 industrialized countries in the 1970s and early 1980s. *Community Dent Oral Epidemiol* 1995;**23**:331–9.

Part Five

Discussion and Conclusion

Conclusions

18

Peter G. Robinson, Sarah R. Baker and Barry Gibson
School of Clinical Dentistry, Claremont Crescent, University of Sheffield, Sheffield, UK

This chapter brings together conclusions from the individual chapters. First, it looks at person-centered care, its value, origins, and manifestations. It proceeds to a review of existing knowledge and identifies outstanding questions about dentine hypersensitivity (DH). The chapter progresses to a contemplation of all the work involving DHEQ, drawing conclusions about its utility and laying out a program of work to finalize the measure and use it. Next, we travel further afield to consider the meaning of DH. Finally, we make an ardent call for greater multidisciplinarity to take all this work forward.

Person-centered oral health care and research

This book set out to present a case for adopting a person-centered approach in oral health care and oral health research.

Such a viewpoint does not ignore biomedical aspects of health. Research in oral pathology has taught us about the nature and causes of disease. Technological development of dental and biomaterials and clinical developments in their application have brought about the oral rehabilitation of millions of people who have been impaired by oral disease. A person-centered approach to health merely requires us to remember the frame of reference for biomedicine. The purpose of dentistry is to make people healthy. It therefore follows that all aspects of health should be considered: the biological aspects of disease; the clinical challenges of treatment; the psychological influences on etiology and consequences of disease; and the social context that determines and frames that experience. In clinical care, this is a case of remembering the person with the teeth. As we have seen, more formal approaches are required to balance biomedical and person-centered perspectives in research. We hope to have adequately represented all viewpoints here.

The notion of a person-centered approach is far from novel or unique to this book, although it is more usually referred to as "patient-centred."

We should not be surprised by this phrase. Most writers in this field (whether as policy-makers, clinicians, or researchers) work within health services or academic departments related to health care institutes, and their focal point will be patients and they will be "patient-centered." However, this phrase can be inaccurate and may be worth reconsidering for other reasons.

Dentine Hypersensitivity. DOI: http://dx.doi.org/10.1016/B978-0-12-801631-2.00023-3

As we have seen, only a minority of people with DH see a dentist about it; therefore, most affected people are not patients. Even people with severe disease will move in and out of being a patient. They will not be patients before they have symptoms or until their symptoms prompt them to seek health care. After their treatment has started, there may be periods when their attendance or registration lapses. Those of us working in public health are concerned with the health of the population, whether or not they are patients. Nonetheless, the sentiment of this book, giving precedence to the lay experience of health, still applies whether or not that lay person is a patient.

There is also something about the word that can place "patients" as passive receivers of treatment, which is far from the active role we would advocate for a person who seeks care, negotiates treatment plans, and assiduously participates in that care. It might not hurt to remember patients are people.

There is a considerable body of evidence supporting the effects of greater patient involvement on better health care outcomes. In an early randomized controlled trial, patients with gastric ulcers who were coached to negotiate treatment decisions with their physicians reported fewer physical limitations, preferred their more active role, and were as satisfied with their care as patients who received a standard educational session.[1]

A second point to bear in mind about person-centered care is that it can apply to a person in the singular, or people in the plural. Clinicians will spotlight individual patients, whereas planners, public health practitioners, and researchers may think about groups of people. The principle of person-centeredness applies in both cases.

The rationale for the centrality of the person in health is a direct consequence of the meaning of health. Definitions as varied as the World Health Organization's "*state of complete physical, mental and social well-being and not merely the absence of disease or infirmity*", Dubos's "*manner in which one responds to the challenges of one's environment*", or Seedhouse's "*foundations for achievement*" share a common idea.[2–4] They see health as something that is active, in which the person acts and uses health for something else, be it getting along with other people, facing the challenges of life, or achieving. These definitions are largely unconcerned with biomedical health. Instead, they highlight the active person. It follows that health care should focus on people and encourage them to be active. To do otherwise may undermine health.

The growing role of the person in health has many manifestations. A marketization of health viewpoint allows people to be consumers acting as informed decision-makers. More vociferous people are also more likely to challenge health care with which they are unhappy, which is evident in more patient complaints, more hearings against health care workers at regulatory organizations, and more litigation.

From a policy viewpoint, the greater role for lay people represents a shift in the focus of health care since the Second World War, away from the dominant and unquestioned doctor or dentist towards a more balanced relationship with more emphasis on the needs and wishes of the patient.

This shift gained prominence in two seminal documents produced by the World Health Organization, which captured this zeitgeist. The World Health Organization

International Conference on Primary Health Care concluded with the Declaration of Alma-Ata.[5] This influential document declared that primary health care "*requires and promotes maximum community and individual self-reliance and participation in the planning, organization, operation and control of primary health care*." In the same vein, the Ottawa Charter for Health Promotion defined health promotion as "*the process of enabling people to increase control over, and to improve, their health*."[6] One of the five key actions of health promotion is strengthening community action, which involves communities in setting priorities, making decisions, planning strategies, and implementing them to achieve better health. Clearly, this process involves empowering communities so that they can control their own endeavors and destinies.

Patient-centeredness is now gaining momentum and is explicit in health care organizations. The US Institute of Medicine set out a comprehensive action plan to redesign health care in the twenty-first century.[7] One of its six aims for improvement was for care to be patient-centered, that is, "*respectful of and responsive to individual patient preferences, needs, and values, and ensuring that patient values guide all clinical decisions*." The document goes on to describe the patients as "*the source of control*" exercising control over their health care in a system that encourages shared decision-making. Likewise, the UK Institute for Innovation and Improvement placed patient-centered care at the heart of its program for improving quality in the National Health Services and exhorted staff to "*listen to your patients; always have them at the centre of your thinking*" which "*becomes a powerful motivator and driver for change*."[8]

The precept is also apparent in the involvement of patients on the boards of hospitals and other organizations. For example, there is greater lay representation on authoritative organizations. In the United Kingdom, the General Dental Council protects patients and the public by maintaining registers of professionals permitted to provide dental treatment, setting and enforcing standards for that treatment and assuring the quality of educational establishments. Its governing body is Council. Six of the 12 members of Council are lay people.[9]

Finally, in relation to health, the same standard is palpable in the control being grasped by, shared with, or given to specific groups of the population. Advocates of people with disabilities have challenged biomedicine's emphasis on "correcting" impairments with medical and other treatments. This movement has argued that disability is caused by the way society is organized and has argued for premises and transport to be made accessible and for resources to be created to reduce the disability related to impairment.[10] The UK National Service Framework for Children applied the same ideas to children's health so that services were "child-centered and look[ed] at the whole child—not just the illness or the problem."[11]

These changes are equally manifest in research. Quantitatively there has been an explosion in the measurement of concepts as diverse as the lay experience of health and disease, satisfaction with health care, attitudes towards other people with disease, or even indifference to oral health. To accompany this interest there has been a commensurate (some might say disproportionate) growth in the literature on the methods and tools to measure these things.

Qualitative research that aims to get right to the heart of what people think, without the distractions of preconceptions of measurement, is also more and more evident in oral and dental research.

More emphasis on the views of lay people is also evident in the design and conduct of research. In the United Kingdom this is termed *Public and Patient Involvement* and many funding agencies expect lay people to be actively involved in the research they support. This application of lay participation is not simply a matter of principle. There is empirical evidence of its effectiveness to increase study quality. A recent systematic review of studies registered with the Mental Health Research Network found that those in which service users were involved in designing or running the trial were 60% more likely to recruit samples on schedule.[12] Studies with greater involvement were four-times more likely to recruit to target.

Local health research organizations within universities and the National Health Services have experts dedicated to fostering this type of involvement. Lay involvement is also more observable in the tightened regulation of research, with lay representation on research ethics committees and greater emphasis placed on the adequacy of patient information sheets and consent forms.

It should go without saying that we endorse this shift towards greater lay participation in health. Moves to encourage lay involvement should be supported. For individual patients this will facilitate greater personal autonomy through specific acts such as the provision of better information or fostering shared decision-making and, more generally, in all approaches that level any hierarchical relationships between health care workers and lay people. At a population level, it may involve community action on the determinants of health and participation in developing and monitoring health services.

In research terms, more work is needed on the psychosocial causes and consequences of oral conditions and how they may be reduced. More research is needed regarding the relationship between clinical status and the impact of the mouth on everyday life. Given the scope of this undertaking, it goes without saying that the work should be multidisciplinary and theory-driven.

Dentine hypersensitivity

In a very traditional way, we started by reviewing the clinical presentation and physiological mechanisms of DH. Katrin Bekes described a condition that caused pain and could disturb people during eating, drinking, and cleaning their teeth. The condition has multifactorial etiology, but there is a fairly consistent understanding of the mechanisms underlying DH based on the hydrodynamic theory. Joana Cunha-Cruz and John Wataha reviewed methods used to diagnose DH and quantified the extent of the burden it causes in a novel and exciting systematic review. Their best estimate of its prevalence was approximately 10%. Under the mantra of "common things commonly occur," this knowledge assists in its diagnosis in individual patients and guides treatment decisions. David Gillam and Elena Talioti's review of the clinical management of DH was also concerned with its diagnosis, especially with the need to tailor management to specific clinical situations. They summarized developments in treatment modalities and brought them together in an algorithm based on the underlying cause of the condition.

All three chapters highlight difficulties in the diagnosis of DH. In some ways, these difficulties are caused by DH being a diagnosis of exclusion, in which all other causes of the symptoms must first be ruled out. However, history may also play a role in this difficulty. The last three chapters of the book revealed DH as a nonproblem problem, where (historically at least) dentistry may not have entirely sanctioned it as a disease, so that it was transposed into the market. As we now know, the prevalence of DH is approximately 10% and attempts to devise and formalize methods to diagnose DH should be commended.

The same arguments may also apply to the treatment of DH. Efforts to rationalize the treatments and a stronger evidence base will assist both clinical and personal decision-making. It should be easier to develop that evidence base now that robust outcome measures are available.

The Dentine Hypersensitivity Experience Questionnaire

Finbarr Allen introduced the notion of measuring lay perspectives of oral disease using measures of oral health-related quality of life. Measures need to be selected carefully depending on the circumstances, the condition undergoing study, and the purpose of the study. The burden placed on people completing the measures must be considered. However, the performance of short measures may be compromised.

The lay perspective of DH was described comprehensively in Chapter 6, and the work described in that chapter is novel, being the first exploration of the experience of DH from the perspective of those affected. Qualitative interviews and analysis identified complex relationships between the symptom of pain and its consequences for the individual, including impacts on everyday activities such as eating, drinking, talking, tooth-brushing, and social interaction. The impacts could be modified by individual and environmental influences. These findings were compatible with the psychological literature on pain.

This first-hand knowledge of DH was used to construct a measure of its effects on everyday life. A rigorous method was deployed to develop the Dentine Hypersensitivity Experience Questionnaire (DHEQ) and preliminary evaluation was promising (Chapter 7). A more practical account of this work forms Chapter 8, where our focus on the effects of one condition allowed us to detail the need for an explicit position and underlying theory, and how our multidisciplinary understanding of DH, pain, and psychometrics allowed us to use the lay accounts to construct the DHEQ. More and more measures of oral health-related quality of life are being developed and we hope our story may be of assistance to colleagues undertaking that work. At least we hope it finally kills the myth that a questionnaire is something one puts together between patients on a quiet afternoon.

Sarah Baker provided an exemplary display of the longitudinal evaluation of an OHQoL measure in Chapter 9. Using data from three randomized controlled trials, she examined change within individuals, differences between people, and change due to treatment. DHEQ proved to be highly responsive to change within individuals with large effect sizes that discriminated between treatments of different efficacy.

DHEQ was distilled into a short form that appears to have practical application in the clinic (Chapter 10). The care taken in the original development of DHEQ and the selection of the items most relevant to people with DH have avoided the floor and ceiling effects discussed by Finbarr Allen. The abbreviated version (DHEQ-15) was sensitive and precise and may be suitable for use with individual patients.

Our review of the developmental work on DHEQ was brought to a close with an important point. We saw how the impact of DH was modified by environmental characteristics in Chapter 6. Therefore, measurement of that impact in new settings must account for those new environments, meaning that measures need to be adapted for use in different languages and cultures. This is not only a case of linguistic translation but also a consideration of how ideas may differ between cultures. Songlin He's careful and rigorous translation in Chapter 11 shows how measures can be made relevant to new target populations.

The work collected in this book demonstrates how DHEQ can measure meaningful and relevant impacts on the everyday life of people with DH. Thus, it is possible and, from a person-centered perspective, desirable to measure the daily impacts of conditions such as DH on people's lives instead of, or in addition to, clinical measurement. Doing so allows us to assess what degree of impacts there are and what impacts are most important from a personal perspective. As a result, we can evaluate the effectiveness of anti-sensitivity treatments from a real-life person's perspective rather than a clinically induced reduction in a pain response. As we have seen, the presence of an accurate measure also gives us the ability to explore questions beyond oral health, such as how individuals adapt to chronic conditions and differ in their response to pain. DHEQ is therefore a promising tool for the study of DH, with applications to wider questions.

The utility of DHEQ can be assessed using tools designed to appraise health-related quality of life measures. The Scientific Advisory Committee of the Medical Outcomes Trust (MOT) compiled eight key attributes of health status and quality of life instruments, along with criteria against which instruments could be reviewed for these attributes (Table 18.1).[13] Chapters 6 and 7 provide the evidence that DHEQ meets the MOT criteria for the conceptual model, reliability and validity. Evidence of responsiveness is present in Chapter 9, and cumulatively all three chapters demonstrate interpretability. Chapter 10 also indicates the utility of alternative modes of administration. Cultural adaptations of DHEQ already exist (see Chapter 11), but we anticipate other authors will provide this information as new versions are translated.

Terwee and colleagues also developed quality criteria for the design, methods, and outcomes of studies of the development and evaluation of health status questionnaires (Table 18.2).[14] DHEQ meets Terwee's criteria for content validity, floor and ceiling effects, internal consistency, and construct validity (see Chapters 6 and 7).

Table 18.1 **Attributes and criteria for reviewing instruments (SAC)**

Attribute	Review criteria
Conceptual and measurement model	Concept to be measured Conceptual and empirical bases for item content and combinations Target population involvement in content derivation Information on dimensionality and distinctiveness of scales Evidence of scale variability Intended level of measurement Rationale for deriving scale scores
Reliability	Methods to collect internal reliability data Reliability estimates and standard errors for all score elements Calculations of reliability coefficients Methods employed to collect reproducibility data Well-argued rationale to support study design and interval between administrations Test retest reliability coefficients Above data for each major population of interest
Validity	Rationale supporting the mix of evidence presented Clear description of the methods used to collect data Composition of the sample Above data for each major population of interest Hypotheses tested and data relating to the tests Clear rationale and support for the choice of criterion measures
Responsiveness	Evidence of changes in scores Longitudinal data comparing groups expected and not expected to change Populations on which responsiveness has been tested, including the time intervals of assessment, the interventions or measures involved in evaluating change, and the populations assumed to be stable
Interpretability	Rationale for selection of external criteria of populations for purposes of comparison and interpretability of data Information regarding the ways in which data from the instrument should be reported and displayed Meaningful 'benchmarks' to facilitate interpretation of the scores
Burden	Average and range of time needed to complete instrument Reading and comprehension level Special requirements made of respondent Evidence of no undue strain on respondent Circumstances when instrument is not suitable Resources required for administration Average and range of time for an interviewer to administer Training required to administer
Alternative modes of administration	Evidence on reliability, validity, responsiveness, interpretability and burden for each mode of administration
Cultural and language adaptations or translations	Methods to achieve conceptual and linguistic equivalence Differences between the original and translated versions How differences were reconciled

Table 18.2 Quality criteria for measurement properties of health status questionnaires

Property	Quality criteria
Content validity	Clear description of the measurement aim, target population, the concepts that are being measured, and the item selection Target population were involved in item selection
Internal consistency	Factor analyses performed on adequate sample size Cronbach's alphas per dimension between 0.70 and 0.95
Criterion validity	Convincing argument that gold standard is "gold" Correlation with gold standard >0.70
Construct validity	Specific hypotheses were formulated and at least 75% of the results are in accordance with these hypotheses
Agreement	Minimal important change <SDC OR MIC outside the LOA OR convincing arguments that agreement is acceptable
Reliability	Intra class correlation coefficient >0.70
Responsiveness	Smallest detectable change < minimal important change
Floor and ceiling effects	<15% of respondents achieved highest or lowest possible scores
Interpretability	Mean and SD scores presented of at least four relevant subgroups of patients and MIC defined

Chapters 7 and 10 demonstrate very good reliability, and responsiveness is demonstrated in Chapter 10. The body of work presented in this book supports the interpretability of DHEQ.

In summary, both DHEQ and DHEQ-15 meet almost all of both sets of criteria, indicating strong support for their utility. Both sets specify information on criterion validity (i.e., comparison against a gold standard measure) but, because no gold standard exists, this assessment is impossible. Descriptive information of the burden associated with the use of DHEQ and DHEQ-15 is required.

Subsequent and future work

The existence of a sensitive, precise, and responsive measure of the effects of a condition on everyday life is of direct benefit to academics studying the condition and, possibly, as we have seen, to people with the condition and those caring for them. However, the same qualities also make DHEQ a useful tool for studying psychological aspects of pain and adaptation that may be applicable to other diseases. In Chapter 12, Marta Krasuska described how she used DHEQ to study adaptation to DH as an example of a chronic disease. In a phenomenon known as response shift, people might adapt to a disease so that its effect on their lives is minimized. Response shift may therefore undermine evaluations of treatment for the condition (Has the treatment worked or has the patient just adapted?). Using DHEQ, Marta was able to compare two ways of modifying a quality of life measure and ended by concluding that one of the currently standard approaches may be not be the best.

Instead of being concerned with the average scores of groups of people, Jenny Porritt's work in Chapter 14 focussed on individual variation in how people experience chronic conditions—another first for dental research. Naturally, our person-centered approach leads us to explore such variation so that we can understand DH in greater detail and interpret individual responses to treatment. Jenny studied factors from psychology and the wider quality of life literature that might explain individual responses to chronic health conditions. Illness beliefs (the perceptions people have toward their health condition) contributed to DH-specific and health-related quality of life outcomes. General health anxiety and types of coping strategies were also important. Again, these findings are compatible with those in general health problems such as arthritis and cancer. Consequently, this knowledge helps inform the development, implementation, and evaluation of interventions for health conditions.

The short form of DHEQ (DHEQ-15) appears to have sufficient psychometric qualities to make it suitable for assessing impacts in individual patients. Its brevity should minimize the burden associated with administering it in clinical services. Although not a diagnostic tool, the measure might be useful when screening for symptoms and impact. Such information could be matched with clinical data to exclude other sources of symptoms. In this way, DHEQ could aid diagnosis of DH and might be helpful in evaluating the response of individual patients to treatments in the clinic. Further evaluation is needed before these applications can be recommended, however.

DH also provided an excellent case study for developing ways to measure a subtle and complex pain. In Chapter 13, Lisa Heaton and colleagues described the development of labeled magnitude scales to quantify the short and episodic pain of DH. Existing pain measurement scales lack the sensitivity and precision to do this. The new approach effectively *amplifies* the minor end of such scales and comprehensively demonstrated superior performance over existing visual analog scales. There is real potential to apply this method to describe relatively minor pain and to use it to assess the effects of treatment.

The meaning of DH

Barry Gibson, Olga Boiko, Ninu Paul, and Melanie Hall considered the social meaning of DH. This work is literally "fundamental" because social meaning is the understanding we share and use to make sense of the world. It is revealed in our language. In a detailed secondary analysis of the language participants used to describe their DH in Chapter 6, Barry and Olga discovered key distinctions in the structure and accounts of DH. Most interestingly, DH could be both a significant problem and not a problem at the same time (a nonproblem problem). It may be that dentistry did not see DH as a legitimate disease in comparison with caries, and so the development of commercial solutions for DH resulted in it being semantically displaced away from dentistry and into the market. Hall and Gibson developed this theme by considering how contemporary advertisements for sensitivity toothpastes produce patterns of meaning that are incorporated into everyday life.

Although much of this book is a study of how we can investigate the impact of the condition on individuals, this section takes us through how society has given DH meaning. In these chapters, DH is a resolutely social phenomenon. This is a much broader view of oral health than either the clinical or the psychosocial perspectives. It starts to ask what *oral health* means, and what it tells us. We have already seen how this work might explain some of the difficulties associated with reaching a diagnosis of DH. What it tells us is the responses to a questionnaire such as the DHEQ are as much social communications as they are responses by individuals. It tells us that we should give more attention to the ways in which society conditions the semantics and meaning of the condition and how this allows individuals to make sense of their symptoms.

Multidisciplinarity

Although concerned with one clinical condition, the scope of this book ranges from the pathophysiology to the social meaning of DH. The chapter authors include (in alphabetical order) chemists, dentists, materials scientists, psychologists, public health practitioners, and sociologists.

DH therefore perfectly exemplifies Popper's statement that "*We are not students of some subject matter, but students of problems. And problems may cut right across the borders of any subject matter or discipline.*"[15] By contrast, disciplines were "*distinguished partly for historical reasons and reasons of administrative convenience (such as the organization of teaching and appointments).*"

The research problems we face bear little relation to our current disciplinary boundaries. To solve those problems, we need to harness the complementary insights and skills of all relevant disciplines. Hopefully, this book demonstrates the benefits of doing that.

References

1. Greenfield S, Kaplan S, Ware Jr JE. Expanding patient involvement in care: effects on patient outcomes. *Ann Intern Med* 1985;**102**:520–8.
2. World Health Organisation. *Constitution of WHO*. Geneva: WHO; 1948.
3. Dubos R. *Mirage of health*. New York, NY: Harper & Row; 1959.
4. Seedhouse D. *Health: the foundations for achievement*. 2nd ed. Chichester: Wiley; 2001.
5. World Health Organisation. *Declaration of Alma-Ata*. <http://www.who.int/publications/almaata_declaration_en.pdf>; 1968 [accessed 11.05.14].
6. World Health Organisation. *The Ottawa Charter for Health Promotion*. <http://www.who.int/healthpromotion/conferences/previous/ottawa/en/>; 1986 [accessed 11.05.14].

7. Institute of Medicine. *Crossing the Quality Chasm: A new health system for the 21st Century*. <http://iom.edu/Reports/2001/Crossing-the-Quality-Chasm-A-New-Health-System-for-the-21st-Century.aspx>; 2001 [accessed 11.05.14].
8. NHS Institute for Innovation and Improvement. *Joined up care*. <http://www.institute.nhs.uk/qipp/joined_up_care/patient_centred_care.html> [accessed 11.05.14].
9. General Dental Council. *The Council*. <http://www.gdc-uk.org/Aboutus/Thecouncil/Pages/council.aspx>; 2014 [accessed 11.05.14].
10. Scope. *What is the social model of disability, and why is it important to us?* <http://www.scope.org.uk/about-us/our-brand/social-model-of-disability>; 2014 [accessed 11.05.14].
11. Department of Health. *National Service Framework for children, young people and maternity services*. London: Department of Health; 2004.
12. Ennis L, Wykes T. Impact of patient involvement in mental health research: longitudinal study. *Br J Psychiatry* 2013;**203**:381.
13. Scientific Advisory Committee of the Medical Outcomes Trust. Assessing health status and quality-of-life instruments: attributes and review criteria. *Qual Life Res* 2002;**11**:193–205.
14. Terwee CB, Bot SDM, de Boer MR, van der Windt DA, Knol DL, Dekker J, et al. Quality criteria were proposed for measurement properties of health status questionnaires. *J Clin Epidemiol* 2007;**60**:34–42.
15. Popper K. *Conjectures and refutations: the growth of scientific knowledge*. London: Routledge; 1963 [Reprinted 2004].

Appendix 1: The dentine hypersensitivity experience questionnaire

This appendix contains the complete version of the DHEQ. Colleagues considering using the questionnaire are advised to read the guidance below beforehand.

The full questionnaire contains 48 items. However, only items 10–43 inclusive form the impact subscales that are used to measure the effects of DH on oral health related quality of life. Table A.1 summarizes the purpose of each part of the questionnaire

Calculating summary scores

A summary score for each impact subscale is calculated as the sum of the item codes for that subscale.

The total DHEQ score is calculated as the sum of the item codes for items 10–43 inclusive.

The score for the effect on life overall is calculated as the sum of the item codes-for items 45–48.

Table A.1 **Format of the DHEQ**

Section	Item Nos	Type of Items	Purpose	Coding
Introductory descriptors	1–6	Closed questions	Describe pain	Each item treated separately
Pain scales Intensity Bothersomeness Tolerability	7–9	Visual analogue scales	Measure pain	Each item treated separately and scaled 0–10
Impact subscales*				
Restrictions	10–13	Likert scale	Measure restrictions in daily activity	7-point Likert scales coded: 1 = "strongly disagree" to 7 = "strongly disagree"
Adaptation	14–25	Likert scale	Measure activities to cope and prevent sensitivity.	
Social impact	26–30	Likert scale	Measure handicap	
Emotional impact	31–38	Likert scale	Measure emotional impact	
Identity	39–43	Likert scale	Measure impact on personal identity	
Global oral health rating	44	Ordinal scale	Measure health perception	1 = "excellent", 2 = "very good", 3 = "good", 4 = "fair", 5 = "poor", 6 = "very poor"
Effect on life overall	45–48	Likert scale	Measure effect on overall quality of life	0 = "not at all" to 4 = "very much"

*Only items 7–43 inclusive form the impact subscales that are used to measure the effects of DH on oral health related quality of life.

The dentine hypersensitivity experience questionnaire

Section one

The following questions are about your sensitive teeth, and the impact it has on your everyday life.

1) Which of the followingbest describe any sensations that you may have felt in your teeth (tick all that apply)

1 ☐ Itchy	2 ☐ Aching	3 ☐ Shooting
4 ☐ Piercing	5 ☐ Tingling	6 ☐ Sharp
7 ☐ Dull	8 ☐ Flashing	9 ☐ Shivery
10 ☐ Lingering	11 ☐ Twinging	12 ☐ Flickering
13 ☐ Stabbing	14 ☐ Shattering	15 ☐ Freezing
16 ☐ Fleeting	17 ☐ Quivering	18 ☐ Pricking
19 ☐ Pain	20 ☐ Discomfort	21 ☐ Twinges
22 ☐ Sensitivity	23 ☐ Other (please Specify)	
24 ☐ None of the Above (go to SECTION TWO)		

From now on in this questionnaire we are going to call what you feel as "***sensations in your teeth***" or "**sensations**".

2) How long have you been experiencing any sensations in your teeth? (tick only one response)

- ☐ Less than six months (1)
- ☐ More than six months but less than a year (2)
- ☐ More than a year but less than five years (3)
- ☐ More than five years but less than 20 years (4)
- ☐ More than 20 years (5)
- ☐ None (0)

3) Which parts of your mouth have been affected? (tick all that apply)

1 ☐ Top front
2 ☐ Top back
3 ☐ Bottom front
4 ☐ Bottom back
5 ☐ None

4) Which of the following cause you to have *sensations*? (tick all that apply)

1 ☐ Cold fluids	2 ☐ Salty foods	3 ☐ Cold foods
4 ☐ Tooth brushing	5 ☐ Hot fluids	6 ☐ Acidy fruits (e.g. oranges)
7 ☐ Hot foods	8 ☐ Sweet things	9 ☐ Having teeth cleaned at the dentist
10 ☐ Hard foods	11 ☐ Sticky foods	12 ☐ Tooth Whitening Products
13 ☐ Cold air	14 ☐ Ice Cream	15 ☐ Metals touching my teeth
16 ☐ Other (Please Specify)		
17 ☐ None		

5) How often do you have any *sensations*? (tick only one response)

- ☐ Several times a day (7)
- ☐ Once a day (6)
- ☐ Several times a week (5)
- ☐ Once a week (4)
- ☐ Several times a month (3)
- ☐ Once a month (2)
- ☐ Less than once a month (1)
- ☐ Never (0)

6) If you have any *sensations*, on average how long do these sensations last? (tick only one response)

- ☐ A few seconds (5)
- ☐ About a minute (4)
- ☐ Several minutes (3)
- ☐ About half an hour (2)
- ☐ Longer than half an hour (Please specify) (1)
- ☐ Don't have them (0)

The following questions are about your sensitive teeth, and the impact they have on your everyday life.

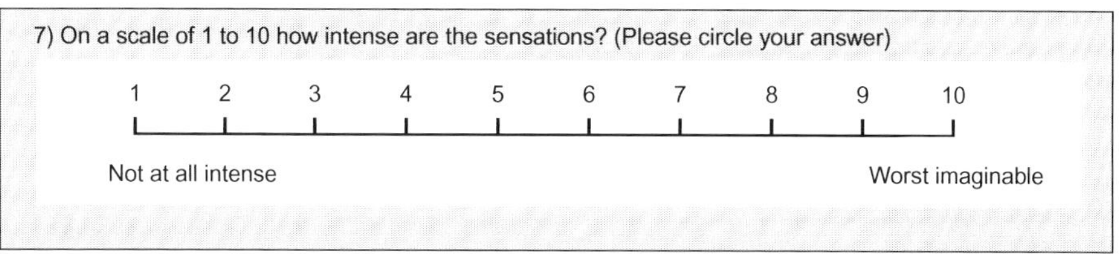
7) On a scale of 1 to 10 how intense are the sensations? (Please circle your answer)

1 2 3 4 5 6 7 8 9 10

Not at all intense — Worst imaginable

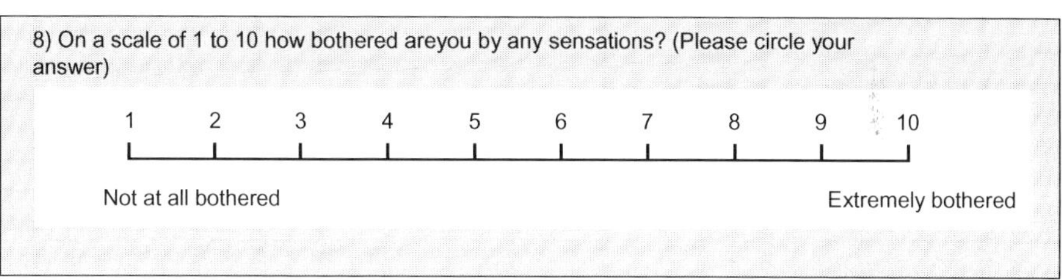
8) On a scale of 1 to 10 how bothered areyou by any sensations? (Please circle your answer)

1 2 3 4 5 6 7 8 9 10

Not at all bothered — Extremely bothered

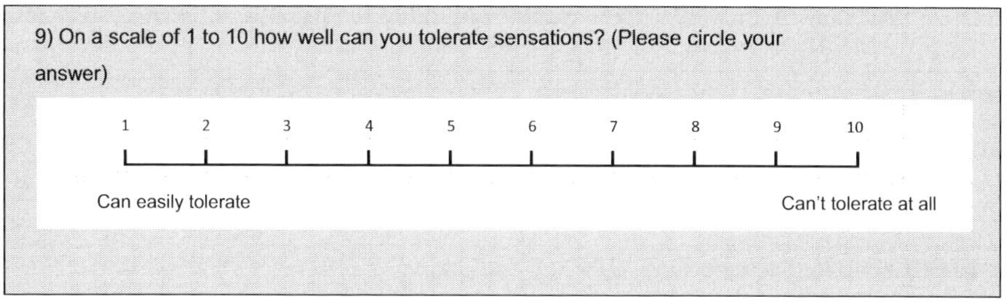
9) On a scale of 1 to 10 how well can you tolerate sensations? (Please circle your answer)

1 2 3 4 5 6 7 8 9 10

Can easily tolerate — Can't tolerate at all

Dentine hypersensitivity experience questionnaire

Section two

The following questions are about **the ways in which any sensations in your teeth affect you in your daily life.** Thinking about yourself ***over the last month*** to what extent would you agree or disagree with the following statements (Please tick only one response for each question)

	Strongly agree (7)	Agree (6)	Agree a little (5)	Neither agree nor disagree (4)	Disagree a little (3)	Disagree (2)	Strongly disagree (1)
10) Having sensations in my teeth takes a lot of the pleasure out of eating and drinking.	☐	☐	☐	☐	☐	☐	☐
11) There have been times when I can't finish my meal because of the sensations.	☐	☐	☐	☐	☐	☐	☐
12) It takes a long time to finish some foods and drinks because of sensations in my teeth.	☐	☐	☐	☐	☐	☐	☐
13) There have been times when I have had problems eating ice cream because of these sensations.	☐	☐	☐	☐	☐	☐	☐

The following questions are about ***the ways in which the sensations in your teeth have forced you to change things in your daily life.*** Thinking about yourself ***over the last month*** to what extent would you agree or disagree with the following statements (Please tick only one response for each question)

	Strongly agree (7)	Agree (6)	Agree a little (5)	Neither agree nor disagree (4)	Disagree a little (3)	Disagree (2)	Strongly disagree (1)
14) I have to change the way I eat or drink certain things.	☐	☐	☐	☐	☐	☐	☐
15) I have to be careful how I breathe on a cold day.	☐	☐	☐	☐	☐	☐	☐
16) I have to leave some cold foods or drinks to warm up before I can have them.	☐	☐	☐	☐	☐	☐	☐
17) I have to cool some foods or drinks down before I can have them.	☐	☐	☐	☐	☐	☐	☐
18) I have to cut up some fruits before being able to eat them.	☐	☐	☐	☐	☐	☐	☐
19) I have to wear a scarf over my mouth on cold days.	☐	☐	☐	☐	☐	☐	☐

The following questions are **about the things you do in your daily life to avoid experiencing the sensations in your teeth**. Thinking about yourself ***over the last month*** to what extent would you agree or disagree with the following statements (Please tick only one response for each question)

	Strongly agree (7)	Agree (6)	Agree a little (5)	Neither agree nor disagree (4)	Disagree a little (3)	Disagree (2)	Strongly disagree (1)
20) I have avoided very cold drinks or foods.	☐	☐	☐	☐	☐	☐	☐
21) I have avoided very hot drinks or foods.	☐	☐	☐	☐	☐	☐	☐
22) When eating some foods I have made sure they don't touch certain teeth.	☐	☐	☐	☐	☐	☐	☐
23) I have changed the way I brush my teeth.	☐	☐	☐	☐	☐	☐	☐
24) When eating some foods I have made sure I bite in small pieces.	☐	☐	☐	☐	☐	☐	☐
25) There are other foods I have avoided.	☐	☐	☐	☐	☐	☐	☐

The following questions are about **the way the sensations affect you when you are with other people or in certain situations**. Thinking about yourself ***over the last month*** to what extent would you agree or disagree with the following statements (Please tick only one response for each question)

	Strongly agree (7)	Agree (6)	Agree a little (5)	Neither agree nor disagree (4)	Disagree a little (3)	Disagree (2)	Strongly disagree (1)
26) Because of the sensations I take longer than others to finish a meal.	☐	☐	☐	☐	☐	☐	☐
27) I have to be careful what I eat when I am with others because of the sensations in my teeth.	☐	☐	☐	☐	☐	☐	☐
28) I hide the way I am eating when I am with others because of the sensations in my teeth.	☐	☐	☐	☐	☐	☐	☐
29) I am unable to fully take part in conversations because of the sensations in my teeth.	☐	☐	☐	☐	☐	☐	☐
30) Going to the dentist is hard for me because I know it is going to be painful as a result of sensations in my teeth.	☐	☐	☐	☐	☐	☐	☐

The following questions are about ***the way the sensations in your teeth make you feel***. Thinking about yourself ***over the last month*** to what extent would you agree or disagree with the following statements (Please tick only one response for each question)

	Strongly agree (7)	Agree (6)	Agree a little (5)	Neither agree nor disagree (4)	Disagree a little (3)	Disagree (2)	Strongly disagree (1)
31) I've been frustrated because I can't find anything that deals with the sensations I have in my teeth.	☐	☐	☐	☐	☐	☐	☐
32) I've been anxious that something I eat or drink might cause sensations in my teeth.	☐	☐	☐	☐	☐	☐	☐
33) The sensations in my teeth have been irritating.	☐	☐	☐	☐	☐	☐	☐
34) I have been annoyed with myself because I did something that I knew caused these sensations.	☐	☐	☐	☐	☐	☐	☐
35) I felt guilty because I might have contributed to the sensations I am having with my teeth.	☐	☐	☐	☐	☐	☐	☐
36) The sensations in my teeth have been annoying.	☐	☐	☐	☐	☐	☐	☐
37) The sensations in my teeth have been embarrassing.	☐	☐	☐	☐	☐	☐	☐
38) I have been anxious because of the sensations in my teeth.	☐	☐	☐	☐	☐	☐	☐

The following questions are about **what the sensations in your teeth mean for you**. Thinking about yourself ***over the last month*** to what extent would you agree or disagree with the following statements (please tick only one response for each question)

	Strongly agree (7)	Agree (6)	Agree a little (5)	Neither agree nor disagree (4)	Disagree a little (3)	Disagree (2)	Strongly disagree (1)
39) I find it difficult to accept that I am a person who has these sensations in my teeth.	☐	☐	☐	☐	☐	☐	☐
40) Having these sensations in my teeth makes me feel different from others.	☐	☐	☐	☐	☐	☐	☐
41) Having these sensations in my teeth makes me feel old.	☐	☐	☐	☐	☐	☐	☐
42) Having these sensations in my teeth makes me feel damaged.	☐	☐	☐	☐	☐	☐	☐
43) Having these sensations in my teeth makes me feels as though I am unhealthy.	☐	☐	☐	☐	☐	☐	☐

The last five questions ask about **how much the sensations in your teeth affect your life overall**.

	Excellent (1)	Very good (2)	Good (3)	Fair (4)	Poor (5)	Very poor (6)
44) Overall how would you rate the health of your mouth, teeth and gums?	☐	☐	☐	☐	☐	☐

	Very Much (4)	Quite a bit (3)	Somewhat (2)	A little (1)	Not at all (0)
45) Overall how much do the sensations in your teeth bother you?	☐	☐	☐	☐	☐
46) Overall, how much do the things you do to manage the sensations bother you?	☐	☐	☐	☐	☐
47) Overall, how much do the sensations in your teeth affect your quality of life?	☐	☐	☐	☐	☐
48) Overall, how much do the things you do to manage the sensations in your teeth affect your quality of life?	☐	☐	☐	☐	☐

Appendix 2: The 15-item dentine hypersensitivity experience questionnaire (DHEQ-15)*

Thinking about yourself over the last month, to what extent would you agree or disagree with the following statements (please check only one response for each question):

1. Having sensations in my teeth takes a lot of the pleasure out of eating and drinking.
2. It takes a long time to finish some foods and drinks because of sensations in my teeth.
3. There have been times when I have had problems eating ice cream because of these sensations.
4. I have to change the way I eat or drink certain things.
5. I have to be careful how I breathe on a cold day.
6. When eating some foods I have made sure they don't touch certain teeth.
7. Because of the sensations I take longer than others to finish a meal.
8. I have to be careful what I eat when I am with others because of the sensations in my teeth.
9. Going to the dentist is hard for me because I know it is going to be painful as a result of sensations in my teeth.
10. I've been anxious that something I eat or drink might cause sensations in my teeth.
11. The sensations in my teeth have been irritating.
12. The sensations in my teeth have been annoying.
13. Having these sensations in my teeth makes me feel old.
14. Having these sensations in my teeth makes me feel damaged.
15. Having these sensations in my teeth makes me feel as though I am unhealthy.

All items should be answered on a 7-point Likert scale labeled and scored as follows: Strongly agree (7), Agree (6), Agree a little (5), Neither agree nor disagree (4), Disagree a little (3), Disagree (2), or Strongly disagree (1).

*The DHEQ-15 total score is calculated as the sum of the item scores

CPI Antony Rowe
Eastbourne, UK
November 06, 2014